Animals in Space

From Research Rockets to the Space Shuttle

Colin Burgess and Chris Dubbs

Animals in Space

From Research Rockets to the Space Shuttle

Published in association with
Praxis Publishing
Chichester, UK

Mr Colin Burgess, BIS
Bonnet Bay
New South Wales
Australia

Mr Chris Dubbs
Edinboro
Pennsylvania
USA

SPRINGER–PRAXIS BOOKS IN SPACE EXPLORATION
SUBJECT *ADVISORY EDITOR*: John Mason, M.Sc., B.Sc., Ph.D.

ISBN 10: 0-387-36053-0 Springer Berlin Heidelberg New York

Springer is part of Springer-Science + Business Media (springer.com)

Library of Congress Control Number: 2006937358

Printed in Germany

Cover design: Jim Wilkie
Project management: Originator Publishing Services, Gt Yarmouth, Norfolk, UK

Printed on acid-free paper

Contents

Authors' preface

Geographically, the two authors who worked on this collaborative effort live half a world apart. Although each had been involved in separate projects involving animal space flights, their paths had never crossed prior to December 2004, when a simple request for information brought them together. Within the space of a few days a collaborative book had been discussed and a proposal sent to Clive Horwood of Praxis.

The proposed book's timing was felt to be perfect by the authors, who both wanted a 2007 release to coincide with the 50th year anniversary of space dog Laika's flight into orbit and the history books. Soon after the proposal was accepted, the authors were quickly thrust into the joys (and tribulations) of research and writing, continually fired and inspired by unearthing little-known or obscure gems of information, in an unflagging spirit of cooperative endeavour and shared discovery.

COLIN BURGESS

It is quite astonishing to realise that the 50th anniversary of the flight into space of a small Russian street dog named Laika is almost upon us. I was 10 years old when Laika's handlers patiently strapped her inside a technological marvel known as Sputnik 2, ready to be hoisted into orbit on a planned one-way journey that would end in universal sadness and condemnation, but forever immortalise her name.

Admittedly the "space bug" had not bitten me at the time – that would occur more than 4 years later with the dramatic orbital flight of astronaut John Glenn – but I did know about Laika. One November night in 1957 our rowdy cub pack had been herded out of the scout hall at a certain time and made to stand under a crystal clear night sky while our cubmaster patiently told us about Laika and Sputnik 2. Suddenly he pointed with excitement above the darkened horizon, and we quickly fell into an awed silence as we watched a small, bright pin-prick of light silently and majestically traverse the star-spangled firmament over the east coast of Australia.

In time, a boyhood fascination with astronauts and space travel became a deep and enduring interest, and while the incredible suspense of the Space Race to the moon was characterised for me by the heroic exploits of American astronauts and Soviet cosmonauts, I also knew, basically, the stories of animalnauts such as Laika, Belka and Strelka, Ham and Enos. They had also been living pioneers of space flight, albeit reluctantly, and they deserved their place in history.

In 1994, at the urging of Australian-born oceanographer and payload specialist astronaut Dr. Paul Scully-Power, I became a volunteer space historian with the newly-formed Australian International Space School (AISS). This remarkable school had been set up to encourage the best science students from all over Australia to look at space science as a career option during a fully-funded, week-long seminar in Sydney.

One of my principal undertakings was to produce the AISS newsletter, and this also led me to write a book specifically for the school on the life and 1984 space flight of Dr. Scully-Power. After this had been published I began work on two other books for the AISS, also aimed at older school students. One was on the life of teacher Christa McAuliffe, who died aboard space shuttle *Challenger* in January 1986. I was given immeasurable help in researching this by Christa's petite but amazingly tireless mother, Grace Corrigan, who kindly hosted my wife Pat and me in her Massachusetts home for several days.

Twenty years after shuttle *Challenger* and her crew of seven astronauts were lost in a launch ascent tragedy, Grace is still incredibly active. Even today she regularly travels across America, giving inspirational talks in schools and other educational institutions, helping to create and fund new Challenger Learning Centres in the name of her daughter Christa and the last crew of shuttle *Challenger*.

Sadly, the AISS began to wind down around this time due to increasing financial and sponsorship difficulties, but through Grace Corrigan's eager support and encouragement the book on her daughter's life was eventually published in the United States, and I pledged every cent of the proceeds to the Challenger Education Fund.

The other book never really wanted to go past manuscript form. Initially it was called "Little Laika: First into Space" and then "Animalnauts: A History of the Pioneering Creatures Who Paved the Way into Space." However, it was never published, with enquiries to publishers producing responses stating, amongst other things, that the deaths of so many animals involved in spaceflight history was probably not a suitable topic for younger readers. The manuscript did, however, provide the basis for two articles written for the British Interplanetary Society's *Spaceflight* magazine – namely, "America's First Astro-Chimps" (July 1996; co-authored with Canadian friend Simon Vaughan) and "Dogs Who Rode Rockets" (December 1996).

In December 2004, a plea for assistance caught my eye in the highly-regarded "collectSPACE" online forum (*www.collectspace.com*). It was written by a Pennsylvania-based enthusiast named Chris Dubbs, who was desperately seeking any further photographs of Russian space dogs for an exhibition he was mounting in New Mexico. I already knew of Chris through my copy of his wonderful children's non-fiction book "Space Dogs," and I was keen to help. In researching my own book on Laika and other space animals I'd accumulated a number of obscure photographs relating to the Russian space dogs, so I sent scans of these to Chris. We quickly began discussing our mutual interest in the subject, and just before Christmas decided to seek out a suitable publisher and put together a proposal for a comprehensive, co-authored book on animal space flights.

I'd been working with Clive Horwood of Praxis for some time in regard to a Springer–Praxis book I'd co-authored with British space writer David Shayler (*NASA's Scientist-Astronauts*) , and I felt this book might work well as part of their superb space science catalogue. To our delight the proposal was accepted, at which time we formulated what each of us would write for the book.

Essentially, Chris worked on the Soviet/CIS side of the story, and fell straight into the horrendously onerous task of investigating the early Soviet ballistic and orbital dog flights – a history replete with misinformation, contradictions, articles written in a language totally foreign to us and a frustrating lack of totally reliable material. But he quickly rose to the challenge, and I am quite confident that no better or more factual account of this facet of spaceflight history will ever be written.

For my part, I worked mostly on the history of animal space flights associated with America and other non-Soviet countries. It was a somewhat guilt-ridden undertaking, as I had a far easier and more travelled path to follow than Chris. These missions and their backgrounds have been extensively and reliably documented, while people associated with the different programmes have been relatively easy to locate and contact.

I harbour no doubts at all that the two of us will one day collaborate on another book on a spaceflight topic, but this effort and its subject has been, and always will be, something very special for both of us.

CHRIS DUBBS

I cannot overstate how indelibly the image of a dog in a satellite burned into my youthful imagination. For me, at the age of 11, there was simply no way to comprehend it. It was too novel, too extraordinary an achievement, that it did not fit within any knowledge base that I possessed. It was mythic.

Like Colin, I waited under the night sky just after sunset for the speck of light to pass overhead. And when it did, I was transfixed. I caught a glimpse on two occasions. I marvelled more for the extraordinary experience given to Laika than I agonized over her fate.

The godfather of Soviet rocketry, Sergei Korolev, once encouraged a reluctant colleague to work for him by explaining the awesome experience of watching a rocket launch. "Once you've seen it," he said, "it will stay with you for the rest of your life." He might just as well have been talking about watching a satellite in those early years. Sputnik and the host of American and Soviet satellites that followed in its wake all left a profound impression on me.

The siren call of this new science led me and some friends to form a rocket club at the dawn of the 1960s. We spoke to community groups about the future of rocketry. We drew considerable crowds to a farm in southeastern Pennsylvania when we launched our rockets. We even had our own animal flight, sending a hapless mouse named Ham to his death when the parachute recovery system failed.

Skip ahead some 45 years. An anniversary of Laika's flight renewed my interest in the subject and motivated me to write a book in 2003 about all of the dogs used in the Soviet space program. Although, I had worked as a writer all of my life, *Space Dogs: Pioneers of Space Travel* was my first venture into writing space history.

The following year, I was invited to serve as guest curator for the New Mexico Museum of Space History in Alamogordo, to create an exhibit, titled "Pupniks," based upon my book. More than anything else, that experience impressed upon me the richness of the subject of animals in space. While at the museum, I was able to view some of the archives from nearby Holloman Air Force Base, home to much of the pioneering work in American rocketry and to the early studies of animals flying in rockets. In fact, the corps of volunteer workers at the museum was rife with retirees who had worked in various phases of those early programmes and were eager to share their reminiscences. A network of collectors also proved wonderfully cooperative in helping me to gather images of the Soviet dogs for that exhibit, including the co-author of this book, Colin Burgess.

Some of these collectors were incredibly knowledgeable and provided many useful suggestions that helped me to unravel a few of the mysteries surrounding the Soviet space dogs. And there were mysteries. Who would have thought that simply compiling a complete list of all of the dog flights and the dogs that flew on them would be such a challenge? Deciphering an ancient language might have been easier. Given the secrecy and the propaganda that surrounded the Soviet space effort, it isn't surprising that the historical record is meagre. But, here and there, the story of the dogs began to emerge. The caption on an old postcard or a TASS photo, the few Russian language memoirs that have seen publication, an oral history project of the Smithsonian Institution, a document translated by the U.S. government in the early 1960s, a newspaper clipping – no individual item was complete and definitive, but all were pieces in the puzzle.

Surprise gifts showed up from around the world, from writers, space enthusiasts, auction houses and museum directors, as if the world had kept these secrets in its collective attic all these years, and now that someone was finally asking questions, they could be given a home.

That wonderful experience has continued in the long preparation of *Animals in Space*. Both Colin and I have been fortunate to have connected with so many people who have given generously of their time and treasures, and their memories. Their enthusiasm for this subject energised us for the task of telling the story of all the animals that made a contribution to the exploration of space.

Colin Burgess
Bonnet Bay
New South Wales
Australia

Chris Dubbs
Edinboro
Pennsylvania
USA

November 2006

Acknowledgements

There are numerous people and organisations to whom we owe sincere thanks for their participation in this book. Some came to us unexpectedly, others responded to appeals for help on a number of online space forums, while still more readily replied to specific enquiries. A mere listing of names can in no way suggest our heartfelt gratitude towards these individuals. Without their interest and cooperation it would have been difficult, if not impossible at times, to collect, transcribe, organise or publish the information, illustrations and anecdotal material contained in this book.

Therefore, in alphabetical order, we wish to acknowledge the kind assistance of:

George Bailey; Tatiana Bogatova; Rick Boos; Jay C. Buckey, Jr., M.D.; Glenn Bugos, Ph.D. (NASA Ames History Office); Cat Carlson; John B. Charles, Ph.D. (NASA JSC); Dr. Gérard Chatelier (CERMA); Deng Ningfeng and Hou Mingliang of the China Astronautic Publishing House; Rosamund Combs-Bachmann (MIT, Mars Gravity Biosatellite Project); Bonnie P. Dalton (NASA Ames); Dwayne Day; Kerrie Dougherty; Edward C. Dittmer, Sr.; Elizabeth Flowers (NASA Wallops Island); Francis and Erin French; Owen Garriott, Ph.D.; Sven Grahn; Duane Graveline, M.D.; John Grazulis; Carol Gums; Rex Hall; Brian Harvey; Ken Havekotte; Ed Hengeveld; Ray Holt; George House (New Mexico Museum of Space History); Institute for Biomedical Problems of the Russian Academy of Sciences; Andrzej Kotarba; Garry Laing; National Air and Space Museum; National Geographic Society; John Locke, Ph.D. (University of Alberta, Canada); Leilani Marshall (NASA Ames History Office); Emily Morey-Holton, Ph.D. (NASA Ames); Roger McCormick; Wayne Meier; Kimberly Merrill (New Mexico Museum of Space History); George Meyer; Mike Myer; Henk Nieuwenhuis; Bill Obenauf; Ralph Papa; Robert Pearlman; Delbert Philpott, Ph.D.; Holloman AFB, Public Affairs Office; Joel Powell; Olivier Sanguy (*Espace* magazine); Chris Scaduto; David Shayler; Asif Siddiqi, Ph.D.; Richard C. Simmonds, D.V.M.; David G. Simons, M.D.; Rob South; Kenneth Souza; Trevor

Sproston; Maciej Stolowski; Jakob Terweij; Bill Thornton, M.D.; Simon Vaughan; François Vigier; Bert Vis; Chuck Vukotich; and Liz Warren.

Special thanks are also due to Rob Kelly for his exceptional work in translating numerous documents from Russian, and to Joseph Bielitzki, D.V.M., for providing the marvellous Foreword to this book.

Foreword

The pursuit of space has presented numerous physical, technical and scientific barriers to those who would leave the surface of the planet to explore in a new environment. That environment is essentially without atmosphere, without gravity, without protection from radiation, without those evolutionary constants that shaped our form and physiology. Of all things associated with being human, our entrance into the environment of space is atypical in every way. Our physiology functions within the certainties of an Earthbound existence; life having evolved within the specific limitations of ambient temperature, atmospheric pressure, atmospheric chemistries and gases, gravitational fields, electromagnetic influences, radiation exposure, light requirements, perceptible sound, the food and water content of our diets, our exposure to other living organisms and our relationship to each other. We exist within defined environmental and social specifications outside of which new science, new technologies and new concepts are needed to sustain our fragile lives.

The exploration of space, perhaps exaggerated by the futurists and science fiction writers of the early 20th century, became a competitive activity following Sputnik's successful flight and was certainly fuelled by the Cold War politics of the 1950s, 1960s and 1970s. The introduction of humans into space became a frenzied activity. Risks were taken to achieve the unimaginable. Behind the high-visibility flight programmes existed steady and cautious life sciences research programmes aimed at understanding the risks that each astronaut and cosmonaut took, from the initial assignment to flight, through launch, flight, recovery and the post-flight period. Much information was gained from those who flew into space; but most of the risk was taken by laboratory animals launched into space to understand those risks, as nations worked to assure the health and safety of the humans who would eventually follow.

The collective memory of the use of animals for life sciences research in space during the modern era spans nearly 60 years since the first rockets carried animals in the late 1940s. During the 1950s and 1960s, the animals that flew into space became part of the Space Race and part of the personality of innovation and progress. The names of some of these animals were part of the space programmes' language but represented the quiet and professional work of hundreds of scientists and technicians. Today some still remember names like Enos, Ham, Able, Baker, Sam, Bonnie, Martine, Pierrette, Lapik, Multik, Tsygan, Dezik, Albert and, of course, Laika. Each flew a mission to help develop telemetry systems, understand physiologic changes, and identify safety and risk factors that humans might face during space flight. Each demonstrated that life in space is hazardous but that these hazards could be overcome. Collective memories sometimes span too many years, too many individuals and too many locations to be of real benefit in recalling the details of the history of space. The details are there, only needing to be linked into a cohesive narrative that appropriately reflects on the culture, the players, the stake holders, the animal subjects and the science.

The use of animals in space research is a unique collection of individual projects coordinated by scientists from many nations linked by common goals and motivations. The competitive nature of the Space Race tended to isolate the different national programmes. However, in the 1970s and 1980s, the Bion programme seemed to transcend political barriers, as the former Soviet Union invited American and European scientists to participate as colleagues in a number of space flights, focusing on animal research and conducted on Soviet space vehicles. Ironically, even during the peak of the Cold War these collaborations continued, providing a scientific bridge between nations. A type of integrated scientific activity occurred among international colleagues separated by language and politics but linked by common problem sets and the common research environment of space. Successes and failures occurred jointly during the Bion programme, forging relationships and friendships that transcend language and politics. Trust and confidence between colleagues from different nations does not come with an academic degree or a research reputation but rather from shared experiences and contributions to common goals and shared responsibility for research progress and programmatic growth.

Life sciences research is complex enough when conducted in the known environments of a terrestrial laboratory, a laboratory where equipment and research subjects remain firmly attached to a surface and governed by the gravity of the Earth. Experimental design and execution become dramatically more difficult in the relatively unknown surroundings of space. Take, for example, the single issue of habitat design.

Significant effort has gone into habitat design in an attempt to meet the welfare needs of animals in space. Most of us go through life never asking how to design an aquatic environment for housing fish, where the water in the habitat, if left unchecked, would float across the room and with little perturbation break into hundreds of smaller, free-floating droplets. Or where, if the fish penetrated the water's surface tension, they could drift away, unable to re-enter their watery home because gravity did not pull them down and through the liquid–air interface. Unless required, nobody wonders how to contain free-floating urine or faeces within an animals' enclosure, or what the risk of inhaling such waste poses to the subject. We never ask, "Where do crumbs from food pellets go if not to the floor?" or "How do you provide water through a sipper when gravity does not assist in moving it to the open end of the sipper tube?" "How will animals sleep?" "Do they nest in a ball, curled up and snuggled, or are they merely floating in free space, startled awake by every contact with every surface?"

These questions go on and on, asked with every animal flight. Habitat design for space flight is a research effort separate from the scientific questions, yet critical to animal welfare and scientific validity. The space agencies conducting animal research design animal habitats through a design system that links life scientists, veterinarians, animal behaviour experts, materials scientists and engineers in a process that will meet the needs of the animal subject, the animal care staff and the scientist, while meeting the design constraints of volume, mass, power and ease of operation.

Many of the animal subjects used in the space programme flew without the companionship of astronaut or cosmonaut. Most were carefully selected for flight based on health, personality, training, ability to perform specific tasks and the ability to tolerate the experimental conditions while in gravity. Training and adaptation of the animal subjects in many cases took months. Technicians and scientists often spent more time in the laboratory working towards flight than at home with friends and family. In many cases the animal subjects became like family. It is possible to have a relationship with a newt, fish, rodent and non-human primate. Laboratory personnel develop commitments not only to the science but also to the experimental subjects. Scientists, astronauts, cosmonauts, laboratory technicians and animal care staff understand the risks, the significance and importance of each research effort and work with total commitment to its success.

The crews that have conducted animal studies in space are unique. Each recognises that these studies are not conducted on rocks, minerals or spectrographic analysis but on living organisms. The use of animals in research is a responsibility and a privilege, one that requires all participants to conduct research in ways that reduce pain and distress, while recognising that the loss of gravity is most likely a stressor. As a way to create a culture of caring for its animal research subjects, NASA has developed bioethical principles for all those individuals who participate in life science research. NASA recognised that, regardless of the importance of the scientific study, the humane treatment and care of the animals involved is equally important. Animals act as surrogates for the human in life science research and as such have a greater moral status. Consequently, each person involved in the research has a greater obligation to provide for their care and to work to minimise any pain and distress that might occur during the research effort. Those who work in life sciences research in the space programmes seem to be continuously aware of these ethical issues.

Animals in Space is an absolutely unique work in the history of science and of space exploration. Several books have been published that outline the essential scientific design of the animal research conducted in space over the last 60 years, since rockets were first invented. This is the first work that summarises the science, the reason for the study, the people and the benefits. It has unified the information that has been published in any number of places into an easy-to-read work that almost makes the reader part of the research team. It personalises the efforts of the research teams; it debunks myths associated with the early space programme; but, most importantly, it tells the truth about the passion and commitment of the individuals involved. It speaks openly of those who opposed the use of animals in research and of those who did not. It describes the practical realities of long periods of research in laboratory spaces far from home, where personnel from many countries worked together to assure the highest quality of science. It does not always capture the fatigue, intensity and angst felt by those working for "mission success". It cannot capture the total passion and commitment of the research community for knowledge and how that information can benefit humanity. The public needs to understand the deep commitment of the space agencies to astronaut/cosmonaut safety and to recognise the consequences of space flight on the human condition.

Time will tell if we can overcome the barrier of space between the Earth and other worlds. Barriers of distance, heat and cold, lack of atmosphere, lack of gravity, radiation storms, lack of available food and water block the way. In our pursuit of space we have asked animals to involuntarily lead the way and demonstrate safety, risks and hazards that should never be first identified by a human space traveller. This book records the animals' contributions, but, more importantly, it recognises the thousands of scientists, engineers, veterinarians, programme directors and managers, technical staff, administrative personnel, animal care staff, flight support personnel, astronauts and cosmonauts, who have made space an attainable goal in an inhospitable environment.

Chimpanzees named Ham and Enos each flew in a Mercury capsule before Alan Shepard and John Glenn, respectively, to reduce the unknown risks of space. We have learned much since then and need to learn much more if we are to go forward; to the moon with self-sustaining habitats; to Mars and beyond with human exploration. Perhaps, we should always remember the words that began John Glenn's orbital flight around the planet, "Godspeed, John Glenn." And ask if it was the briefest of prayers to protect against unknown risks of space, unidentified during Enos's brief flight, or just good wishes to begin a new period of discovery.

Joseph T. Bielitzki, M.S., D.V.M.
NASA, Chief Veterinary Officer, 1996–2001

To those countless unknown and unnamed animals whose lives were sacrificed in paving a safer path for humans to follow into space.

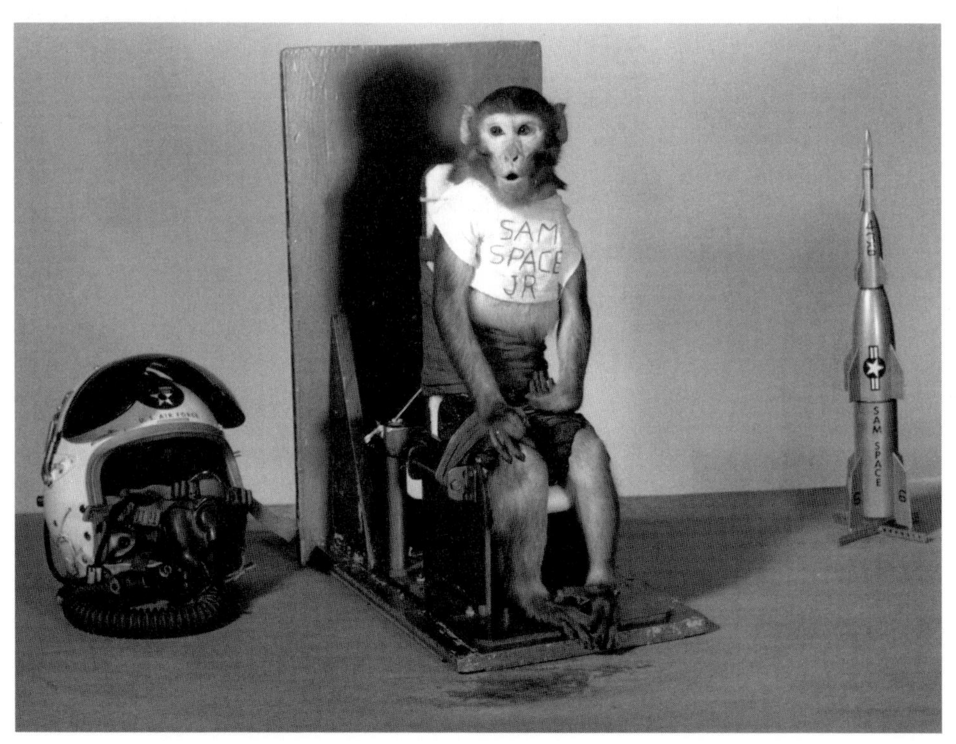

Sam Space Jr., one of a small group of monkeys training for space flight circa 1959 at the USAF School of Aviation Medicine, then located at Randolph Air Force Base, Texas. (Photo: USAF National Archives)

Figures

DEDICATION

PROLOGUE

CHAPTER 1

CHAPTER 2

CHAPTER 3

CHAPTER 4

CHAPTER 5

CHAPTER 6

CHAPTER 7

CHAPTER 8

CHAPTER 9

CHAPTER 10

CHAPTER 11

CHAPTER 12

Abbreviations and acronyms

A-1	Aggregate-1 or Prototype-1
AAEU	Aquatic Animal Experiment Unit
ABMA	Army Ballistic Missile Agency
AEM	Animal Enclosure Module
AFB	Air Force Base
AGI	International Geophysical Year (French)
AIDS	Acquired Immune Deficiency Syndrome
AISS	Australian International Space School
AKA (a.k.a.)	Also Known As
AMC	Amphenol Micro Coaxial
AMFL	AeroMedical Field Laboratory
AMRL	AeroMedical Research Laboratory; Army Medical Research Laboratory
AP	Advanced Projects (Lockheed's Advanced Development Projects Unit – known also as the "Skunk Works")
ARDC	Air Research and Development Command
ARPA	Advanced Research Projects Agency
ASCS	Automatic Stabilisation and Control System
ASPCA	American Society for the Prevention of Cruelty to Animals
BEM	Bee Enclosure Module
BIOCORE	BIOlogical COsmic Ray Experiment
BION	BIOlogical Investigations of Space (Project)
BIOS	BIOlogical Satellite
BOSS	Bioastronautics Orbital Space System
BOTEX	BOTany EXperiment
BRIC	Biological Research in Canisters
BTS	BioTelemetry System
CAAHS	Chongqing Academy of Animal Husbandry Science

CalTech	California Institute of Technology
CAS	China Academy of Sciences
CASC	China Aerospace Science and Technical Consortium
CASI	Center for Aerospace Information (NASA)
CCCC	Center for Captive Chimpanzees Care
CEBA	*Centre d'Etudes de Biologie* (Centre for Biological Studies)
CEBAS	Closed Equilibrated Biological Aquatic System
CERMA	*Centre d'Enseignement et de Recherches de Médecine Aéronautique* (Research and Studies Centre of Aerospace Medicine)
CIA	Central Intelligence Agency
CIEES	*Centre Interarmées d'Essais d'Engins Spéciaux* (Inter-arms Special Weapons Test Centre)
CIS	Commonwealth of Independent States
CNES	*Centre National d'Etudes Spatiales* (French National Space Agency)
Convair	Consolidated Vultee Aircraft Corporation
CPE	Circadian Periodicity Experiment
DEFA	*Direction des Etudes et Fabrications d'Armés* (French Directorate of Armaments)
DVM	Doctor of Veterinary Medicine
ECG	Electrocardiogram (also known as EKG)
ECS	Environmental Control System
EKG	*See* ECG
ESA	European Space Agency
EST	Eastern Standard Time
EVA	Extra-Vehicular Activity (spacewalks)
FOE	Frog Otolith Experiment
FOEP	Frog Otolith Experiment Package
FSW	Fanhui Shei Weixing
GDR	German Democratic Republic
HIV	Human Immunodeficiency Virus
HZE	High-ioniZing high-Energy particles
IAF	International Astronautical Federation
IAM	Institute of Aviation Medicine
IASM	Institute of Aviation and Space Medicine
IBMP	Institute of Biomedical Problems
ICBM	InterContinental Ballistic Missile
IEPT	Institute of Experimental Pathology and Toxicology
IGY	International Geophysical Year
IKI	Institute of Space Research
IML-2	Second International Microgravity Laboratory
IRBM	Intermediate Range Ballistic Missile
IRE	Institute of Radiotechnology and Electronics
JATO	Jet Assisted Take-Off

JSC	Johnson Space Center
JSLC	Jiuquan Satellite Launch Centre
KH	KEYHOLE
KSC	Kennedy Space Center
LSD	Dock Landing Ship
MHz	Megahertz
MIA	Mouse In Able (Project)
Mirak-1	Minimum Rocket-1
MR-BD	Mercury–Redstone Booster Development
MSFC	Marshall Space Flight Center
NACA	National Advisory Committee for Astronautics; National Advisory Committee for Aeronautics
NAS	Naval Air Station
NASA	National Aeronautics and Space Administration
NIH	National Institutes of Health
NPIC	National Photographic Interpretation Center (pronounced as "en-pick")
OART	Office of Advanced Research and Technology
OFO	Orbiting Frog Otolith
OKB	*Opytnoe Konstructorskoe Byuro* (Experimental Design Bureau)
PARE-1	Physiological and Anatomical Rodent Experiment 1
PETA	People for the Ethical Treatment of Animals
PSE	Physiological Systems Experiment
RAE	Royal Aircraft Establishment
RAF	Royal Air Force
RAHF	Research Animal Holding Facility
RASKO	*Raketa Stucznej Kondenzacii* (Experimental Science Rocket)
SAA	South Atlantic Anomaly
SAM	School of Aviation Medicine; Sally, Amy and Moe (three black mice)
SARAH	Search And Rescue And Homing (Beacon)
SLS-1	Spacelab Life Sciences 1
SMPAA	*Section Medico-Physiologique de l'Armée de l'Air* (Medico-Physiological Section of the [French] Air Force)
SPAF	Single Pass Auxiliary Fan
SPOG	Special Projectile Operations Group
SPURT	Small Primate UnRestrained Test
SRV	Satellite Recovery Vehicle
STARS	Space Technology and Research Students
STG	Space Task Group
STS	Shuttle Transportation System; Space Transportation System
TBS	Tokyo Broadcasting Service
TCF	The Coulston Foundation

UFO	Unidentified Flying Object
USAAF	United States Army Air Force
USAF	United States Air Force
USC	University of Southern California
USDA	United States Department of Agriculture
USSR	Union of Soviet Socialist Republics
USSTAF	United States Strategic Air Forces (Europe)
V-2	*Vergeltungswaffe 2* (Retaliation Weapon 2); at first, this stood for *Versuchmuster 2* (Prototype 2)
Veronique	*VERnon electrONIQUE*
VFEU	Vestibular Function Experimental Unit
VfR	*Verein für Raumschiffahrt*
VHF	Very High Frequency
WAC	Without Any Control
WD-40	Water Displacement formula 40
WSPG	White Sands Proving Ground
ZIB	Russian acronym for "substitute for missing dog Bobik"

Also by Colin Burgess in this series

NASA's Scientist-Astronauts (2006), ISBN 10: 0-387-21897-1

Prologue

Remarkably, in his classic 1865 book *From the Earth to the Moon*, French science-fiction writer Jules Verne told of a lunar trip successfully undertaken by three men. It was a journey that would have many striking similarities to the first lunar landing by human beings more than a century later.

Verne's three-man spacecraft had been named the *Columbiad*, while the Apollo 11 spacecraft was called *Columbia*. The *Columbiad* craft was fired towards the moon from Florida, prophetically just 120 miles from the present-day Cape Kennedy, and just like Apollo 11 in 1969, it splashed down in the ocean. There are many other amazing similarities: for instance, one of Verne's intrepid explorers had the surname Aldan, while one of the Apollo 11 crewmembers had the surname Aldrin. But unlike Apollo 11, Verne's crew carried along some non-human passengers, as seen in a contemporary wood-cut illustration featuring a pair of cockerels and a small dog, which curiously bears a striking resemblance to yet another small canine that would achieve immortality in 1957 as the first animal ever to orbit the Earth – a dog named Laika. Jules Verne's name for his spacefaring dog was also a most prophetic choice – it was called Satellite.

He may rightfully be regarded as a true doyen of science fiction, but not even a writer as wonderfully imaginative and prophetic as Verne could have conjured up the rich and dramatic history that involved animals and their part in the exploration of space. The unfolding of that great adventure awaited only the appearance of vehicles capable of broaching our atmosphere and the dark, mysterious envelope of space beyond.

Throughout history our unexplored frontiers – jungles, seas, deserts, the North and South Poles – have been conquered by a procession of intrepid explorers. There is, however, one powerfully seductive frontier we have only just begun to penetrate, and it can be found just a few miles from each of us, no matter where we stand on this great planet. It is a vertical frontier, more challenging and hostile than any we have ever faced before. It is space.

A contemporary illustration from Jules Verne's prophetic novel shows the three explorers travelling to the moon aboard their ship *Columbiad*, accompanied by a small dog named Satellite. (Illustration: U.S. Library of Congress)

Spanning every decade of the 20th century, legions of men and women have climbed ever higher in an unstoppable thirst for knowledge and adventure. The tallest trees, the Tower of Babel, the summit of Mount Everest have all been ascended in that compelling quest to go beyond what we knew before. Balloonists ascended silently into the frigid regions of the upper atmosphere, where the great, gas-filled sphere above them froze to the fragility of paper-thin glass. Post-war aviators pushed their supersonic aircraft to the limits of human and mechanical endurance, clawing their way to heights where no air passed over the wings to create lift: into an unforgiving environment where an aircraft can surrender to the laws of high-altitude nothingness by suddenly tumbling out of control.

Pressurised cabins, breathing masks and high-altitude suits had to be devised and developed that would allow humans to survive in the near vacuum of the upper atmosphere. Barely 11 miles above the Earth's surface, the air pressure is so weak that pilots without suitable protection would be unable to breathe, and their blood would bubble and fizz like uncorked champagne.

It took the impetus of war to advance the technology that finally prepared humankind for the first major assault on the frontier of space. As the Second World War drew inevitably to an end, conquering armies swept in and plundered equipment and personnel from the once-formidable and terrifying German V-2 rocket programme. Having safely landed their valuable charges on American soil, the military quickly despatched dozens of these captured scientists and rocket parts to a remote desert area of the American southwest. Meanwhile, the Soviet Union, working with a more modest share of the V-2 booty and captured German technicians, focused its rocket development efforts on a stark and forbidding outpost named Kapustin Yar, on the empty steppes north of the Caspian Sea.

If mankind's future is defined by the challenges it undertakes, then it began to write a new future in the decades immediately following the war. This remarkable concept of space travel – the future – was happening in two remote parts of the world. Two groups of scientists; two programmes. In both places so much of the focus was on the development of rockets for military use, but some insiders were unambiguously taken with the vision of manned space travel.

In the Soviet Union it might be said, without too much exaggeration, that one man carried the vision of this future, along with the force of will to make it happen: Sergei Korolev. No one burned with the same urgency as Korolev. In the United States the enormous energy, influence, foresight and work of Wernher von Braun is undeniable, but the American effort was not centred quite so much in one man. But, without any doubt, their passionate involvement in rocketry and in creating a human presence in space was an integral part of the vast, rapidly-developing story of astronautics. It was a totally new concept, which captivated people from all nations during the tense Cold War years of what has become known as the Space Race. National and international prestige was at stake, and these two men stood firmly at the helm as an awed and anxious world followed the news of ever-bolder triumphs. In this era of superpower propaganda and political chest-beating, humankind's reach into space was always tempered by the worrisome reality of an accelerating military–space tension, bringing an added urgency to the effort.

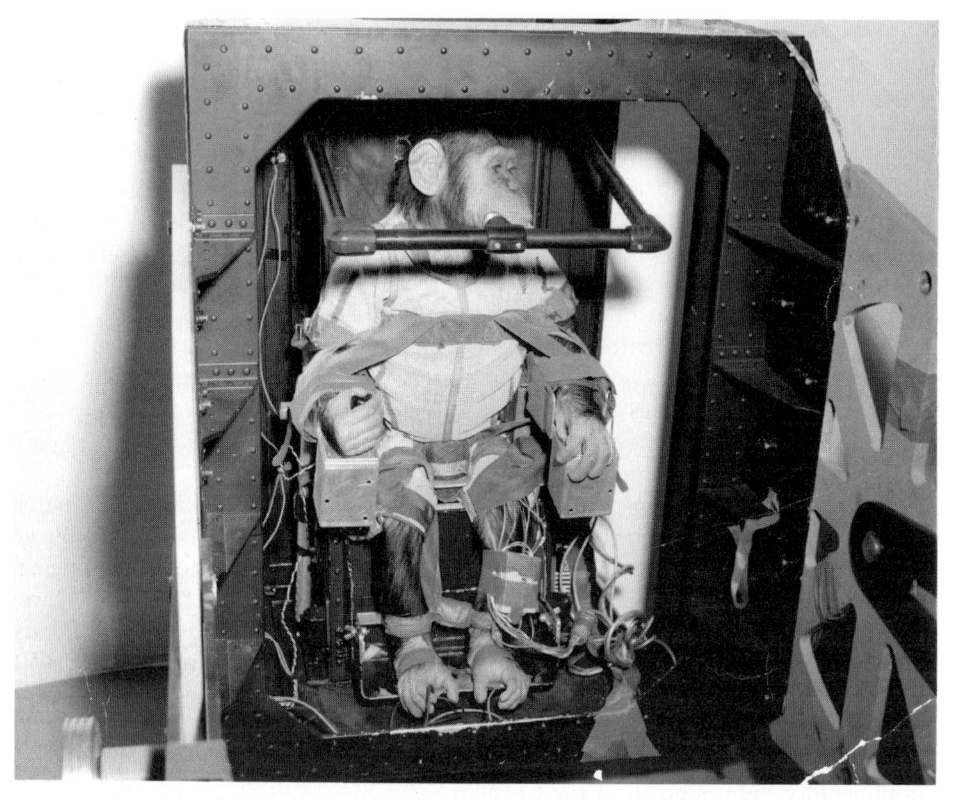

Chimpanzees such as this one undergoing psychomotor repetition training at Holloman Air Force Base helped pave the way for early manned space missions. (Photo: USAF, New Mexico Museum of Space History)

Our first fledgling steps on a path to the stars may have unfolded at some of the most inhospitable locations on the planet, the work shrouded in a dark curtain of secrecy, but it was driven by a resolute scramble to identify and overcome each obstacle to that future. And those obstacles and challenges were legion. Until aeronauts and scientists discovered the means to overcome the tremendous pull of gravity and ways to survive in the airless upper atmosphere, humans could not venture very far from the home planet. Ultimately, it came down to a matter of physics and physiology.

The physics problem was simple, if not the solution. It meant increasing the velocity of rockets from around 3,100 miles per hour, the highest speed reached by the V-2 rocket, to nearly 18,000 miles an hour, the velocity required to escape Earth's gravity and travel into orbit.

The physiology problem, on the other hand, was even more daunting. In 1946, when the first launch of a captured V-2 rocket from American soil took place at Holloman Air Force Base in New Mexico, the fastest any human had ever travelled in an aircraft was 606 miles per hour, a record set by RAF Group Captain Hugh Wilson in a Gloster Meteor jet aircraft on 17 November 1945. With little more to go on than basic physics and biology, scientists were not even certain that a person could survive a brief ballistic flight in a rocket, given the crushing force of acceleration and the disorienting period of weightlessness that would follow. Just how would the human body react to these unfamiliar forces and to the malevolent dangers of cosmic radiation that silently waited in the upper atmosphere?

To answer these questions, and to avoid sacrificing would-be astronauts, researchers began looking to the use of warm-blooded animals, whose physiology closely resembled that of humans.

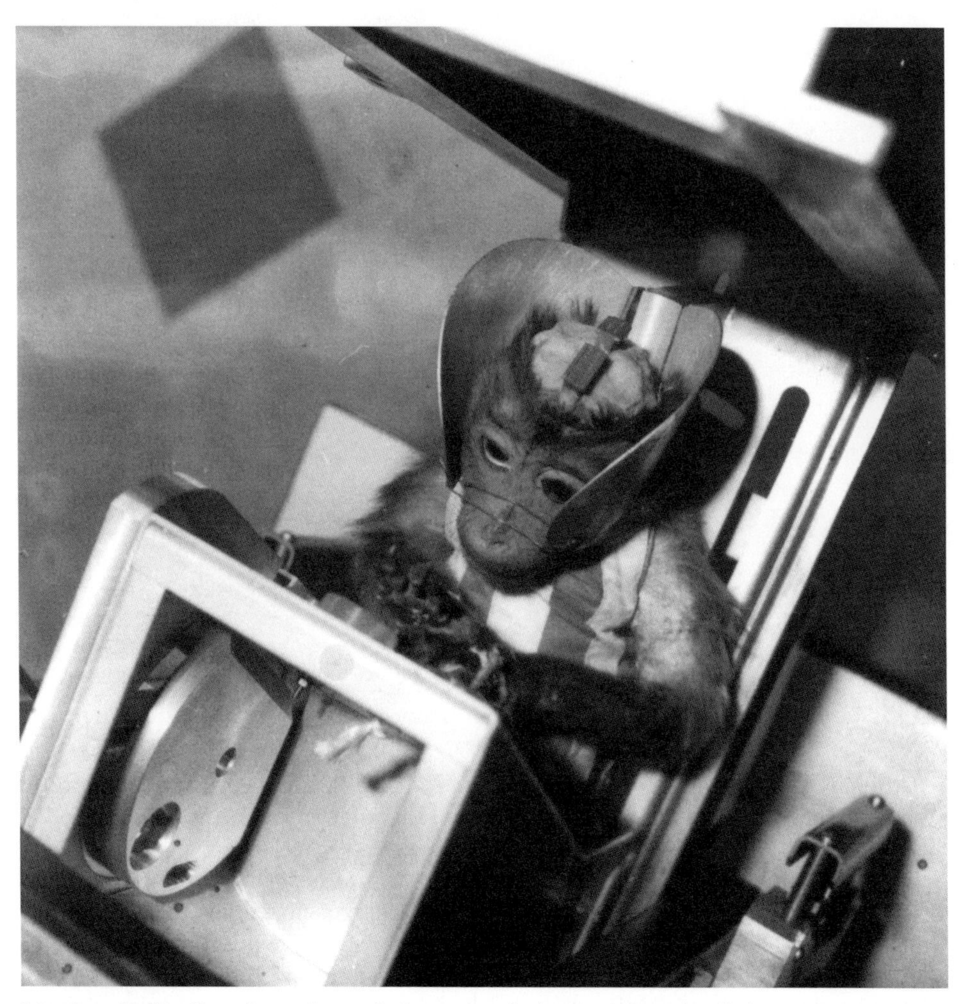

Martine, the first French monkey to fly into space, is shown undergoing flight psychomotor tests in preparation for her historic 1967 mission aboard a Vesta rocket. (Photo: Dr. G. Chatelier, CERMA)

Russian street dogs such as Malyshka were the true pioneers of the Soviet space programme. (Photo: authors' collections)

Animals would prove to be our bridge to the future. They gave substance to the vision. If humans could not yet travel in space, animals could. Their contribution has been profound and continuous. In a biological sense, from 1948 to 1960 they *were* the space programmes. They were writing the future. In a host of laboratory and field tests, in high-altitude balloons and occupying the cramped nose cones of early rockets, animals became the test subjects for the U.S. and Soviet space programmes. To settle questions about acceleration, vibration, noise, radiation exposure, extremes of pressure and temperature, weightlessness and basic survival in sealed capsules, animals stood in for humans. To test the rockets, capsules, parachutes, recovery systems, environmental systems and satellites, animals gave their lives. We saw ourselves, and what we could accomplish, reflected in their successes and failures. Bonds developed between American scientists and the monkeys, and the Soviet scientists and the dogs. It was an incredible collaboration. When dogs and monkeys died, the future seemed more remote; whereas, their accomplishments brought tomorrow to our doorstep.

On 12 April 1961, Soviet cosmonaut Yuri Gagarin became the first person to rocket into space. This was a truly historic mission, paving the way for other men and women to follow. On his return, Gagarin made a speech in which he acknowledged the contributions of those who made the mission such a success. He specifically mentioned the earlier flight of a small dog named Laika, whose name will be forever remembered in history books as the first creature to make a true space flight.

This book is the story of Laika and other dogs, monkeys, chimpanzees and assorted Earthly creatures whose pioneering flights proved that men and women could be launched with relative safety aboard powerful rockets and survive the harsh environment of space. Many of these biological flights, and particularly that of Laika, brought strong protests from animal lovers around the world. But each of the animals, whether they survived or not, gave us information and experience so desperately needed if we were to leave this planet in our quest for knowledge about our universe and ourselves.

This book is a tribute to Laika and the other animals that pioneered space flight and is gratefully dedicated to their memory.

Laika, prior to her history-making flight. (Photo: authors' collections)

1

Taming the rockets: From wrath to research

In order to understand the story of the first animals in space, it is important to place them in their context within the advances of rocketry and space flight. The rockets that would eventually fly them had fascinating histories, as did the people who designed them. As rockets grew bigger and more powerful, and as their uses diversified from weapons to possible deliverers of humans into orbit, the need to see if living creatures could survive their flights grew stronger. First, however, the rockets had to fly.

THE MAN BEHIND THE VISION

Inevitably, discussions on the genesis of the spaceflight programmes of different nations always seem to involve the curious paradox that exists in the history of American rocketry and space flight. The passage of time brings a host of innovations and changes, but it does not diminish one irrefutable fact – that the man most responsible for giving the United States an unassailable supremacy in the all-out race to land the first person on the moon was a former, and formidable, enemy of the Allies.

Even though the person in question found the use of his rockets as weapons of war unpalatable, he nevertheless took advantage of the opportunity to work for his Nazi overlords to create a series of lethal projectiles, crammed with high explosives, which included variants specifically designed to be launched on civilian targets.

During the Second World War this family of ballistic missiles would become universally known, feared and despised as the V-weapons. Ironically, neutered versions of these same rockets would one day carry a host of animals into the heavens on peaceful, science-gathering missions.

Dreams of tomorrow

The man principally charged with exploiting the potential of this innovative propulsion system was Wernher von Braun, born 23 March 1912, the son of a Prussian baron. He would later become an engineering graduate of the Zurich Institute of Technology, Berlin's Charlottenburg Institute of Technology and the University of Berlin.

As a young boy, his mother Emmy had given Wernher a precious gift on the day of his confirmation in the Lutheran Church. Not a watch and long pants like most Lutheran boys, but a telescope. "So, I became an amateur astronomer," he would reflect in later years, "which led to my interest in the universe, which led to my curiosity about the vehicle which will one day carry a man to the moon" [1].

The vehicle to which he was alluding was the rocket.

Rocket man

There was an inkling of things to come when von Braun was just 11 or 12 years old. Curious about the effects of propulsion, he strapped a cluster of gunpowder rockets to a wagon and ignited them, sending it on a perilous path along a crowded street toward Tiergarten Strasse. "I was ecstatic," he later recalled. "The wagon was wholly out of control and trailing a comet's tail of fire, but my rockets were performing beyond my wildest dreams" [2]. A police officer soon put an end to his excitement, grabbing hold of the boy's arm and threatening to arrest him. He was later released in the charge of his father.

While this escapade was an interesting experiment, and his telescope had certainly given him a boyhood fascination with the stars and space travel, von Braun first became truly immersed in rocket technology when he was 17 years old, while reading Hermann Oberth's authoritative 1923 book, *By Rocket Into Interplanetary Space.* He later became an enthusiastic member of the German Rocket Society, or *Verein für Raumschiffahrt (VfR)*, which had formed in July 1927.

Made up mostly of keen, young scientists, and with Oberth as its first president, VfR eventually set up a base in the Berlin borough of Reinickendorf. Soon afterwards they began carrying out a series of experimental rocket propulsion tests at their new base, which was basically an abandoned ammunition dump they had gleefully endowed with the nickname *Raketenflugplatz*, or Rocket Airfield.

One of the founding members of the Rocket Society was the prolific science writer and space advocate Willy Ley, later recognised for his gifted ability to accurately frame complex technical material in a form easily understood by the lay reader. He once wrote that the Rocket Society had to make considerable efforts to try and raise funds "to convert Professor Oberth's theoretical work into practical reality. It succeeded to some extent and built and fired quite a number of successful liquid-propellant rockets; but it must be borne in mind that in those days, from 1930 to 1932, a rocket was considered 'successful' if it worked at all."

A group of German rocket pioneers in 1930. At centre, Hermann Oberth stands beside one of his rockets. Wernher von Braun (in short pants) is second from the right. (Photo: NASA Marshall Space Flight Center)

A minimum rocket

The group had soon developed a small, liquid-fuelled rocket called the Mirak-1 (Minimum Rocket-1), which was propelled by a highly-volatile mixture of liquid oxygen and gasoline. The first successful static test of a Mirak-1 took place at Bernstadt, Saxony, giving the team tremendous encouragement to continue. During a second static test-firing the following month, however, the liquid-oxygen tank split open and their diminutive rocket exploded.

Despite these setbacks, the development programme continued. In 1932, von Braun's impressive work with the VfR managed to attract the early attention of a former artilleryman, Captain Walter Dornberger.

Acting on behalf of the Ordnance Department of the German Army, Dornberger had been assigned the task of heading a covert investigation into the feasibility of rockets and their potential for military use. At this time, the Army was vitally interested in the development and production of a weapon that did not contravene the many strictures attached to the Treaty of Versailles, and one that could provide Germany with future defence capabilities. Signed at the end of the First World War, this treaty forbade Germany from manufacturing a whole raft of armaments that could potentially be used in warfare, although the army's military analysts had noted that it only precluded solid-fuel rocket research.

Dornberger convinced von Braun he had a future beyond amateur rocketry, and Colonel Karl Becker, chief of ballistics and ammunition for the German Army, concurred. He told von Braun that once he had received his bachelor's degree the Army would underwrite his graduate study at the University of Berlin, where Becker was also a professor. The only proviso was that von Braun's principal study had to be on liquid-fuelled rocket engines.

The first recorded experience of von Braun's involvement in animal experiments relating to acceleration forces came about in 1931, while he was in Zurich completing a semester of study. Together with a fellow student named Constantine Generales (who would later become an eminent biomedical investigator) he rigged up a rudimentary centrifuge with a bicycle wheel as its centrepiece, in order to determine the effects of acceleration on a living creature. The crude machine worked, and in time several mice had been strapped to the inner circumference of the wheel and spun at great speed. One day their unappreciative landlady found sprays of blood on her wallpaper, and further experiments involving mice were abruptly curtailed.

The work begins

Once he had received his bachelor's degree and begun his doctoral research, von Braun was put to work by Dornberger (by then promoted to Colonel) in an official capacity at the Army's Ordnance Research and Development Office in Kummersdorf, 17 miles south of Berlin. A static test site for ballistic missiles had been established there, in a clearing deep within the Brandenburg Forest.

In his post-war book *V2*, Dornberger wrote: "Our nineteen-year-old 'student', Wernher von Braun, had come to us fresh from his work on the rocket airfield . . . That enterprise was slowly dying of chronic lack of money, so he had joined the Army Weapons Department on the 1st October 1932. He now belonged to my specialist staff" [3].

Von Braun was swept up in the excitement of this new challenge, which he actually saw as a way of inducing the German Army to give much-needed funding to the Rocket Society for their experiments, and claimed that he was not overly concerned about the possible military consequences of his work. "Besides, in 1932 the idea of another war was absurd," he would stress many years later. "The Nazis were not then in power. There was no reason for moral scruples over the use to which our researches might be put in the future. We were interested in only one thing – the exploration of space" [1].

CREATING THE ROCKETS

With little to guide them other than their own experience and the published works of American rocket pioneer Dr. Robert H. Goddard, combined with the theories of Romanian–German scientist Hermann Oberth, von Braun's team began to design and develop a rocket motor that would eventually produce an amazing 650 pounds of thrust. Known as the A-1 (*Aggregate-1*, or Prototype-1), it embodied a liquid-oxygen fibreglass tank located inside a larger tank filled with alcohol, and was equipped for flight stability with a 70-pound flywheel gyroscope secured in the nose section. It was a problematic beast, however, with the awkward fuel arrangement making it prone to exploding without warning.

Dornberger later wrote that the A-1 was too top-heavy. "The centre of gravity lay too far ahead of the centre of pressure," he reflected. Undaunted, the team then worked on a new design. "So far as the motor is concerned, it was a replica of the A-1, but the gyroscope had been moved from the nose of the missile to the middle, between the oxygen and alcohol tanks" [3].

Higher and heavier

By early December 1934, two modified 3,300-pound A-2 rockets nicknamed Max and Moritz were ready for launch-testing. These were fired separately from the North Sea island of Borkum, and, while both were successful, one managed to soar over 7,000 feet into the sky to the delight of the development team. Still, Dornberger could not convince Adolf Hitler that this was a viable weapons programme, and the facility had a difficult time scraping together sufficient materials and resources.

Work soon began on the A-3 rocket, which required a vastly more ambitious design than the earlier missiles. The most immediate problem they faced was that there seemed to be no way to stabilise and control this much larger and heavier rocket's trajectory during the critical first few seconds after lift-off, when the test missiles were sluggish, unstable, and in the early throes of acceleration. Work continued on the rocket's internal guidance system and movable surface controls were added to the fins, but to von Braun's frustration these initially proved of little post-launch benefit. When the motor ignited it created incredible dynamics that were difficult to harness. The slightest instability would cause the rocket to keel over and explode before it could gather sufficient speed to claw its way into the sky.

Von Braun and Peenemünde

In April 1937, following an extensive search for a suitable, remote test-firing range, von Braun recommended the entire top-secret operation be relocated to the small fishing village of Peenemünde, situated beyond the mouth of the Peene River in the northwest peninsula of Usedon, which forms part of the German Baltic coast. The peninsula was also densely forested, which would help conceal test stands and workshops from prying Allied aircraft. Von Braun would become the facility's technical director.

Through patient development and testing, the rocket team slowly began to overcome several problems associated with the A-3, at the same time setting to work creating the even more ambitious *Fernraketen* (long-range rocket) A-4, a single-stage rocket that weighed more than 12 tons and stood over 46 feet high. This rocket came about as a result of the Ordnance Department instructing von Braun and his team to produce a ballistic weapon capable of carrying a 1-ton payload over a distance of between 150 and 200 miles. To achieve these goals the A-4's motor, burning the same mixture of liquid oxygen and alcohol, would have to develop 60,000 pounds of thrust. Solving the vexing guidance control problems now became even more of a priority.

Developing the A-5

In a somewhat novel approach, von Braun's team constructed a smaller-scale test model of the A-4, known in turn as the A-5. Similar in size to the underachieving A-3, it employed the same rigid but lightweight geodetic framework used in the massive Zeppelin airships. Changes to the missile's control system also meant that it could be steered in the first few seconds of flight by movable graphite vanes located in the rocket exhaust. By 1939, a series of the A-5s had been produced and flight-tested over the Baltic, reaching heights of around 35,000 feet. After extensive tests on these smaller prototypes, Von Braun's rocket team went into full-time development of the A-4, although project engineers would soon realise that creating an operational, long-range rocket was a far more difficult task than they had realised. Despite numerous setbacks, the first models were ready for test-firing by the spring of 1942.

Birth of the "terror weapon"

The first test flight of a *Fernrakete* A-4 would take place from Test Stand VII at Peenemünde on 13 June 1942. Meanwhile the rocket had been given an alternative designation: V-2. At first this stood for *Versuchmuster 2*, or Prototype 2; however, for propaganda purposes this would later be modified to *Vergeltungswaffe 2*, or Retaliation Weapon 2 [4]. To their chagrin, the test flight ended in abject failure 3 miles high when the rocket became unstable, rolled around its long axis and exploded, sending the shattered remains cartwheeling into the Baltic Sea a mile from the launch site.

On 16 August a second rocket also exploded at an altitude of 7 miles after its guidance system froze. Only on 3 October, with the launch of the third V-2, did the rocket come close to achieving the desired optimum performance. Although flying on a steeper trajectory than planned, the missile managed to stagger to an altitude of 50 miles and achieve a satisfactory range of around 120 miles before crashing into the sea 5 minutes after lift-off.

On 7 July 1943, Adolf Hitler ordered Dornberger, now a Major-General, to a meeting at his headquarters in the East Prussian town of Rastenberg to discuss the V-2 rocket. Dornberger took von Braun with him, as well as his head of the Department of Instruments, Guidance and Measurements, Ernst Steinhoff. Hitler was accordingly briefed on the successful test flight and potential of the V-2, together with his chief of staff of the armed forces, Field Marshall Wilhelm Keitel; the army chief of armaments, General Walter Buhle; and minister of armaments, Albert Speer. Previously indifferent and unreceptive to the potential of a guided ballistic missile programme, Hitler ordered that the rocket go into immediate full production, now giving the launch centre a top-priority ranking. "Manpower and material streamed in," wrote Dornberger. "We accomplished in weeks what would have taken months and years" [3].

The launch of a V-2 from Peenemünde while other missiles are being prepared for firing. (Photo: U.S. Army)

However, optimism at Peenemünde would soon turn to dismay when the next 28 test-firings failed for a variety of reasons: premature engine cut-offs, tumbling, flying too shallow, falling back to Earth – it was not until 12 August 1943 that the first truly successful launch was recorded.

Severe setbacks and relocation

Just as the Peenemünde team had begun to taste success, the Allies identified the launch facility as a priority target. On the evening of 17/18 August they sent in heavy bombers to flatten the launch centre. Under a full moon, 596 aircraft of Bomber Command dropped nearly 1,600 tons of bombs on the experimental factory at Peenemünde, following up with another heavy bomber raid on 25 August. The factory production of the V-2 was quickly relocated to a secret underground plant at Mittelwerk, near Nordhausen in Thuringia. Here slave labour from Dora, a sub-branch of Buchenwald concentration camp, was set to work at the end of August 1943. Their task was to extend an abandoned, hillside gypsum mine a mile into the mountain, and carve out ample room for a new underground missile plant.

A wooden mock-up of a V-2 rocket is shown loaded on a *meillerwagen* hydraulic transporter in February 1942. (Photo: U.S. Army)

Other V-2 production plants were soon undergoing construction at sites near Berlin, Friedrischshafen and Vienna, but Allied intelligence gathering was one step ahead by now, and persistent aerial bombing eventually meant that the Nordhausen Mittelwerk factory became the sole location for V-2 production. By 1945, nearly 900 of the deadly missiles were being constructed every month at the plant, mainly by slave labourers literally forced to work to death in appalling circumstances.

Once completed, the rockets were transported by rail car to secret storage areas. When called upon they were towed behind half-track vehicles to a mobile launch site together with fuel trucks and other support vehicles. The rockets had been transported on a *meillerwagen*, a piston-driven hydraulic vehicle capable of raising the rockets to a vertical position. Once the rockets were ready, a launching platform would also be raised, assuming the full weight of the missile. Fuelling would then take place, after which the cradle clamps would be released, while the *meillerwagen* and the other support vehicles were driven a short distance away for safety. The launch platform was simple in design and easily transportable, featuring a 10-foot rotatable ring housed in a square, angle-iron framework that was supported at its corners by jacks. With its convoy of support vehicles, the *meillerwagen* was capable of delivering a V-2 to a selected site and launching the missile within a 4- to 6-hour timeframe.

From Peenemünde to prison

But von Braun's brilliant career in rocketry was about to take a very unexpected turn. In February 1944, he was summoned to appear at Gestapo Headquarters in East Prussia for urgent talks with Heinrich Himmler, who controlled the SS and Gestapo and was one of the most powerful and dangerous men in Germany at that time. Himmler, who had extensively toured the facilities at Peenemünde and knew of the missile's potential, asked von Braun if he might consider abandoning the Army's V-2 programme in favour of joining his staff and developing similar weapons for the Gestapo. In a gutsy move, von Braun declined the invitation.

Three weeks later von Braun and several of his most important engineers were forcibly arrested and thrown into a Stettin prison, charged with sabotage of the V-2 project and a possible defection to British hands. A furious Walter Dornberger spoke with Himmler, but his complaints were curtly dismissed. He then contacted Field Marshall Wilhelm Keitel, asking the reason behind von Braun's arrest. Keitel told him that von Braun and some of the other men were reported to have been overheard at a function stating it had never been their intention to make a weapon of war out of the rocket. According to Dornberger's memoirs, Keitel said it seemed they had worked "at the whole business of development only in order to obtain money for their experiments and the confirmation of their theories." He added that "their object all along had been space travel ... and consequently have not applied their whole energy and ability to production of the A-4 as a weapon of war" [3].

Armed with this information, Dornberger then made a direct appeal to SS General Heinrich Müller, chief of the Gestapo, and an active participant in some of the most heinous crimes committed by the Nazi regime. Dornberger stated that the charges against von Braun and the others of sympathising with the British to the point of treason were entirely false, and their expertise, especially that of von Braun, was absolutely crucial to the ongoing development of the V-2. He went so far as to personally vouch for their loyalty – a dangerous move. Müller finally agreed that the charges might have been maliciously fabricated, and 2 weeks after their arrests the men were released from prison for a 3-month probationary period in order to continue their vital work.

LEARNING THE SECRETS

As Allied and Soviet troops forged towards Nazi Germany in the weeks following the Normandy landings, it became increasingly evident that the war was coming swiftly to an end. With Germany's home defences showing increasing signs of crumbling in the face of these overwhelming forces, Hitler began looking in desperation to the V-2 as a last-ditch means of defiance.

A rocket named Waterfall

Although the V-2 was a single-stage missile with an integrated system of guidance, work had been continuing on refinements and variants to make it an even more potent weapon. One such mutation was the *Wasserfall* [Waterfall], an anti-aircraft missile capable of carrying a 200-pound warhead, guided by ground-based radio signals that would also be used to detonate the warhead. While the *Wasserfall*, and variants such as the *Schmettering*, *Enzian* and X-4 would all be flight-tested, several factors such as materials and fuel availability fortunately combined to keep these missiles from fully operational status.

In one stroke of good fortune in mid-1944, the Allies gained some vital information on Hitler's rockets through a wayward *Wasserfall* development V-2. Launched from Peenemünde on 13 June, it had veered completely off course and strayed towards the Swedish coast, ending its troubled flight in an airburst over the Kalmar area. Parts of this missile were recovered and quickly flown across to England for expert evaluation.

That same year, after the missile base at Peenemünde had been bombed by Allied forces, the Polish Resistance reported that a new launch facility had been set up at Blinza in eastern Poland, where tests of the V-2 would continue. Whenever a V-2 launch took place, members of the Resistance would immediately rush to the impact site. However, their recovery plans were generally thwarted by heavily-armed German troops who would reach the landing zone first, secure the area and remove the rocket's essential components for later examination by guidance technicians.

The Resistance and the rocket

Finally, on 20 May 1944, a V-2 fell from the skies and plunged into the muddy river Bug near Sarnaki nad Bugiem, a village 80 miles east of Warsaw. This time members of the Resistance were the first to arrive. They found the rocket sticking upright in the muddy water. With only minutes in which to act, the Poles hurriedly hauled the partially-submerged rocket down and rolled it deeper into the muddy river, after which they herded in some nearby cows into the river, effectively muddying the water even more and effectively screening the missile from view. Following an extensive but fruitless hunt, the Germans finally abandoned their search, allowing the Poles to retrieve and dismantle the missile without interruption. An expert team of engineers working under partisan Jerzy Chmielewski removed all the missile's crucial parts and later made some detailed drawings [5].

Once they had learned of this operation, and in cooperation with the Polish Resistance, the Allies set in motion an audacious plan to fly an unarmed DC-3 Dakota transport plane from Brindisi, Italy to German-occupied Poland. Here it would make a night landing on the evening of 25/26 July at a makeshift airfield, where the four-man crew would pick up the parts and drawings and fly them straight back to Brindisi.

On the evening set for Operation *Trzeci Most* (Third Bridge), a large Resistance task force was standing by the field, accompanied by 20 suitcases packed with 920 pounds of V-2 parts, technical drawings and other vital intelligence on the rocket, waiting for the Dakota. Suddenly, the operation took an unwelcome turn when two German FW190 fighters swept in and landed on the field. The waiting Resistance fighters could only nervously watch and hope. They did not know if they had been betrayed, and the Germans were waiting for the transport aircraft to arrive. Then, with only a few hours remaining before the Dakota was scheduled to land, there was a sudden flurry of activity at the field. To everyone's relief the engines on the German fighters sputtered into life, and they soon took off, quickly vanishing in the growing darkness.

Just before midnight the sound of another aircraft could be heard approaching the field, and torches were switched on at each corner of the landing strip as the Dakota roared in. Once it had pulled up, the V-2 parts and paraphernalia were hastily loaded on board together with five Polish Resistance operatives, including Chmielewski, and the pilot prepared to take off. Then, all the carefully-laid plans went awry as the aircraft's tyres became bogged down in mud from some recent heavy rain. The pilot desperately ran the DC-3's two engines at full throttle, creating a hideous din that echoed across the countryside. There was a German post located just over a mile away, which had to have been alerted.

The audacious plan now seemed doomed, and there was even talk of burning the Dakota so it would not fall into German hands. Then, there was sudden salvation. Some local farmers had heard the racket and made their way to the field. Wanting to help, they tore wooden palings from nearby fences and laid them on the field, scrabbling away at the muddy ground with shovels and bare hands, hoping they would take the weight of the aircraft. Once a number of planks had been laid in front of the two main wheels the pilot tried to coax the aircraft forward once again, and this time it worked. Sixty-five minutes after landing, the transport plane had gathered sufficient momentum to take off. It vanished into the night sky just as the German forces arrived, running into a Resistance ambush. Within 3 days the parts and the drawings were safely in London [6].

Through this and other gathered intelligence, the Royal Aircraft Establishment (RAE) was able to interpret the information and assess the likely maximum range of the rocket, the warhead type and weight. They passed this information along to the Big Ben Committee, which was directly involved in disseminating intelligence on the existing and potential threat of Hitler's V-weapons. The information also went to the Bodyline and Crossbow Organisations, both of which supported counter-strike operations against V-weapon sites. It was a desperate race against time.

The devastation begins

On 6 September the first combat V-2 carrying a warhead struck Paris. Then, just two nights later at 6:43 p.m., another V-2 fired on London exploded in suburban Chiswick. Sixteen seconds later another fell and detonated in Epping. Six months of terror and destruction had begun.

The Germans now unleashed a massive bombardment on London and Paris with the first of 4,000 missiles they would eventually fire against Allied targets. With a range of around 260 miles, and flying faster than sound, the V-2 was a terrifying weapon.

The devastation caused by an exploding V-2 can clearly be seen in this photograph taken in Antwerp, Belgium, on 13 October 1944. (Photo: A-4/V-2 Resource Site)

For those in their path, there seemed to be no form of defence against these silent intruders, which fell unpredictably and without any warning to those on the ground. The only real saving grace for the Allies was one of logistics; the V-2 programme was not only limited by the rate at which the missiles could be produced, but fuel was also becoming an increasingly scarce commodity due to Allied bombing and blockades.

OPERATION PAPERCLIP

By early 1945 the battered German Army was retreating, and the V-2 arsenals, together with their mobile launchers, had likewise been driven farther inland. Soon, they were so far out of range of their intended targets they could no longer reach the prime cities of Paris, London and Antwerp, and had become totally ineffectual as a means of bombardment.

A brutal bombardment

At 4:45 p.m. on 27 March 1945, the final V-2 fell at Orpington in Kent. British records later indicated that 1,115 of these missiles had exploded on British soil, but fortunately a greater precision of trajectory had eluded the German scientists and engineers. Although the city of London was a large target, fewer than half the missiles – 501 in all – actually fell within the London Civil Defence Region. According to a 1948 report tabled by Air Chief Marshal Sir Roderic Hill: "2,511 people had been killed and 5,869 seriously wounded in London and 213 killed and 598 seriously injured elsewhere" [7]. Nevertheless, of the 1,359 V-2s fired at London, German records indicate that only 169 actually failed en route. Under vastly different circumstances that would have been regarded as an impressive engineering record.

Means of escape

During January and February of 1945, Peenemünde was being evacuated in stages as Russian forces moved ever closer. "The Russian Army was approaching from the east," von Braun would later write. "It was about 100 or 80 miles from Peenemünde; so close that we could already hear the artillery fire at night. It was very obvious to me and my associates that the war was lost, and the decision whether we wanted to wind up on the East side or on the West side had to be made before the Russian Army arrived" [8].

Not surprisingly, the Peenemünde personnel were terrified of being captured by the Russians, whose reputation for retaliatory cruelty preceded them. As one member of the rocket team stated at the time: "We despise the French; we are mortally afraid of the Soviets; we do not believe the British can afford us; so that leaves the Americans" [9]. They unanimously opted for surrender to the U.S. forces.

When one of several confusing orders came through ordering the evacuation or dispersal of nearly 3,000 Peenemünde associates and their families, von Braun was well prepared. He chose the order that best suited his plans, and then assembled 525 evacuees together with tons of sensitive documents relating to rocket design, construction and testing, ready to travel by truck, train and any other means to Bleicherode, a town in the Harz Mountains a short distance from the underground Mittelwerk V-2 production plant.

In a hazardous exodus carrying tenuous evacuation orders through the quickly-crumbling Reich, von Braun headed south with his group and a 13-year accumulation of missile documentation. Towards the end of their journey, and while all but a couple of his people headed for Bleicherode, von Braun quickly made arrangements for the precious documentation to be secretly buried several miles to the north, in an abandoned mine shaft near Goslar. Following this, he completed the journey to Bleicherode through chaos and a number of air raids.

Meanwhile, knowing of the rapidly-escalating collapse of the Third Reich, the SS commander at Bleicherode, General Hans Kammler, was desperately hatching his own plans for survival. Several concentration camps fell under his administration, and he realised he would become accountable if he was captured. Once the vast convoy of

Peenemünde evacuees had arrived, Kammler ordered 450 key personnel, including von Braun, to relocate to an empty army camp situated at Oberammergau in the foothills of the Bavarian Alps, where they would be relatively safe from Allied bombing.

Von Braun was rightly suspicious of this move, as Kammler's hidden agenda was to hold these valuable men hostage and then personally surrender them as a means of

A broken left arm was no deterrent for von Braun in his dramatic bid to surrender to Allied forces. Walter Dornberger (in hat) is standing next to him. (Photo: U.S. Army)

self-preservation. Accordingly, the men were transported to Oberammergau where, instead of living in the camp, they took up lodgings in several local villages.

By this time von Braun was hampered by a badly-broken left arm, suffered in a car accident in mid-March, which had been set in a cast and which he wore in a sling. Soon after their arrival he received a welcome surprise when General Dornberger joined them.

Death in a Berlin bunker

In his post-war book, *V2*, Dornberger wrote that his Peenemünde executive staff had received orders to evacuate to Oberammergau on 3 April, "as the American tanks advanced through Bleicherode towards Bad Sachsa. From that time onwards we were accompanied by Security Service men. I suspected what that meant. Were we to be used as hostages in armistice negotiations? Or were we to be prevented from falling into enemy hands?" [3].

Several days later, on 30 April, word reached the group that Hitler had committed suicide in his Berlin bunker, and von Braun set plans in motion for an immediate surrender to the Allies.

On the rainy afternoon of 2 May, von Braun and six other men met outside the Haus Ingeborg, an inn located in the town of Oberjoch, ready to conduct the surrender on behalf of the entire group. His engineer brother Magnus, who had been in charge of gyroscope production at Mittelwerk, was one of the seven, as was Walter Dornberger. The seven men then clambered into three cars and drove toward the Austrian village of Schattwald, which they knew was under American occupation. As they approached Schattwald on the morning of 2 May, Private First Class Fred Schneikert stepped out in front of the small convoy and waved them down, his M-1 rifle at the ready. Magnus von Braun spoke English, and once he had convinced the soldier of their intention they were taken to the local Allied commander.

Once officials became fully aware of the importance of these key rocket personnel, arrangements were quickly made to move them to a captured German army barracks at Garmisch-Partenken, although several months would pass while they awaited their fate. Unfortunately for Dornberger he was regarded as a person of interest beyond being a rocket scientist, and he was transferred to British forces. They imprisoned him for 2 years before his ultimate release to the United States.

Beating the Russians to the spoils of war

Major General Hugh Knerr was the deputy commander of the United States Air Force in Europe, and once he had established the identity and importance of the Peenemünde group, he made a recommendation to Lieutenant General Carl Spaatz, USSTAF commander, that the Army Air Force "make full use of established German technical facilities and personnel before they are destroyed and disorganised" [10].

With Hitler dead and conquering troops pouring into Germany, the Allies knew that German rocket development would be a trump card in post-war competition.

Allied commanders also realised that only the Americans and the Russians had the resources and the determination to mount a major rocket development programme.

In light of what von Braun revealed to him, Knerr hurriedly organised a covert mission group which was despatched to the mine near Goslar, 200 miles to the north, where the Peenemünde documents were disinterred and rushed back to the American zone. They were just ahead of Soviet forces, which then swept into the Goslar area soon after.

Knerr then wanted to send the German scientists and their families to the United States, where they would be able to continue their work in rocketry under the auspices of the U.S. government, without the constant fear that they or their families could be kidnapped by the Russians. It was a far-sighted strategy on Knerr's part that would soon prove of immense importance. He would later write that the occupation of German scientific and industrial establishments "has revealed the fact that we have been alarmingly backward in many fields of research. If we do not take the opportunity to seize the apparatus and the brains that developed it, and put the combination back to work promptly, we will remain several years behind while we attempt to cover a field already exploited" [11].

A covert operation

A purposeful but desperate scramble was already well under way as the conquering Allies sought out many of Nazi Germany's technological secrets, plundering as much equipment and expertise as they could round up ahead of the Soviet thrust into Germany. What they unearthed from the rubble of the Reich would astonish them. Hidden away in laboratories, underground factories and workshops were once top-secret guided missiles and supersonic rockets, jet aircraft and a number of deadly, cutting-edge technical and biological weapons.

Of paramount importance, and under the direction of Colonel Holger Toftoy, who would later play a large role in von Braun's life, they narrowly beat the Russians in the race to a mammoth technological coup. Following his directions they located the vast subterranean complex at Nordhausen, and in a massive undertaking removed all the V-2 missiles and tons of parts from an area that Soviet forces would eventually claim as being in their zone of occupation.

Behind General Knerr's words was the genesis of what became known as Operation Paperclip, which would deny the Soviet Union these intellectual resources by quickly spiriting von Braun and more than 700 others out of Germany. The unusual name for this operation originated from the simple fact that those engineers and technicians selected to travel to the United States were distinguished by paperclips on their files, which combined their scientific papers with regular immigration forms.

The aim of Operation Paperclip was simply stated: "To exploit German scientists for American research and to deny these intellectual resources to the Soviet Union." Moving quickly in the national interest, President Truman authorised Operation Paperclip that August, although it must be recorded that he expressly forbade the participation of anyone found to "have been a member of the Nazi party and more than a nominal participant in its activities, or as an active supporter of Nazism or militarism" [11].

Operation Backfire

Shortly after the German surrender in 1945, and while procedures for sending von Braun and the other German experts to the United States were being sorted out, the British hurriedly arranged for a demonstration of the captured rocket technology. A number of reassembled V-2 rockets would be test-launched from the site of a former German Navy artillery range, located 10 miles southwest of the harbour town of Cuxhaven, as part of a military research project code-named Operation Backfire.

The operations site for Operation Backfire was located within a small birch forest at Altenwalde, near Arensch. It was here that a host of captured German scientists, engineers and other Peenemünde rocket personnel had been assembled in two groups for detailed interrogation on the rockets.

A captured V-2 is prepared for an Operation Backfire launch from a road-based motorised unit at Cuxhaven, September 1945. (Photo: U.S. Army)

Thirty of these rockets and assorted components for dozens of others had been captured by liberating forces and transported to the artillery range for storage and handling. Most were destined for overseas transportation and later scrutiny by rocket engineers and technicians. Others were undergoing preparation for a limited but comprehensive evaluation of the rocket's entire systems. This British-run operation would oversee controlled, radar-tracked test-firings of the V-2 out over the Baltic and North Seas [12].

The rockets to be used in these test-firings were painted in a black-and-white checkerboard pattern for photographic analysis of their flight characteristics, while test launch facilities were hastily constructed or upgraded by a large contingent of British troops and engineers.

V-2 rockets launched by the Allies

Captured Peenemünde rocket engineer and designer Wernher von Braun was one of those brought to Cuxhaven to assist with these tests on behalf of the British Special Projectile Operations Group (SPOG), an adjunct to the Air Defence Division.

General Walter Dornberger was also transported to the area. However, despite their involvement in the German rocket programme and these post-war test launches, both men were still under close scrutiny, and neither would be permitted to attend the actual firings of the missiles.

The first attempted test launch on 1 October 1945 fell victim to a faulty igniter unit and was scrubbed. The following day, witnessed by British, American and Russian military observers, another V-2 was successfully launched and arced out over the Baltic Sea. Two more launches would be conducted, with the last of the test rockets plunging into the planned target area in the North Sea on 15 October [13]. Having encountered very few problems of significance, Operation Backfire was considered a success and subsequently shut down. All of the data and photographs were later sent for analysis by rocketry and weapons experts. Operation Backfire, for all its problems, successfully completed a comprehensive technical documentation of the formidable rocket, its fuel systems and composition, and support procedures.

An agreement with the Allies

Still in Germany, von Braun was sent to a U.S. Army detention centre where he was interrogated for several weeks about his involvement in the missile programme. Eventually, Colonel Toftoy explained that their questioning was at an end and asked if von Braun would consider becoming a loyal citizen of the United States in order to continue his work. Some researchers would even suggest that it was von Braun himself who had proposed this to his captors. "I said I would try," von Braun later observed [1].

In his book, *To a Distant Day: The Rocket Pioneers*, author Chris Gainor sums up the life and overall career of this extraordinary man. "It is impossible to imagine the space age of the twentieth century without Wernher von Braun," he writes. "It was his unique fate to overcome his participation in the century's darkest deed to make possible its most sublime achievement" [14].

WORKING FOR THE AMERICANS

On 18 November 1945, the first Germans landed on American soil. Included in their number were most of the V-2 control and guidance specialists, who had been offered contracts to work in America.

A close-won race

Wernher von Braun's mentor, Professor Hermann Oberth, was another of those who would enter the United States under the auspices of Operation Paperclip, and he would assist von Braun's Peenemünde team in assembling V-2 rockets from captured components. Three years after entering the United States he was permitted to return to Germany and resume his teaching profession, but he would later serve as an advisor to the U.S. Army Redstone Arsenal in Huntsville, Alabama, from 1955 to 1958.

When reflecting on the success of Operation Paperclip, and the surrender of von Braun's Peenemünde team, one fact remains paramount: had von Braun and almost all of his top scientists and engineers been captured by Soviet forces and taken to Russia, that country would almost certainly have won the later Space Race, and perhaps changed the entire outcome of the Cold War.

Sent to the steppes

Meanwhile, in the last months of the war, the Soviet Union had been equally eager to appropriate German personnel and technology for their own purposes. Soviet "trophy battalions" combed through a ruined Germany in search of industrial equipment and materials. When Soviet forces captured the German rocket facility at Peenemünde in May 1945, they were discouraged by how thoroughly the Allies had cleared it of materials and personnel. However, in quick succession, they uncovered the underground rocket production facility at Nordhausen, a rocket engine test site at Lehesten, and the last headquarters of Wernher von Braun at Bleicherode.

Efforts to gather rocket parts and equipment from these and other sites met with considerable success, and a rigorous programme was begun to test and replicate critical components and recreate complete engineering documentation for the German rockets. Some 4,000 German rocket and technical personnel would eventually be rounded up in the sweep of enemy territory. They would play a critical role in helping the Russians to rebuild the A-4 rocket and explore advances on that design.

One major rocket designer claimed by the Soviets was Helmut Gröttrup, the assistant to Dr. Ernst Steinhoff, director of the Guidance, Control and Telemetry Laboratory. The previous year he had been one of those placed under arrest along with von Braun, and was later released from prison following Dornberger's representations to Heinrich Müller.

A U.S. Army photo of an A-4 (V-2) scrap heap found outside the vast underground rocket production facility at Nordhausen, Germany. Although the U.S. captured the bulk of the German rocket personnel and materials, the Soviets uncovered repositories of parts, unlooted facilities, and a cadre of engineers and technicians who would play an important role in launching the Soviet rocket programme.

Gröttrup was unhappy about the possibility of being sent to the United States for an extended period without his wife, so they decided to stay in Germany, making their way back into the Russian zone. Later, working under the command of Lieutenant-General Lev Gaidukov, who headed a Soviet group responsible for acquiring German rocket technology and engineers in the Russian zone after the war, Gröttrup would lead the Soviet-affiliated missile programme. By mid-1946 he was in charge of over 5,000 workers at the Zentralwerke factory in Mittelwerk, charged with reconstituting the V-2 programme [7].

Some 30 rockets had been assembled by September and were ready for test-firing at a German launch test facility at Lehesten, a small mountain town in the heart of Thuringia. These engine tests would be carried out under the direction of Colonel Valentin Glushko, who would later become a key figure in the Soviet manned spaceflight programme.

An engineer named Korolev

One Russian rocket expert Gröttrup particularly enjoyed working alongside was Gaidukov's knowledgeable deputy on the Soviet Special Commission, Sergei

In this 1946 photo, Sergei Korolev (right) is pictured in Germany with Georgi Tyulin. The effort to seize German rocket assets, recruit German personnel and recreate the A-4 rocket brought the best and brightest Soviet engineers and specialists to Germany immediately following the war. Many of these individuals would later play leading roles in the Soviet rocket development and space programme. (Photo: NASA)

Korolev. Impressed with the man's professionalism and expertise, Gröttrup also noted the deep respect in which Korolev was held by his fellow engineers.

Within a few years, Korolev would be serving as chief designer of long-range missile development. But in 1945 he was just one of a number of Soviet engineers and technicians who had been working with aviation and rocket design since the 1920s. Earlier that year, Korolev had been awarded a Badge of Honour for his work on the development of rocket motors for military aircraft. Then, hastily commissioned as a colonel in the Russian army, he was shipped off to Germany to join in the effort to study and test the V-2 rocket.

On 21 October, Gröttrup attended a meeting in Bleicherode headed by Gaidukov, which would discuss possible improvements to the V-2. The following morning, to his complete surprise, he received a hysterical call from his wife Irmgard. She told him that several thousand German technicians, along with all 200 Soviet rocket engineers and their families, had just been woken by troops banging on their doors and bluntly informed that they were to prepare for evacuation to the Soviet Union on 12 hours' notice. They were required to be on a train to begin their journey the following day. Ahead of them lay a 7-year assignment on the Russian steppes.

According to the book *The Rocket Team* by Frederick Ordway III and Mitchell Sharpe, the Germans received a surprisingly good reception when they finally arrived in Moscow on 28 October, although some would soon find themselves working under poor and sometimes volatile conditions.

On the arrival in the USSR, the Germans from Zentralwerke were split into two approximately equal groups. One was sent on to the island of Gorodomlya in Lake Seliger, some 150 miles northwest of Moscow. Conditions on the island were incredibly primitive, and the island was in a region that had seen some of the bitterest fighting on the Russian front only four years earlier, in January 1942. As a result, the Nemets (Germans) were received with outright hostility by the local Russians. The other group, the Gröttrups among them, was settled in the northeastern section of Moscow, near Datschen, in relative comfort [7].

By early 1947, all rocket personnel and technology had been transferred from Germany to locations in the Soviet Union.

Russia and the V-2

Two years later, on 30 October 1947, the Soviet Union would launch its first V-2 rocket, designated by them as the R-1, from the new test range at Kapustin Yar. It flew 185 miles downrange before plummeting back to Earth, smashing down right on target. The missile race – and the subsequent Space Race – had begun. Soon, many captured V-2s would be launched from newly-constructed launch sites on both sides of the world. And many would carry animals as unwitting, exploratory precursors to the human flights that would follow.

These fledgling biological flights would give rise to the new science of space medicine, ushering in an era when animals would take the lead in establishing the physiological threshold of rocket flight.

REFERENCES

[1] Dennis Piszkiewicz, *Wernher von Braun: The Man Who Sold the Moon*, Praeger, Westport, CT, 1998
[2] Erik Bergaust, *Reaching for the Stars*, Doubleday, New York, 1960.
[3] Walter Dornberger, *V2* (translated from the German edition, *V2: Der Schuss ins Westall*), Hurst & Blackett, London, 1954.
[4] Wikipedia, on-line encyclopaedia, *V-2 Rocket*. Website: *http://en.wikipedia.org/wiki/V2_rocket*

[5] Yahoo Groups, *Polish Forces during World War II: Operation Wildhorn III.* No author given. Website: *http://groups.yahoo.com/group/polishforces*

[6] Janusz Zurakowski, *Kazimierz Szrajer and the German Rocket, WW II, July 1944.* Website: *http://www.zurakowskiavroarrow.homestead.com/KazimierzSzrajerEng~ns4.html*

[7] Frederick Ordway III and Mitchell R. Sharpe, *The Rocket Team*, Apogee Books, Burlington, Ontario, 2003.

[8] Hearings, Senate Preparedness Subcommittee, *Inquiry into Satellite and Missile Programs*, chaired by Senator Lyndon B. Johnson, 25 November 1957.

[9] Walter A. McDougall, *The Heavens and the Earth: A Political History of the Space Age*, Basic Books, New York, 1985.

[10] David S. Akens, Historical Account, *Origins, Marshall Space Flight Center*, NASA, MSFC Huntsville, AL, 1960.

[11] David Darling, *The Encyclopedia of Astrobiology, Astronomy and Spaceflight: Operation Paperclip.* Website: *http://www.daviddarling.info/encyclopedia/P/Paperclip.html*

[12] Michael Grube, *Operation Backfire: Versushskommando Altenwalde* (translated from the German). Website: *http://www.lostplaces.de/backfire/index.html*

[13] A-4/V-2 Resource Site, article *Operation Backfire* (no author given). Website: *http://www.v2rocket.com/start/makeup/backfire.html*

[14] Chris Gainor, *To a Distant Day: The Rocket Pioneers* (to be published), University of Nebraska Press, Lincoln, NE, 2007.

2

Holloman and the Albert Hall of Fame

Just to the south of Trinity Site, where the first nuclear explosion was triggered, lies Holloman Air Force Base, and the White Sands Proving Ground from which the United States took its first feeble steps into space. Here the rumbling of rocket motors, the piercing high-pitched scream of jet engines, the cacophonous background noises of the space age became ordinary.

It is a forlorn waste, but across the vast, unfertile expanse there exists a sense of space and grandeur matched only by the deep blue of the sky above which seems to beckon man into the greater cosmic wilderness. Under that sky, in the dust-blasted cinder-block buildings of the Aeromedical Field Laboratory at Holloman, we were answering that call, slowly, tentatively edging upward, away from the earth. And our longest step was close at hand.

Lt. Colonel David G. Simons with Don A. Schanche, *Man High*

The inaugural firing of a V-2 rocket from American soil took place on 16 April 1946 at a remote military site near Alamogordo, New Mexico. Newly painted, the rocket looked surprisingly striking in a gaudy white-and-black checkerboard pattern that would enable technicians to study launch film post-flight to determine the amount of spin generated during the first 20 seconds of the ascent. The missile had been meticulously checked and prepared for what should prove to be a defining moment in American rocket research.

AN INAUSPICIOUS START

An anxious wait was nearly at an end for the assembled technicians, military brass and project scientists as the countdown dwindled to zero. A sudden flurry of sparks erupted from the foot of the rocket, dancing outwards as they hit the blast deflector. Within moments these ignition sparks had coalesced with burning liquid fuel to

Loading alcohol fuel aboard a V-2 before firing from the U.S. Army Ordnance Proving Ground at White Sands, New Mexico. Note the use of a ladder to top up the fuel. (Photo: U.S. Army)

produce a dazzling, golden torrent of flame. Thick clouds of smoke began billowing across the concrete launch pad, while the sound of rolling thunder filled the air. Full combustion took 3 seconds, after which the pad restraints fell away, unleashing the rocket to begin its ascent into the bright blue sky. After seeming to balance itself momentarily, the V-2 rose and began surging away from the pad, consuming 33 gallons of liquid oxygen and ethyl alcohol fuel for every second of flight.

Punching a hole in the sand

Accelerating rapidly, the 40-foot missile tore a passage through the sky, leaving a thick, white vapour trail in its wake. Then, when it was 3 miles up, the collective exultation of those on the ground turned to dismay as a stabilising fin broke loose and the rocket began lurching off course. A hurried signal transmitted from the ground halted the flow of fuel to the engine to prevent an explosion. Now, completely lacking control and thrust, the errant missile tumbled around wildly before finally plunging back to earth, where it slammed into the gleaming white desert sand.

Although successful in some ways, the maiden flight of a V-2 from the American mainland was an inauspicious start to a programme that would quickly recover to set the United States on a magnificent but difficult course for the stars.

THE COMING OF THE MISSILES

On 6 February 1942, 2 months after the United States had been suddenly thrust into the Second World War, work began on converting vast tracts of south–central New Mexico wasteland into a sprawling construction site. No one could know it then, but within 5 years this unwanted stretch of scrubby desert would be transformed from a military training zone into a vital launch and operations centre for the burgeoning science of rocket development and biomedical research.

America takes over

The Alamogordo Bombing and Gunnery Range had originally been conceived as a fully-operational wartime service base for the British Overseas Training Program. This programme, as the name suggests, would have allowed uninterrupted training for British forces aircrew well away from the tumult of the European conflict. The catastrophic events at Pearl Harbor were followed by the immediate entry of the United States into the conflict, and this caused many plans to change. One resulted in Britain's military leaders abandoning any thoughts of conducting overseas aircrew training in New Mexico. Their U.S. counterparts subsequently realised the potential of the Tularosa Basin site as an ideal place to train American military personnel, and construction work began on a bombing and gunnery range.

Within months the range would become home to the Alamogordo Army Air Field. Here, over the next 3 years, intensive training would be undertaken by American crews flying heavy bombers such as the B-17, B-24 and B-29, prior to their combat deployment in either the Pacific or European theatres of war [1].

Thunder across the desert

Alamogordo, whose Spanish name translates to "fat cottonwood", was once little more than a remote ranching and farming community. When America found itself thrust into the Second World War, the majority of local farms were huddled around what had begun life as a railroad junction important to the transport of mountain lumber, located 10 miles east of the Tularosa Basin and nestled at the sprawling feet of the towering, blue–grey Sacramento and San Andreas Mountains.

The most notable geological feature of the Tularosa Basin remains the vast, ever-shifting deposit of glistening gypsum dunes. Sparsely punctuated with stub growths and cactus, the place is inhospitably hot and infested with rattlesnakes, while the glare from the gypsum is almost blinding. It was here, in a remote corner of the Alamogordo Bombing and Gunnery Range, that a momentous event took place at 5:29 a.m. on 16 July 1945.

Operating under the code name Project Trinity and working in unparalleled secrecy, a cloistered group of scientists and top military personnel detonated the world's first atomic bomb in the north–central part of the range at a place called Point Zero (later Trinity Site). According to author George Meeter in his book *The Holloman Story* [1], no information was released on the Trinity test until similar atomic bombs had fallen on the unsuspecting Japanese cities of Hiroshima and Nagasaki. The locals, however, were all too aware that something had happened.

"The light flash and the thunder clap made a vivid impression over a radius of roughly 160 miles," Meeter wrote. "Glass was shattered 120 miles away in Silver City, New Mexico. And Chicago newspapers carried an item on the explosion of a munitions dump somewhere in New Mexico." The site would remain off limits until September 1953. It is still only open to the public for 2 days a year. Enough radioactivity lingers to fog photographic film [1].

Holloman is born

Post-war, both the immediate and long-term future of the military base were uncertain. Rather than simply shut it down, the U.S. Air Materiel Command concluded that the bombing range would provide an ideally remote site for crucial testing and development of missiles and pilotless aircraft, and the facilities were transformed into an Air Force guided missile test range.

On 13 January 1948, 4 months after the Air Force had split from the Army and become a separate branch of the U.S. armed forces, the Alamogordo Army Air Field underwent a name change to the Holloman Air Development Center. It now became a permanent Air Force installation forming part of the Air Research and Development Command (ARDC), named in honour of the late Colonel George V. Holloman, a noted pioneer in guided-missile development, killed in the crash of a B-17 Flying Fortress on 19 March 1946.

Meanwhile, just to the south of the Holloman range, the drifting dunes of the Tularosa Basin provided an appropriate name for the Army's White Sands Proving Ground (WSPG), which was established on 10 July 1945, just 6 days before the detonation of the first atomic bomb.

A PLACE KNOWN AS WHITE SANDS

The White Sands story actually began in 1944, shortly after Germany had unleashed the first of its deadly V-2 missiles on England. Senior U.S. Army officers realised that their country had made no real provision for research and development of an active missile programme, and a meeting was organised in Washington, D.C. to formulate plans. It was decided that a secure missile base should be set up somewhere in the United States, dedicated to the development of retaliatory weapons that might be used against Germany.

A team of U.S. Army Ordnance and Engineers officers began scouring maps and scouting suitably remote sites, and their investigations soon led them to White Sands in New Mexico, adjacent to the Alamogordo Bombing and Gunnery Range. Moving quickly, U.S. Secretary of War Henry L. Stimson signed paperwork establishing what would become White Sands Proving Ground. Soon after, work began on developing suitable missile test facilities and living quarters on 4,000 square miles of the Tularosa Basin. The facilities at WSPG would even include a formidable blockhouse with walls some 10 feet thick.

A prime testing facility

Naming the base White Sands was actually something of a geographic misnomer, as the proving grounds were actually located some 30 miles south of the gypsum desert. But that mattered little in the scheme of things. WSPG would quickly develop into a prime testing facility for the United States' post-war rocket research programmes.

Demonstrating substantially more foresight and initiative than their British military counterparts, the United States Army shipped around 300 freight carloads of captured V-2 components to White Sands for assembly and test firings. Here the vast, uninhabited terrain and clear skies would provide ideal conditions for launching rockets.

Colonel (later Major General) Holger N. Toftoy was the officer-in-charge who had not only organised the transfer of 127 captured German rocket scientists to the United States, but also arranged the shipment of components. Toftoy, then the acting chief of the Rocket Branch of the Army Research and Development Division, had already begun formulating plans for the most effective use of the rockets once they had been assembled.

In one amusing slip of the pen, a later report co-authored by Toftoy and Major (later Colonel) James Hamill, who selected the parts of the V-2 rockets that would be shipped to the United States, made several important suggestions. One of these read: "From these discussions came the plan for giving a private a ride in a V-2 at White Sands Proving Ground." A line has been drawn through "private" and the handwritten word "primate" substituted.

Putting the pieces together

After the 30,000 parts and sections had been transported to New Mexico, the Peenemünde rocket scientists and technicians were also assigned to WSPG. Prior to assembly, the myriad components were carefully checked under the supervision of specialists, including Wernher von Braun. In carrying out this task, the Americans received some particularly valuable training, not only in the handling and assembly of the missile's components, but also in making any necessary modifications. It led to an understanding of the systems and flight characteristics that would greatly assist them in ballistic missile design.

Although crude in many ways, a fully-assembled V-2 was nevertheless an impressive vehicle. It stood a little over 46 feet in height with a girth of 5.5 feet, weighing in at a massive 28,300 pounds when fully fuelled. The propellant used was a mixture of 2,500 gallons of ethyl alcohol and liquid oxygen, which it consumed at a tremendous rate of 275 pounds per second. The rocket would achieve a staggering 56,000 pounds of thrust, and reach a maximum velocity of 3,600 miles per hour. This equated to a mile per second, or five times the speed of sound [2].

A smaller sounding rocket

At the same time, experimental firings of liquid-fuelled American rockets were also taking place at White Sands. One promising vehicle undergoing tests was the WAC-Corporal sounding rocket. Measuring just 16 feet in length and a foot in diameter, it was appreciably smaller than the V-2, but as a test missile it was comparatively more powerful than the German rocket. In the fall of 1945, launched as a second stage atop a solid-fuel rocket generically known as "Tiny Tim", the WAC-Corporal was already attaining a height of 40 miles.

A ROCKET FOR SCIENCE, NOT WAR

After an inglorious end to the first V-2 firing 3 weeks earlier at White Sands, a second launch took place on 10 May, and this one would prove to be almost flawless. On this occasion the rocket managed to soar to an altitude of just under 70 miles.

Expanding the scope

Although the firings were carried out by military personnel to gain valuable experience in the handling and launching of rockets, the former warheads of the V-2 rockets were shipped to various universities and research laboratories around the United States, where they would be packed instead with instruments dedicated to high-altitude research. This initiative was the brainchild of Colonel Toftoy, who had decided to expand the scope of the firings beyond military applications by establishing a far-sighted policy of reserving space in the noses of these missiles for scientific instrumentation.

A group of scientists known as the Upper Atmosphere Research Panel had been formed to take advantage of the space aboard these rockets, which could make soundings of the upper atmosphere well beyond the altitudes then achievable by research balloons. This group drew its members from the Army Signal Corps, the Applied Physics Laboratory of Johns Hopkins University, the U.S. Air Force, the Naval Research Laboratory and universities such as Harvard, Princeton, the California Institute of Technology (CalTech) and the University of Michigan.

Following the first successful test-firings, further launches with instruments tucked in the nose cones followed, and they soon revealed some interesting challenges. The V-2 was rapidly proving to be a largely unsatisfactory vehicle for such research. In simple terms, it had been designed as a weapon; this meant there was precious little room for all the instruments scientists wanted to cram inside the nose. The V-2 was also proving to be notoriously unstable once its fuel was exhausted and the engine had subsequently shut down during the ascent. Twenty miles above the ground the atmosphere became far too thin to act as any sort of stabilising force on the non-controllable rocket, which would begin to gyrate and go off course. Some tumbled end over end before exploding and plummeting back to Earth.

There were other things to consider in setting out research programmes: principal among these was the reality that the number of reconstructed rockets was limited, and their firings were both difficult and costly.

Jumping on the bandwagon

Despite many problems associated with the use of V-2 rockets, a growing number of scientists were impressed by different facets of the preliminary data and wanted to be involved. They began submitting ideas for improving the payload instruments and gathering additional data.

The first successful transmission of telemetry from a V-2 in ballistic flight occurred on the 17th firing at WSPG on 23 January 1947. During this breakthrough test, performance data on the missile's entire operating system was transmitted to ground recording stations. Useful as this would prove, the physical recovery of certain equipment in the nose section had now become a very real priority for researchers.

To develop this capability, a series of flights known as "Project Blossom" was inaugurated under the direction of the Air Materiel Command, and would begin with the 20th flight of a V-2 at WSPG. Project Blossom (so named because of the way parachutes opened on descent) would employ seven flights, each testing the possibilities of ejecting a nose-cone canister equipped with a parachute recovery system at the very zenith of the rocket's climb.

Project Blossom

The first of these flights took place on 20 February 1947. A parachute system was packed aboard the rocket, but it was not a conventional design for that time. There were actually two parachutes: the first was 8 feet in diameter and comprised a series of ribbons sewn together, each panel covered with a metallic mesh to aid in radar tracking. Its primary purpose was to reduce the sudden shock created when the 14-foot main parachute was deployed, opened and engaged.

On this 20th V-2 firing, a canister attached to the parachute system contained fruit flies and a variety of seeds. This was an experiment aimed at determining the effects of cosmic rays at the upper atmosphere, including the possibility of biological mutations. Cameras and photo-electric cells had been secured in the canister to assist in monitoring other experiments on board.

Initially the launch went well. Then, 27 seconds into the ascent, a disturbance was noticed in the rocket's pitch motion – later traced to the loss of a jet vane. Ten seconds later the missile began to roll, at which time the canister and its parachute system were purposely jettisoned. Despite the circumstances this deployment went well, and the parachute system operated satisfactorily. The canister touched down 50 minutes after being released from the wayward carrier rocket, but not before breaking the altitude record for a parachute drop. Radar had been used to track the parachute descent, and the canister was soon located by a recovery team.

ANIMALS TO RIDE THE ROCKETS

As part of the Blossom experiments, it was determined that a limited number of biological flights would also be conducted at WSPG under the auspices of the Air Force's Cambridge Field Station (later renamed the Air Force Cambridge Research Center) at Wright Air Development Field in Dayton, Ohio. It was proposed that live animals be used in these V-2 firings to investigate the hazards and any limiting factors associated with space flight.

Unwilling but essential test subjects

While it has long been, and continues to be, a subject that can invoke the expression of strongly-held opinions, it seems to be an unavoidable reality that without animal research we would know far less today about biological systems and the many illnesses that afflict human beings and other animals. While animals should only ever be used when there is realistically no other known alternative, and should be treated with humane respect to create a minimum of avoidable suffering, medical researchers have stated that the historical use of animals in medical tests has enabled them to find new and ever-better treatments for diseases such as cancer and antibiotics for infections. The scientific community also believes that these tests have helped radically improve surgical techniques, and aided in the development of vaccines for a vast number of deadly and debilitating viruses. Over many years, animals have unwillingly served as surrogates for human beings in procuring vital information that could not be obtained in any other way. In recent decades, as well, valid arguments about minimising unnecessary suffering have led to great improvements in their care and use.

In the late 18th century and early 19th century, experimenters such as Humphrey Davies and Henry Hickman discovered – after tests involving animals – that pain-free surgical and dental procedures could be carried out following the inhalation of nitrous oxide by patients, rendering them temporarily unconscious during the operation. Further work on the safe use of anaesthetics such as ether would continue, using animal subjects such as rabbits and dogs.

Corneal transplants, safer methods of blood transfusion, anticoagulants, diphtheria vaccines and kidney dialysis all came about in the early 20th century as a result of medical research involving animals. Prior to this time, Type 1 diabetes was regarded as nothing less than a death sentence, but life-saving insulin was developed through experiments on dogs, rabbits and mice. While countless benefits have obviously been derived for humans, it must also be pointed out that even today's pets live longer and far healthier lives due to the parallel development of vaccines and treatments that are now standard in veterinary medicine.

Following the end of the Second World War, and in much the same way as these early medical breakthroughs came about, scientists from a variety of disciplines became aware of imminent high-altitude and space flights involving rockets. When this research related to the question of flying humans into space, there were unresolved questions concerning the effects of rocket flight and space travel on the human body – notably acceleration, deceleration, heat, cold, extreme noise, eating and drinking in weightlessness, cosmic radiation and claustrophobia to be evaluated and overcome.

Researchers began to look at the very real prospect of conducting initial physiological experiments using these primitive rockets, and it was felt to be a natural and logical step to propose the use of live animals in their experiments. Not only was this because animals were seen as less of a sacrifice to send than a human, but the rockets could also carry the much lighter payload of an animal far easier.

The programme's guiding spirit

The proposed series of V-2 animal flights would require a much larger and heavier payload capsule. It was suspected, however, that the dual-parachute system recently used so successfully might not be able to cope with the massive forces of deceleration and stress associated with the heavier payload. Despite this they would try, and they would fail.

About this time Dr. James Henry came into the picture, soon becoming the guiding spirit behind an audacious programme of biological space research. Even though his laboratory would never enjoy the luxury of a research rocket to itself, he and his team would ultimately send living creatures into the upper atmosphere and return them from the very fringes of space.

Earlier in his career, as an assistant professor of aviation medicine at the University of Southern California (USC), Dr. Henry's principal assignment had involved a study of blood action under heavy gravity weights. As a spare-time project while waiting for a centrifuge to be built at the university, he helped design a partial-pressure suit for emergency extreme altitude protection, which would later become standard issue for military pilots. For a time he would work on acceleration physics with USC's Human Centrifuge facility, before leaving the university to serve in the Air Force as an environmental physiologist. Because of his pioneering work with high-altitude protective clothing, he was appointed chief of the Acceleration Section at Wright Aero

Dr. James Henry (left) and an unidentified handler with two spaceflight candidates. (Photo: U.S. Army)

Medical Laboratory in Ohio, studying cardiovascular problems related to altitude and acceleration [3].

A unique proposal

In April 1948, Dr. Henry's department was approached by a parachute specialist from a laboratory also based at Wright Field. He had a rather unusual proposition. His group had been working on a project for the Cambridge Field Station, aimed at the development of electronic and photographic equipment for gathering information on the physics of the upper atmosphere. One of the laboratory's specific tasks was to design and manufacture parachutes of sufficient strength to permit the safe recovery of valuable instruments that were being sent up in the V-2 rockets on behalf of the Cambridge centre. Parachute viability would prove to be a critical factor in these and all the early animal flights.

The Cambridge centre had been allocated a total of six V-2 rockets from the precious remaining supply, but after the first two firings at White Sands the parachutes had failed to deploy properly and valuable instrumentation had been lost. Due to incredibly high speeds and the wind resistance encountered, conventional parachutes were simply not up to the task. After the second failure it became apparent to the head of the research programme at White Sands and Holloman Air Force Base, Dr. Marcus O'Day, that tremendous aerodynamic forces were at work, despite the incredibly thin atmosphere through which the rocket was plunging during the early phase of its return.

Dr. Henry was informed that the third Project Blossom rocket would feature a detachable nose cone, similar to one undergoing development as a possible escape capsule for pilots of the proposed X-2 piloted rocket plane. The rocket's nose cone would be filled with recording and telemetry equipment, but the safe recovery of this equipment was secondary to the military's concerns about the aerodynamic and acceleration forces at work on the human body at such altitudes and speeds. It would help immeasurably if the problem of an adequate parachute system could be resolved, and quickly.

Then came the rub; as the third Blossom V-2 would have a nose cone shaped like the X-2's escape capsule, Henry's Aero Medical Laboratory was asked if it could supply and train suitable "simulated pilots" for this and later launches. It would be a unique opportunity to adequately determine whether a living creature could survive crushing acceleration and other forces at work during an actual rocket flight [4].

Exploring the possibilities

James Henry was intrigued by the prospect, so he in turn approached Major David Simons, a young medical officer who had been working alongside him. After discussing the project, they agreed it presented the perfect opportunity, through experimentation, to investigate certain theories about the effects of acceleration, gravity and weightlessness, especially those propounded by a former colleague from the laboratory, Dr. Otto Gauer.

Dr. Gauer, who had once headed the acceleration laboratory at the Wright Air Development Center, was one of the cadre of captured German scientists brought to the United States at the end of the Second World War. He had theorised that multiple-g acceleration followed by weightlessness might result in profound physiological effects on humans, and specifically that acceleration forces encountered during a rocket launch and re-entry would depress circulatory function. This, he said, could lead to a number of conditions that had been previously observed during experiments in high-altitude aircraft flights, such as blood pooling in the extremities and the brain (known as "red-out") or an insufficient supply of blood to the brain ("blackout").

According to Gauer [5], the lack of gravity would greatly compound the problem, as blood vessels might relax and fail to carry out the capillary action needed to assist the heart in the circulation of blood. This would place even further strain on an organ already overworked by the stresses of rapid acceleration and the resultant g-forces. In turn this could lead to circulatory dysfunction, heart failure, severe muscle cramps and even pneumonia, as well as a disruption of the normal processes of the nervous system.

Furthermore, Gauer and another German physician named Heinz Haber had determined that the brain receives signals on the position, direction and support of the body from four mechanisms – pressure on the nerves and organs, muscle tone, posture, and the labyrinth of the inner ear. The two physicians felt that these four mechanisms might offer conflicting signals in a weightless state and wrote that such disturbances "may deeply affect the autonomic nervous functions and ultimately produce a very severe sensation of succumbence [*sic*] associated with an absolute incapacity to act" [5].

An irresistible challenge

Henry and Simons discussed these issues, as well as the timing involved. If they accepted the challenge, the Aero Medical Laboratory would be given barely 2 months in which to prepare a compact, airtight container and also train suitable animals for the ballistic firings. Like Henry, Simons was swept up in the excitement, knowing that this opportunity presented an irresistible challenge. He was ready to make an immediate start.

Lt. Colonel David G. Simons during his Project Man High balloon research days. (Photo: D.G. Simons)

THE MEN, THE MISSION AND THE MONKEYS

David Simons was born in Lancaster, Pennsylvania in 1922. His father, a physician, had always encouraged and guided him into taking on a similar career in medicine. Fortunately this is where he also saw his future and was subsequently enrolled at his father's old school, Jefferson Medical College in Philadelphia.

Simons entered Jefferson Medical College as a private in the wartime Army Specialist Training Program, and would later serve as an intern at Lancaster Hospital, where he completed a 15-month rotating scholarship. He then owed some military time and would serve 2 years in the Army Air Corps.

Right place, right time

There had always been an interest in astronomy and rockets for Simons as he grew up, but he was also deeply fascinated by electronics and radio and had become a radio ham. He decided to combine his medical training with electronics and asked the Army personnel staff if he could somehow be assigned to an electronic research facility. To his delight he was sent to the Wright Field Aero Medical Laboratory's electronics section, which serviced the laboratory's acceleration unit. Here he would meet and work under the direction of Dr. James Henry.

"They were physiologists concerned with the physiology of both positive and negative human acceleration tolerances and protective devices," Simons told Kennedy Space Center director Gregory Kennedy during a 1987 interview for the International Space Hall of Fame's Oral History programme. He said the laboratory was working on "ways of increasing man's ability to withstand more than man was built for, that could be dished out by airplanes. So, I was having a great time working with them at the centrifuge, designing and making electronic gadgets for them" [6].

A wonderful opportunity

During his interview, Simons recalled the day that Dr. Henry, then his project director at Wright Field, had told him about the proposed biological rocket flights.

"He said, 'Well, what would you think of having an opportunity to help us put a monkey in a captured V-2 rocket that would be exposed to about two minutes of weightlessness and measure the physiological responses to weightlessness?'

"I said, 'Oh! What a wonderful opportunity! When do we start?' "

Like Henry, he knew that this first effort to send living creatures beyond Earth's atmosphere might help define some of the problems humans would face later on in true space flight.

"They wanted us to put a simulated pilot in the Blossoms," Simons recalled. "That was their reason for the experiment. In our minds – that is, in mine and Dr. Gauer's and Dr. Henry's at least – the main point of the experiment lay in solving a few of the problems connected with travel beyond the atmosphere. That was a prime target for us. And 'beyond the atmosphere', by definition, means the total atmosphere" [6].

Project Albert

Simons, who had elected to stay in the U.S. Army Air Corps when it split from the Army, found himself assigned to the role of project officer for what had begun to be referred to as Project Albert. Simons has no idea where the name came from, although he felt it may have originated in the parachute branch. "They had some very imaginative people over there," he said, "and our project was rather a pet of theirs. So they would be very prone to dub it some sort of name like that, and once you start calling something like that a name, why of course it rapidly sticks if there's no contra-indication" [6]. He was billeted at Holloman AFB and would travel across to White Sands Proving Ground to conduct set-up work on the exciting new project.

"The research group Jim Henry organized was not a large one," Simons jokingly recalled in a recent interview with the authors. "It consisted initially of Captain James Henry, First Lieutenant David Simons, and two monkeys" [7]. They would later be joined in the task by Major P.J. Maher, Captain E.R. Ballinger and other medical researchers.

Finding a way

At first, the program team of Henry and Simons thought that designing a sealed capsule capable of containing a monkey, with sufficient oxygen to last for about 2 hours, some means of measuring and storing physiological data relating to the animal's heartbeat and respiration, as well as chemicals to absorb carbon dioxide exhalations, was not the most difficult task on their plate. They were wrong – it was actually a very complex engineering job, and neither of them were engineers. As pilot and author Lloyd Mallan would state in his 1958 book, *Men, Rockets and Space Rats*, the two men learned the hard way, after work had already begun on the capsule.

It was planned according to specifications: one side was to be convex so that it would fit snugly against the concave inner wall of the V-2's nose-cone; the remaining three sides were to be squared off like a box; the whole thing would then slide neatly into the rocket warhead, permitting other instruments to be packed against it [8].

When Simons visited the Franklin Institute Laboratory of Research and Development outside Philadelphia, the first problems began to emerge. The Institute's people had no idea how to fit an anaesthetised monkey inside a capsule into an instrument-packed nose cone. Likewise, as Mallan noted, Simon's team had no idea about instrumentation, but they simply had to work it out, and quickly.

Major Dave Simons went back to the Aero Medical Laboratory in Ohio with a new set of measurements and a desperate determination not to give up. The capsule was redesigned. It now had only three sides, two of which looked like the gabled roof of a rickety shack about to tumble over in a wind storm. The remaining side, still convex, resembled a section of an old milk bottle that had been partially welded in a fire. The odd form was built for them out of aluminium with welding at the edges to assure its being airtight [8].

An unidentified pair of hands holding the prototype capsule that would contain the first Albert monkey. (Photo: D.G. Simons)

A SUITABLE FLIGHT SUBJECT

Finally, the rudimentary capsule was ready. Meanwhile, Henry and Simons had been given the added job of preparing the animals. They had chosen to work with *Macaca mulatta* rhesus monkeys, due to their high intelligence and many physiological similarities to humans.

Training – with a note of caution

It was felt that an ability to learn simple tasks quickly, combined with their natural docility, would enable the rhesus monkeys to perform complex tasks in restricted situations. There was another important factor: the animals' reaction time in such tests – seven-tenths of a second – had proved to be only marginally slower than that of humans. Although they were ordinarily easy to keep and train in captivity, Simons nevertheless felt the ones he worked with were far from friendly. "I learned to always wear gloves when handling them," he recalled [6].

Simons's May 1949 report, *Use of V-2 Rocket to Convey Primate to Upper Atmosphere* (Air Force Technical Report 5821), is rich in details of the preparations for this endeavour.

"For a monkey to survive a V-2 rocket flight uninjured," he wrote, "he must, first of all, be mechanically secured so the oppositely directed warhead separation and parachute opening shocks of the order of 10 to 20 g's will not injure him. At the same time, he must be provided with means for existence in a virtual vacuum. The simplest and most direct approach to the latter problem is to enclose the animal in a pressurized container.

"Since the rocket firing may be delayed several hours after the animal had been placed in the nose-section, and since recovery might extend well into the next day, means must be provided for sustaining the animal in this container for 12 to 24 hours. This means that within the pressurized container, there must be an oxygen supply and an adequate material to absorb the carbon-dioxide and water vapour for 24 hours, plus the instrumentation for recording physiological reactions under conditions which do not exceed the animal's capacity for heat regulation.

"The size and shape of the Albert I capsule was determined by the space available in the rocket after all the other components were in position.

"By utilizing all the space available in the capsule, it was possible with difficulty to place a 9-pound rhesus monkey in it after the installation of associated equipment. The capsule was made as large as possible and still fit into the rocket test section. To conserve weight, it was constructed of aluminum. Serious difficulty was met in pressurizing the odd-shaped container and servicing the component parts. These components were difficult to reach because the access door was as small as possible to minimize pressurization problems, also, because of the internal bulkheads" [9].

The Albert capsule

The capsule that would hold Albert I had been carefully manufactured to slide into a small section of the nose cone. It would also be carefully pressure-tested to ensure there were no leaks, but early tests demonstrated that there was very little integrity in the makeshift capsule; air spewed out everywhere.

There were no exact precedents to work by, but Henry's team knew that if their capsule was not fully sealed and pressurised the occupant would rapidly expire from a lack of oxygen once it flew into the thinner atmosphere. Reinforcing strip-metal bulkheads were added to overcome a ballooning problem with the capsule's walls, extra riveting was installed, and a thick rubber compound was applied to areas around the leaks. This all seemed solid and satisfactory until a final test was carried out. Once again multiple streams of air hissed from the capsule as soon as it was pressurised. Some extra welding and caulking seemed to remedy the problem.

The final appearance of the crude capsule may have been far from attractive, but it was now satisfactorily airtight and ready for the primate passenger.

Too much monkey business

While Dr Henry was engaged in work associated with the V-2 flight, Major Simons flew down to McDill Air Force Base accompanied by the two flight-ready monkeys in order to conduct some performance tests. To his annoyance, one of the animals later managed to escape from his cage at the base. Even though the exits were all guarded the monkey somehow slipped out through a broken window and dashed away. He was subsequently sighted several times, but no one was quick enough to catch the elusive runaway.

Three weeks later Simons received a letter, informing him that a captured monkey was being held at the Tampa City Police Station, and asking if it might belong to him.

"It seems this particular monkey was tired of the Air Force Base and had wandered off into town a few miles away," Simons said of the mischievous animal. "It stuck its nose into a lady's kitchen one morning and began to snoop around. The lady happened to be a meticulous housekeeper, especially in the kitchen. Her concept of neatness naturally did not include having a monkey crawling around among her saucers and teacups. So she made the mistake of trying to remove the monkey from her cupboard by force. The monkey took exception to her attack; he started throwing teacups and saucers in her direction. She and the monkey then began running round and around her kitchen, until it became rather the worse for wear!" [8].

The mischievous escapee was retrieved from the police, while the woman was finally placated and compensated for her broken crockery.

"It seems that the base was tendered a bill for quite a few hundred dollars by the lady," Simons recalled. "She included emotional damages to her personality, lost teacups, irreplaceable china and the general besmirchment of her clean kitchen in the bill. This taught me a lesson: never trust a monkey who is not anaesthetised!" [8].

THE ALBERT FLIGHTS BEGIN

It was 11 June 1948, and a gleaming V-2 stood prepped and poised to deliver a Holloman monkey into the upper atmosphere.

Three hours before the launch Henry and Simons clambered into a Jeep with a small passenger and slowly made their way out to the launch gantry. They pulled up close to the launch gantry and spent a few moments watching the sleek rocket being fuelled before they clambered out of the vehicle. Simons was nursing and gently stroking the docile rhesus monkey now known by the name of Albert, while Simons was checking once again that he had everything he needed in his sterile medical kit.

Preparing Albert for flight

Less than an hour before launch, as Henry and Simons prepared the monkey for the flight, a pad photographer was buzzing around the launch gantry and he noticed something hurriedly scrawled in chalk on a fin of the V-2. He quickly passed on word of his find. Some unidentified wit had added a well-wisher's invocation to the occasion by writing (and slightly misquoting) a line from Shakespeare's *Hamlet* on the vehicle, much as an actor would be invited to "break a leg" before their first performance. "Alas, poor Yorick," it read. "I knew him well" [8]. They never did find out who the mystery culprit was, but a few of the launch team soon began to jokingly refer to the monkey as Yorick. In a light-hearted example of serendipity giving way to fact, the animal is still identified as Yorick in some post-flight histories of the mission.

Events now began to rapidly tumble over each other as launch time drew near. Forty-five minutes before the scheduled lift-off James Henry opened his medical kit and carefully anaesthetised Albert with 10–15 mg of sodium pentobarbital, administered intravenously, before the electrocardiograph needles and a respiration unit lever arm were to be sutured into place.

Simons said an anaesthetic was administered for a number of reasons. "I think the most important one was that if the parachute opened properly, the monkey would get what was, at that point, an unknown opening shock that was calculated to be pretty impressive. And, depending on what altitude and what the position of the nose cone was and its orientation, things can go wrong ... like if the parachute streams and it comes in either half open or tangles – well, the monkey could land hard but not fatally. And we wouldn't want the monkey to be lying there suffering until we could get there. It can take a while to track down where the nose cone landed and get to it.

"I think one other thing, too, from the monkey's point of view – the thing we put them in was a seat that they were strapped into. And they didn't have room to wiggle at all. They were totally restrained because our space was as cramped as it conceivably could be . . . so was the weight limitation. If they were awake and conscious, this would be a cruel thing to do to the animal if it wasn't necessary.

"For those reasons, we did anaesthetise the animal. It meant that we had to try to extrapolate what reactions one would get *conscious* and *awake* in this situation, compared to asleep and anaesthetised. But, if there were any really gross, serious physiologically disturbing factors, they should show up, even under those circumstances. It was a first cut; no animals had ever experienced anything like this before" [9].

Straps, supports and steel springs

If the firing of the V-2 carrying Albert went as planned, nothing was expected to cause the animal any untoward pain during or after the flight. To further protect the animal from injury due to any unexpected difficulties, he was also given an intramuscular injection of 2 cc of Luminal, a long-acting barbiturate used as a sedative. Albert was then placed on a felt-padded aluminium seat which was hinged at hip level to allow for his easy insertion into the narrow confines of the capsule.

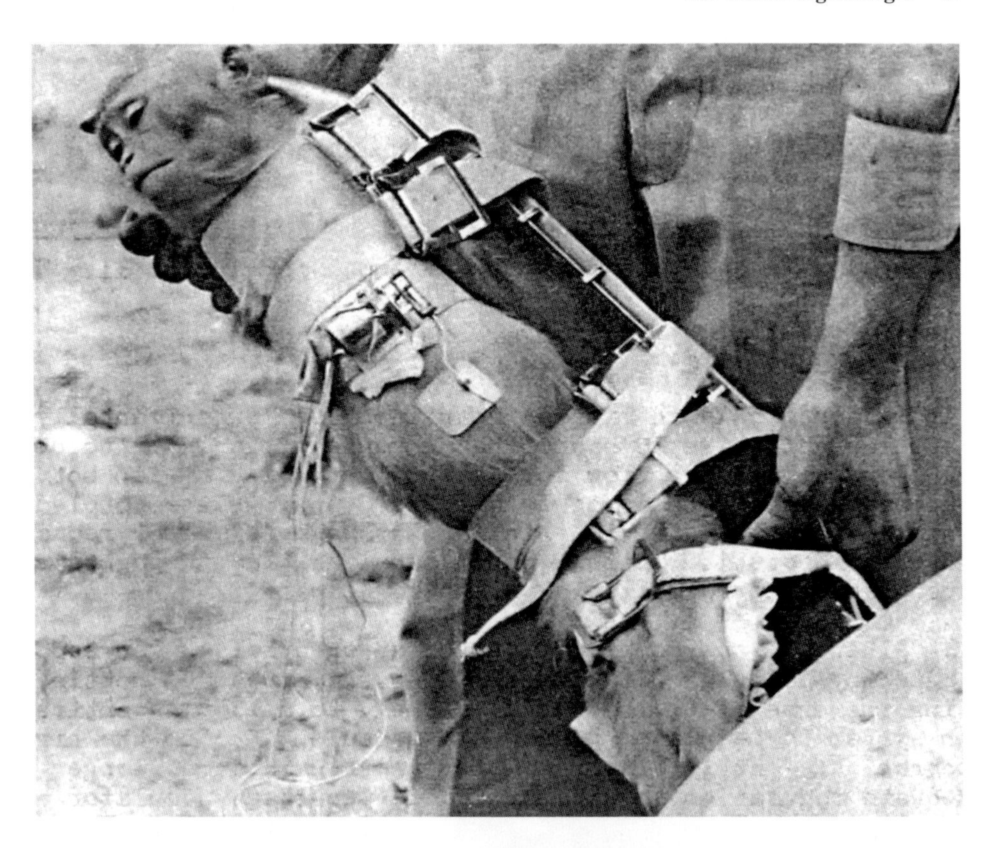

In both of these photos, a forlorn-looking Albert I is inserted into his makeshift capsule. (Photos: D.G. Simons)

The cylindrical capsule into which he would be placed was 3 feet long and a foot in diameter, constructed of sheet iron welded to steel end rods, which were sealed to the aluminium end rings by 24 bolts. A rubber gasket was cemented to each end plate, seated between 1/64″ beads on the edges of the steel rings. It was also fitted with a

Project Blossom III rocket with Albert I aboard is readied for Launch. (Photo: NASM, Smithsonian Institution)

tubular rack that slid in and out of the capsule. All components were mounted on this rack where they would be readily available for manipulation.

With the anaesthetised monkey now seated, restraining straps were threaded beneath the rods on both sides of a platform welded to the rack. A head support would be bolted onto this platform once it had been installed in the capsule. A net-type support was placed around the monkey and then he was secured to the platform by tying the straps in pairs beneath it. Other straps also ran longitudinally over Albert's shoulders and through the crotch area, and then transversely along the length of his body. Great care had been taken in the design of this restraint device to prevent any obstruction of the monkey's airway. Albert's respiration unit was then pinned to the straps before his insertion into the capsule.

Although the 9-pound monkey was the smallest available for the flight, Simons reported that the space inside the capsule "was so limited that his head had to be placed in a cramped, forward position with the neck acutely flexed" [9].

An end before a beginning

Once the monkey had been installed in his seat it was carefully squeezed into the narrow capsule, which was then flushed through with oxygen. Thirty minutes before launch time, the capsule was transported to the gantry's upper level for insertion into the nose section of the V-2. This was achieved by sliding the monkey-bearing capsule through a circular opening. Once the capsule was secured in place the opening was covered by a plate fitted with Zeus fasteners. Then, as Simons explains, there was a major problem.

"After installation of the Albert I capsule in the test section, no indication either of heart action or of respiration could be obtained. The filament supply voltage was all that could be checked outside the capsule and time did not permit opening of the capsule to determine the cause of the difficulty. The record, subsequently obtained from the Cook recorder, was equally void of evidence of physiological activity. Either the monkey died or there had been a failure of the electrocardiographic apparatus" [9].

It would prove to be a crushing setback.

"Disturbed about the whole thing"

The launch countdown would continue despite some apparent problems with Albert, eventually winding down to zero. The V-2 roared into lusty life and blasted a path into the heavens, reaching an altitude of 37 miles. The rocket's velocity was measured at some 18% below general average, and the burn-out took place prematurely 57.5 seconds from lift-off.

Separation of the nose section took place at the very crest of the rocket's flight, but the parachute bulkhead system failed to operate as planned. The parachute popped out and streamed behind but could not open fully in the thin air until it reached 25,000 feet, when it finally bit into the thickening air and blossomed. Unable to withstand the sudden weight and velocity of the plummeting nose section, it exploded into shreds. Soon after the nose cone slammed into the ground at high speed and shattered as it skidded and tumbled across the desert sands. The capsule containing Albert was so badly deformed that even if he had survived the actual flight, death would have been an inevitable conclusion.

"We were disturbed about the whole thing," an unhappy Dr Henry said later. "We had allowed barely enough capsule space and had in all probability lost the animal due to breathing difficulties in its cramped posture even before the rocket was fired. Further, our recording system was inadequate. We got no information or data from the test – but we had been initiated into rocketeering and now knew it was a problem of achieving foolproof reliable instrumentation" [8].

Identifying the problems

Despite the loss of Albert and the failure of the parachute system to operate at extreme speeds and altitude, project scientists were determined to do better on the next test. Parachute experts worked hard to identify and solve the problems associated with the earlier flight, while the capsule was redesigned to provide a less cramped position for the next monkey. Additionally, a stand-by capsule and monkey would be prepared in the event of a problem occurring after the first animal had been inserted into the nose section.

Trying to find some answers

Further studies carried out by Henry and Simons gave an indication that the test monkeys had a poor tolerance to heightened body temperatures, even when lightly anaesthetised. Furthermore, they were subject to shock from stasis (a slowing or stopping of blood flow) when immobilised for prolonged periods in a hot environment.

As the temperature in the nose section could reach up to 20 degrees higher than the ambient local temperature, some means of cooling the capsule's interior after its insertion into the nose cone had to be devised. An ideal solution was found by encircling part of the capsule in a wool-insulated canvas bag filled with dry ice. This lowered the temperature inside the capsule by around 20–30 degrees, and the cooling would last for more than 4 hours. This would be more than sufficient to cover the launch and recovery period.

As a further measure of controlling the animal's body temperature, rubber diapers would be fitted over the monkeys' gauze diapers to eliminate the possibility of urine evaporating. This evaporation had been found to cause relatively high humidity of the capsule during preliminary test runs.

ANOTHER MONKEY CALLED ALBERT

It would be another year before the second V-2 launch in the Blossom series took place. Using very little imagination, a smaller, $6\frac{1}{4}$-pound monkey that had been selected was promptly named Albert II. This time, the animal's capsule had been redesigned and professionally manufactured in a workshop by fully-qualified machinists. The finished capsule was 3 feet long and 13 inches in diameter.

Henry and Simons try again

On the day of the launch, 14 June 1949, Albert II and a second anaesthetised monkey were prepared to the point where both were ready for insertion into the capsule. Then, once the fuelling of the V-2 had begun, the rack holding the prime test animal was given a final, thorough check and placed inside the capsule.

As before, the lid of the capsule was sealed, the plugs in the relief valve were removed and the capsule was flushed with oxygen. Next the experimental dry ice bag was filled and the edges sealed around a third of the length of the capsule with adhesive tape. Finally, the capsule was carried to the top of the gantry where it was inserted into the nose section of the waiting rocket. On this occasion the installation of Albert II's capsule was completed just 45 minutes before scheduled lift-off.

Some modifications had been made to the rocket after the first Blossom flight. This V-2, and those that followed in the programme, now sported a slightly elongated nose cone, which not only provided a considerably larger instrumentation area, but offered a less cramped environment for the monkey.

With greatly improved instrumentation now installed, no problems were encountered in monitoring the monkey's respiration and heart rate. Both showed a slight increase after he had been inserted into the nose section, but this was attributed to the dry ice being slow to take cooling effect. There was certainly no evidence of physiological distress recorded right up to 45 seconds before launch, when the solenoid-powered AMC plug, connecting the rocket to recorders in the blockhouse, dropped out as scheduled. From that time on, Cook tape recorders in the rocket took over monitoring duties.

If at first ...

At 3:30 p.m. local time, the second Blossom rocket roared into life and soared skywards. The rate of acceleration climbed rapidly, peaking at around 5.5 g's. Sixty-six seconds into the flight all the fuel had been exhausted. The rocket continued to soar upwards some 7° from the vertical until it reached a peak altitude of 85 miles $3\frac{1}{2}$ minutes after lift-off. Five seconds later, right on schedule, the nose section separated from the booster as the return journey back to Earth began.

Five minutes and 17 seconds after the V-2 had left the launch pad the main parachute blossomed out with a shock of around 12–13 g's and remained open for several seconds. Just as things looked promising, the parachute suddenly streamed and the nose section began to drop at an alarming rate. It is believed the parachute may have begun to billow again, as the nose section unexpectedly tore loose from its shroud lines, eventually smashing into the ground at high speed. The impact was so severe it created a crater 10 feet wide and 5 deep. Amazingly enough, despite the devastation, the Cook recorder tapes were recovered, allowing for an analysis of the monkey's respiration and electrocardiogram data.

Despite the forces of acceleration and separation, the two jolts and free-fall, all the indications were that the monkey had survived in relatively good condition right up to the moment of hitting the ground.

Simons managed to remain philosophical: "Since the animal remained alive until impact, it is likely that it would have been recovered, had the parachute operated properly," [9] he reflected.

Although Albert II died at the end of a troubled flight, he nevertheless holds a relatively unknown but unique place in spaceflight history as the very first living creature ever known to have made a rocket flight exceeding 60 miles.

A later report on the Albert flights includes an electrocardiographic record of Albert II throughout his flight. As David Simons told the authors in 2006: "The heart rate was clearly disturbed by autonomic input from the sudden change in G forces both at burn-out (6 g's to 0 g), going to altitude and again when it impacted the atmosphere on return, which at that velocity presents an appreciable resistance rather suddenly. That is the first record made of an animal response to space flight and of its response to several minutes of pure weightlessness (no change from normal in an anaesthetised animal). I think it deserves recognition as such – several years ahead of the Russians" [10].

Simons departs, and solutions are sought

Despite being heavily involved in the Albert project, David Simons would leave Holloman after this second primate flight. He was bound for the Air Force's School of Aerospace Medicine, where he would undertake training as a flight surgeon. From

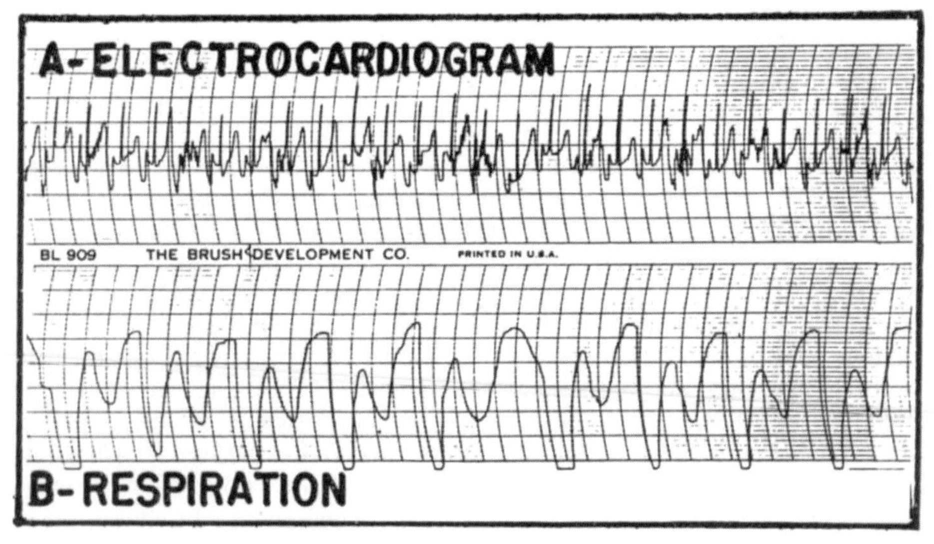

Taken from a post-flight technical report on the use of V-2 rockets to convey primates to the upper atmosphere, this historic recording of electrocardiogram and respiration data from Albert II was taken at approximately 85 miles altitude. (Courtesy: D.G. Simons)

The late Brigadier-General Homer Boushey was heavily involved in early rocket research, and he drew this whimsical cartoon to illustrate how the Holloman monkeys must have felt. (Courtesy: Homer Boushey)

there he was assigned back to Wright Field just as the Korean War broke out and he would spend the next 30 months on duty in the Far East. Eventually he would return to Holloman as chief of the Space Biology Laboratory, but his involvement this time would be in the area of cosmic radiation exposure during high-altitude balloon flights.

Meanwhile, the failure of the parachute system on the latest V-2 flight and the loss now of two test animals were both a disappointment and frustration for Henry. When questions were raised by journalists about the continued use of monkeys in the flights, he argued that the animals were involved in proving humans could survive rocket flights into space, which had profound future implications of both a civilian and military nature.

While the undamaged return of the scientific instruments related to the animals' physical well-being was desirable, albeit not essential, the Aeromedical experts needed to assess whether the flight animals experienced any undue strains that might result in harmful after-effects. Henry wanted bigger and better parachutes, and there were many who concurred with this argument, according to author Lloyd Mallan:

If a monkey could survive a V-2 flight, then so too would a man. The Air Research and Development Command was in full agreement with this attitude. But apparently it was an impossible task to develop a successful parachute for a V-2. Because of the load and altitude conditions of the Blossom series, the problem was never solved [8].

THE FRUSTRATION OF FAILURE

Two more test flights were conducted in September and December of 1949, but both flights would end in failure. The V-2 carrying Albert III left the launch pad as scheduled on 16 September, but the rocket's velocity was below optimum levels by approximately 15% from the time of lift-off. A little over 10 seconds into the flight there was a sudden explosion in the tail section, although the thrust continued. Fourteen seconds later there was an even more violent explosion, and this time the rocket disintegrated 3 miles above White Sands. A later examination seemed to indicate there had been a breach of significant size in the alcohol fuel system.

The final animal flight

Launched on 12 December 1949, the rocket carrying Albert IV began its flight smoothly enough, ascending on a good trajectory. Then, as before, the test flight climaxed in a problem with the parachute system. Despite surviving the rigours of ballistic flight, the monkey was killed when the nose cone hurtled down and slammed into the desert. Once again all heart and respiratory data gathered during the flight was recovered from the nose cone and gave no signs of any serious disturbance in the animal's functions. Nor were the forces associated with acceleration and deceleration of a magnitude sufficient to raise any concerns with this physiological aspect of rocket flight.

One final V-2 test flight remained, and this time James Henry decided a mouse would take the place of a monkey. A movie camera had been developed at the Technical Photographic Laboratory at Wright Air Development Center, which would be installed in the nose cone to record the mouse's movements at fixed intervals during weightlessness. Knowing the probable fate of the nose section at the end of the mission, the camera was securely anchored inside a solid steel box.

On this occasion, Dr. Henry's team would make no attempt to monitor the animal's heart action or respiration. As the intention of this flight was to record the conscious reactions of an animal during fluctuating gravity conditions, the mouse would be neither anaesthetised nor restrained. A special transparent Plexiglas container was developed, which – somewhat ironically – resembled a large wedge of cheese. The floor of the fixed cage in which the mouse would travel was made of wire gauze, allowing the animal to maintain a firm foothold during the flight, and giving it some degree of reliance on tactile and visual senses.

"The V-2 clobbered in"

The launch of a V-2 containing the unrestrained mouse took place on 31 August 1950, and the missile flew satisfactorily to an altitude slightly in excess of 85 miles. Anxious project scientists from the Aeromedical Laboratory had been fervently hoping for better results this time, but once again the failure of the parachute system let them down.

"The V-2 clobbered in, killing the mouse," said Dr Henry. "But the camera was all right and the pictures showed the mouse floating around, with no apparent air of confusion" [8]. It was also noted that allowing the mouse a foothold on the wire-gauze floor meant it did not appear to be seriously disturbed. Henry would later report that the mouse "no longer had a preference for any particular direction, and was as much at ease when inverted as when upright relative to the control starting position" [4].

That would not be the end of Project Albert, as a new programme was already under way that would see a continuation of the biological rocket flights. Even as project scientists had been preparing for the final V-2 flight, plans were already taking shape to continue their experiments aboard a newly-developed sounding rocket known as the Aerobee. Unlike the cumbersome V-2, however, this slick rocket had been purpose-built to conduct high-altitude research and experiments.

AEROBEE FLIES

The first phase of Project Blossom was at an end, but other programmes using V-2 rockets continued. Earlier that same year, on 24 February 1949, a WAC-Corporal sounding rocket had been successfully launched atop the nose of a reconstructed V-2 at WSPG. This combination rocket was known as the Bumper-WAC, and the co-operative amalgamation of German and American technology foreshadowed a momentous day in January 1958 when the same international effort would result in sending America's first satellite, Explorer I, into orbit.

Overcoming the difficulties

During earlier test flights the WAC-Corporal had proved to be a reliable vehicle, leading Colonel Holger Toftoy to suggest in 1946 the possibility of combining the V-2 rocket and the WAC-Corporal to create a two-stage vehicle. As the smaller rocket had no guidance mechanism of its own (hence the acronym WAC – Without Any Control), Toftoy felt it could be launched as a second stage at high speed during the ascent of the V-2 in order to achieve stable high-altitude flight. Once again his judgement would prove to be impeccable.

The flight on 24 February would be the fifth dual-stage launch in the Bumper series after highly disappointing results and booster losses in the first four attempts. On this occasion the V-2 struggled to a height of 20 miles before a trigger mechanism was actuated, at which time the smaller WAC-Corporal lit up and disengaged from the carrier rocket, racing into the skies at 5,150 miles per hour. Just six and a half minutes after the mated rockets had lifted off from WSPG, the WAC-Corporal upper stage had reached a height of 250 miles.

Following an unfortunate international incident on the next Bumper flight, when the combination rocket skewed off course and tumbled into a Mexican cemetery south of Juarez, the entire Bumper programme was shifted overland to a new test facility at the Cape Canaveral Auxiliary Air Force Base in Florida.

A successor vehicle

Meanwhile, the potential created by the WAC-Corporal programme led to the development of a successor vehicle that could reach greater heights, and carry a far heavier payload. The end result was the Aerobee, 18.8 feet in length and just 1.25 feet in diameter.

Initial tests of the two-stage sounding rocket took place in November 1947, while the far heavier V-2s were still being launched on research flights. The Aerobee would rapidly prove to be a far more reliable and highly versatile research vehicle, capable of carrying 154 pounds of instruments to altitudes in excess of 70 miles [2].

Partial success

Determined to continue with their biological programme, project scientists at the Aeromedical Laboratory were granted permission to design modified animal capsules for use within the nose cone of the Aerobee sounding rockets.

The first of these flights, carrying a monkey called Albert V, was launched on 18 April 1951 and reached an altitude of 36 miles. Physiological data were successfully transmitted from the capsule, and there were no signs of the monkey suffering any

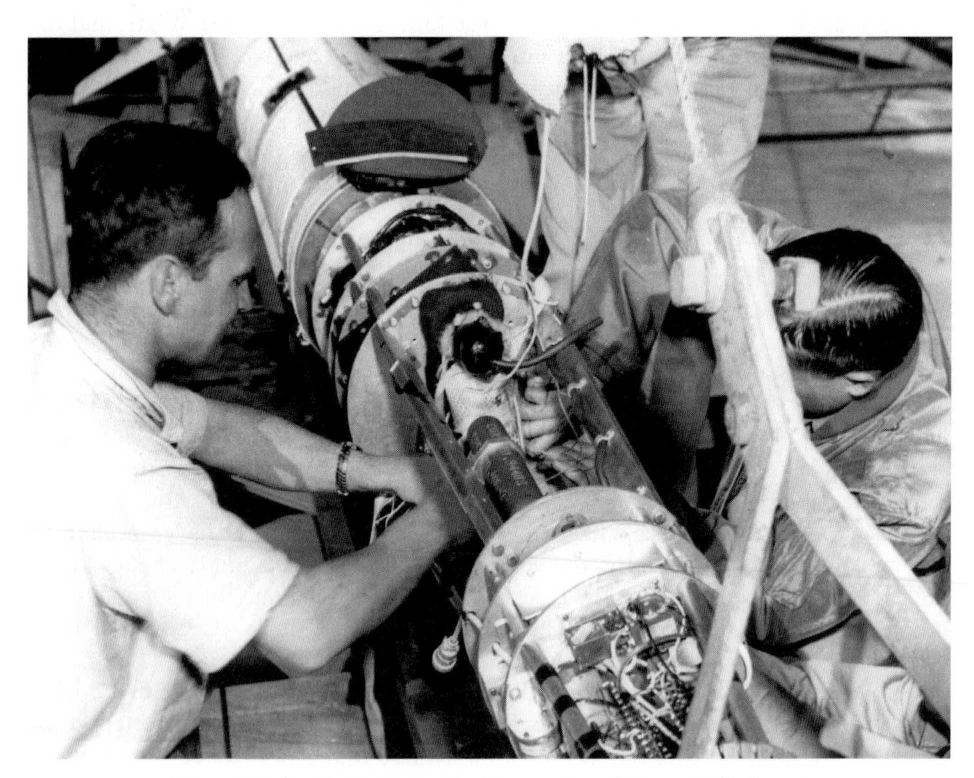

Albert V is loaded into the Aerobee rocket. (Photo: U.S. Army)

Albert VI is recovered after landing, but the hapless monkey would not survive due to heat prostration. (Photo: U.S. Army)

gross disturbance during the rocket's ascent. To the scientists' dismay, however, this latest Project Blossom test flight ended in similar fashion to earlier attempts when the parachutes were unable to cope despite several modifications. The nose cone crashed heavily to Earth in hilly desert country somewhere near Holloman, and despite an extensive search was never located.

A long-awaited breakthrough was just around the corner. During the next flight on 20 September that year, an Aerobee rocket carrying a monkey named Albert VI and 11 mice climbed to an altitude of nearly 45 miles.

Nine of the mice were aboard specifically for exposure tests related to cosmic radiation, while two others had been placed in separate Plexiglas compartments of a slowly rotating drum in order to film their reactions to gravity-free conditions. The drum, mounted with its axle across the long axis of the rocket, would rotate at a steady 4 revolutions per minute. Some time prior to the flight, one of the mice had undergone a surgical procedure on the labyrinth of the middle ear, depriving it of the orienting vestibular function, which is responsive to gravitational forces.

This time the single parachute worked adequately, bringing the nose cone to ground with a solid thump. All 11 mice on board survived the flight and landing, as did Albert VI. Unfortunately it took the recovery team longer than expected to arrive at the touchdown area, while the metallic nose cone was baking under the fierce midday sun of southern New Mexico.

Despite some immediate attention the hapless monkey passed away in a Medical Corps station wagon on the way to the air base hospital, just 2 hours after the landing. The cause of death was given as heat prostration or landing shock, and was most likely a combination of both. Two of the mice would also die post-flight, although it was later determined that none of the survivors had suffered any detrimental after-effects related to their exposure to cosmic radiation.

Breakthroughs at last

The Albert VI flight in September would be the final Aerobee biological launch conducted in 1951. As with primates on the earlier ballistic flights, the monkey had displayed no obvious signs of disturbance in his heartbeat or breathing, although there was a slight but almost insignificant decrease in his arterial blood pressure during the sub-gravity period.

Film recovered from the nose cone showed useful disparities in the activities of the two mice in the rotating drum. The mouse which had undergone surgery managed to cling onto a small crack it had found inside the compartment's wall. It seemed quite composed, undisturbed by weightlessness and appeared to be moving about at will. By comparison, the untreated mouse did not seem to enjoy the sensation and continually clawed at the air. For the 2 minutes of weightlessness this mouse found orientation difficult and was unable to control its actions, appearing quite agitated.

By analysing film of the mice, the Air Force concluded that little or no loss of physical or mental prowess would occur for subjects while experiencing weightlessness during brief rocket flights. The value of having a fixed reference, such as the crack one mouse found in his container, was also more than adequately established [11].

Finally, the parachute designers came up with what they hoped was the solution to their problem. The first, smaller drogue was designed as a circular series of cloth strips separated by air gaps, rather than being sewn completely together as before. It would pop out to slow and stabilise the nose cone as it fell through the thinner air of the upper atmosphere. The second and larger parachute would then deploy and open closer to the ground.

PATRICIA AND MICHAEL

The first test of the new parachute system occurred on 21 May 1952, when two Philippine *macaque* monkeys named Patricia and Michael were scheduled to be launched on the third Aerobee biological flight, together with two mice named Mildred and Albert – the latter named as a final salute to the series of flights.

A great milestone achieved

Unlike the previous flight, neither mouse underwent surgery. They would be individually contained in a two-section rotating drum similar to that used on the previous attempt. The only difference would be the inclusion of a stable, notched perch that one of the mice could cling onto during weightlessness, while the other had nothing to grasp within its smooth-walled drum.

As with all previous monkeys, each of the two primate candidates was given an anaesthetic before lift-off to keep them calm during the flight. Following this they were strapped into sponge rubber couches using nylon webbing, then carefully secured into a pressurised capsule 3 feet long and 8 inches wide. For the sake of physiological comparison during high acceleration, Patricia was seated upright, with Michael supine. Before the capsule was sealed and inserted into the nose cone, each monkey was provided with a facemask through which it received a supply of oxygen.

Everything worked well; the launch was successful, the rocket achieved a height of 36 miles while travelling at close to 2,000 mph, and to everyone's jubilation the new parachute system worked without a hitch. All of the animals were recovered in good health, and a great milestone had been achieved in American space flight.

Once again, physiological data recorded on the flight indicated that none of the animals involved had suffered harm during any period of their journey into the upper atmosphere.

According to David Simons, the recorded data from this and the earlier flights unquestionably indicated "that the weightless state itself produces no disturbance of circulation in terms of heart rate or arterial and venous blood pressures. This does not mean that the circulation might not be involved secondarily, due to emotional and autonomic reactions to weightlessness. Such secondary reactions are essentially the same whether caused by weightlessness, a rough sea, or an obnoxious mother-in-law" [12].

Analysing the results

Post-flight, Dr. Henry would observe that film footage of the two mice in the rotating drum clearly demonstrated that the one provided with a perch had remained "oriented and quiet" [4] through the period of sub-gravity, while the second mouse without any

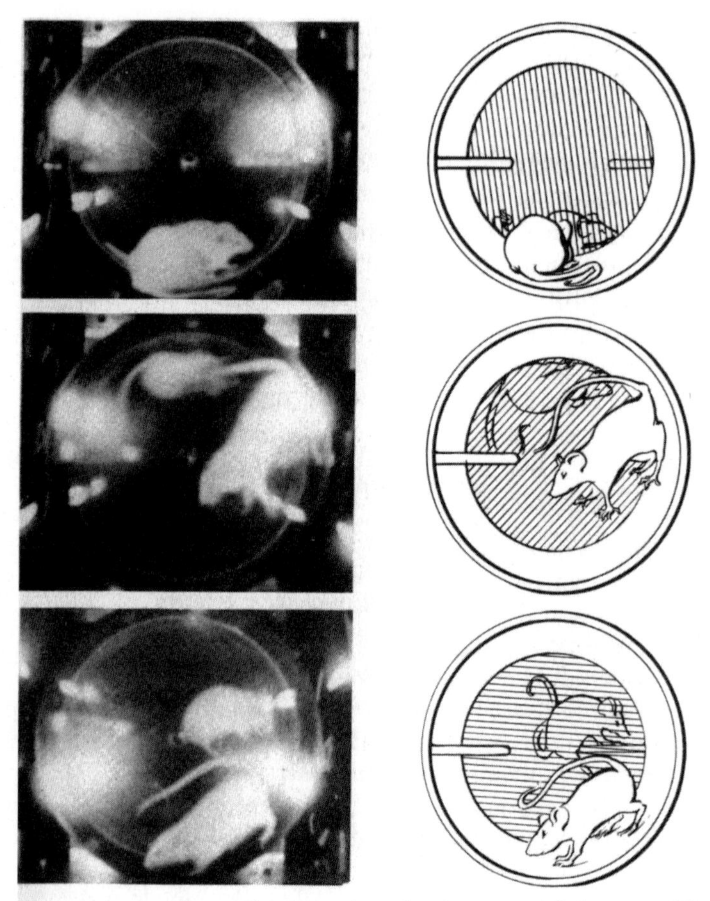

This US Air Force montage shows the two mice adapting to weightlessness. (Photos: U.S. Army)

foothold also seemed to be relatively calm. As the Aerobee was coasting during this phase of the flight, it was also rotating at around 1 revolution per second (together with a little pitch and yaw motion), and this action produced some transverse acceleration of about 0.5 g. Dr. Simons would later relate this to the actions of the mice. In his report he stated the experiment "suggests that a little G can go a long way in supplying helpful orientation" [13].

Once the nose cone had separated from the carrier rocket and begun to fall back to Earth, it had achieved pure weightlessness for about 15 seconds before it finally encountered some atmospheric drag. Simons later wrote that, during this brief period, the mouse without the perch "hopped disconcertedly back and forth" [12].

According to Dr. Henry, "The entire series of experiments showed that the stresses imposed by a brief rocket flight into the ionosphere, and by the operation of this particular escape-capsule system, are well within the range of tolerance of the animals used, and probably of man as well. More important still, the work showed

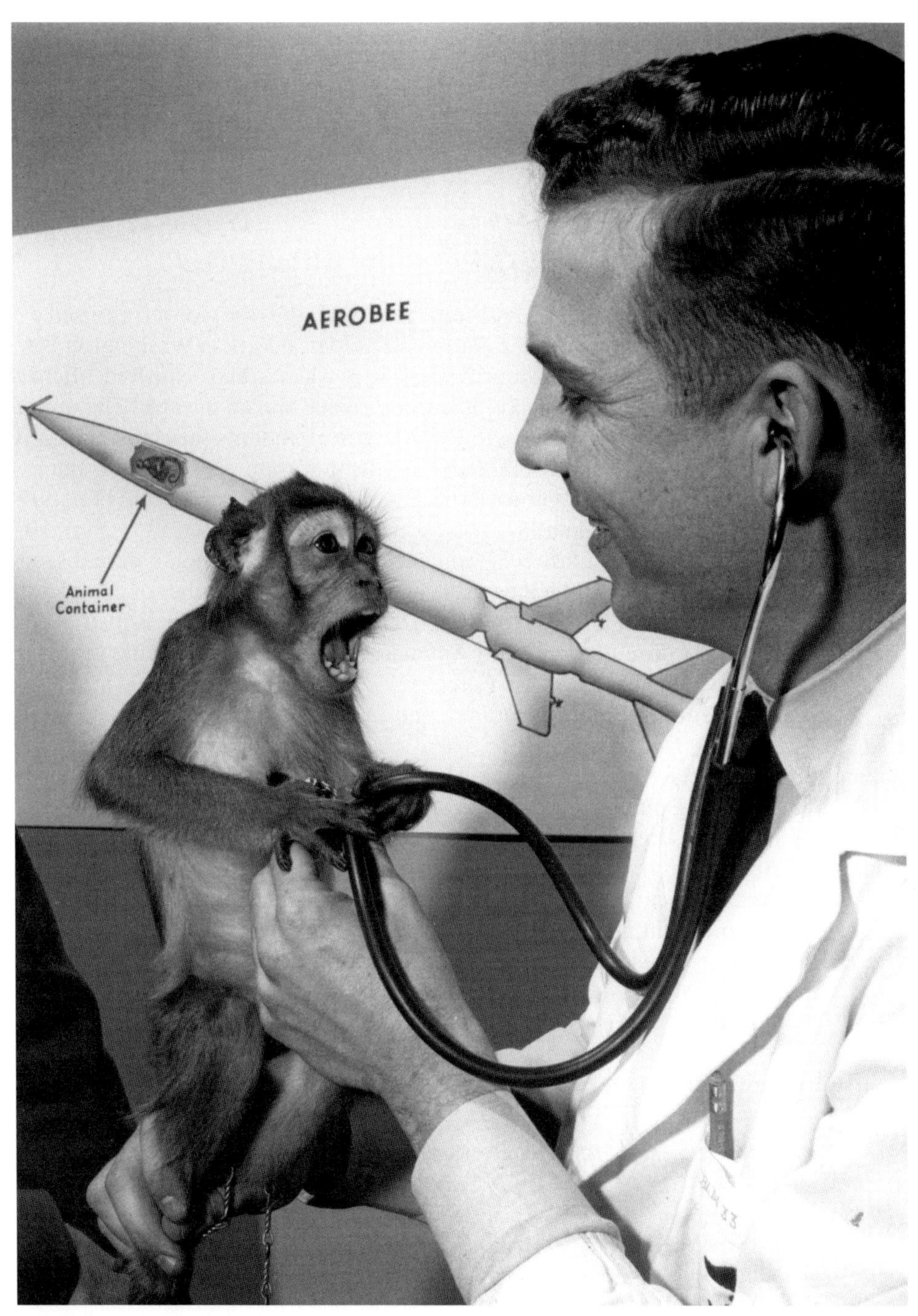

Michael is examined after his successful space flight. A cutaway of the Aerobee rocket is in the background. (Photo by Robert Sisson, used with permission of the National Geographic Society)

that tests of physiological and even psychological reactions can be conducted by proxy and in miniature scale by using animals.

"The next step awaits the development of rockets of significantly greater performance. Were such vehicles to become available it should be possible to telemeter physiological information in spite of a fourfold or even greater increase in distance from the Earth."

The sad saga of an ill-tempered monkey

When all the examinations and tests had come to an end, the two pioneering monkeys were given a home in a large cage at the National Zoological Park in Washington D.C. Both lived happily until one day about 2 years later when a bad-tempered Michael became annoyed with his mate and gave Patricia a savage bite on the arm. The wound became infected, and despite the best efforts of the zoo attendants they could not save her. A spokesman would later state that Patricia's death was purely accidental and not connected in any way to her flight aboard the Aerobee rocket. Michael died of natural causes at the zoological park in 1967.

From this flight in 1952 until 1957, animal research rocket flights were discontinued in the United States. While these tests had resulted in a number of complaints from animal lovers in the United States and abroad, including the British Royal Society for the Prevention of Cruelty to Animals, there was an interesting aspect to other letters received by the U.S. Air Force.

According to authors James Hanrahan and David Bushnell in their 1960 book, *Space Biology: The Human Factors in Space Flight*, some people were more than willing to take the place of the animals. The authors wrote that the Aerobee flights "inspired a surprising number of human volunteers to write and offer themselves as passengers in the next rocket. Such offers arrived at Holloman from as far away as the Philippines. Sometimes they were made by persons hoping to pay some debt to society by gathering scientific data at considerable risk and inconvenience to themselves; one offer, in fact, came from a resident in Washington State Penitentiary.

"So far, all the offers have been declined with thanks" [14].

REFERENCES

[1] George F. Meeter, *The Holloman Story*, University of New Mexico Press, Albuquerque, NM, 1967.

[2] Gordon J. Vaeth, *200 Miles Up: The Conquest of the Upper Air*, The Ronald Press Company, New York, 1956.

[3] Monika Guttman, *James Paget Henry, Developer of the Pressure Suit, Dies*. University of Southern California Public Relations. Website: *http://www.usc.edu/uscnews/stories/2403.html*

[4] Clyde R. Bergwin and William T. Coleman, *Animal Astronauts: They Opened the Way to the Stars*, Prentice-Hall, Englewood Cliffs, NJ, 1963.

[5] Christian Drummer, Massimo Cirillo and Natale De Santo, *Journal of Nephrology*, online article, "History of Fluid Balance and Kidney Function in Space," 2004. Website *http://www.sin-italy.org/jnonline/vol17nl/180.html*
[6] Gregory P. Kennedy, Oral History programme for the Kennedy Space Center International Space Hall of Fame, interview with David G. Simons, M.D., conducted 30 September 1987.
[7] David G. Simons, M.D., email correspondence with Colin Burgess, 10 June 2005.
[8] Lloyd Mallan, *Men, Rockets and Space Rats*, Julian Messner, New York, 1958.
[9] David G. Simons, M.D., technical report: *Use of V-2 Rocket to Convey Primate to Upper Atmosphere*, United States Air Force Air Material Command, May 1949. Permission to quote extensively from this report given by Dr. Simons.
[10] David G. Simons, M.D., email correspondence with Colin Burgess, 7 April 2006.
[11] *Journal of the British Interplanetary Society*, "Rocket flights of mammals to 200,000 feet," London, 1953, author not identified.
[12] David G. Simons, *Review of Biological Effects of Subgravity*, USAF report, May 1955, p. 211.
[13] Erik Bergaust and William Beller, *Satellite!* Lutterworth Press, London, 1957.
[14] James S. Hanrahan and David Bushnell, *Space Biology: The Human Factors in Space Flight*. Basic Books New York, 1960.

3

Pioneers of destiny: The suborbital dog flights

If you had visited Moscow's Institute of Aviation Medicine (IAM) in the summer of 1950, you might have been surprised to hear the clamour of dogs, to see them trotting about in pressure suits, and being spun in centrifuges. A cluster of small, light-furred dogs had taken up residence in that post-war summer, when the Institute was just beginning to turn its focus from the physiology of airplane to rocket flight. In the years following the Second World War, the major research thrust of the IAM was on the biological problems associated with flight.

The Institute, first established in 1934, had been closed during the war, but reopened in 1947 to investigate the problems related to high altitude, acceleration, the protection of pilots and rescue procedures.

HAVE YOU EVER SEEN A ROCKET BEING LAUNCHED?

Busy with such research, IAM physician, Dr. Vladimir Yazdovskiy, had certainly not expected the phone call he received in the winter of 1949. The caller introduced himself as Sergei Korolev, in charge of the development of "special equipment". Korolev had already taken the lead role in Soviet rocket development, but even he was forbidden to mention the exact nature of his research on the telephone.

As Yazdovskiy recalled in a 1995 interview, Korolev needed someone to take the lead on biomedical research in preparation for future manned space flight. He was already busy with research in aviation medicine, Yazdovskiy explained. But, Korolev, whose persuasive personality was legend, brushed aside any objection.

"Have you ever watched a rocket being launched?" Korolev asked. When Yazdovskiy's response came back in the negative, Korolev added, "Well then, if you've seen it once, it will stay with you for the rest of your life" [1].

A programme in its infancy

By the end of 1949, biomedical research in rockets had already become more than a theoretical science. The U.S. had conducted four animal flights using rhesus monkeys loaded into the nose cones of captured German V-2 rockets. Unfortunately, none of the animals had survived the flight, but the effort must surely have caught Korolev's attention. At the same time, the development of large rockets in both the Soviet Union and the United States was still very much in its infancy. Other than a handful of scientists and technicians, very few people had in fact seen a launch.

Beginning with the help of German technicians after the war, the Soviet Union had aggressively pushed rocket development, with the ultimate goal of developing a rocket powerful enough to deliver a military warhead to another continent. By 1948 they had tested their own modified version of the German V-2 rocket, the R-1, at their new launch facility at Kapustin Yar, a semi-arid region about 75 miles east of Stalingrad.

Concurrent with this, development proceeded on rockets that would be used for scientific purposes. As well, Soviet researchers would be steadfastly involved in a systematic programme aimed at testing factors that could prove harmful to human organisms in attaining ever greater velocities and higher altitudes. This would not only aid in the short-term development of insulated pressure suits and hermetically sealed cabins, but might one day extend to placing a Soviet person into space. Many biological questions needed to be resolved, and soon.

The first group of scientific – or geophysical – rockets, derivatives of the R-1, included the R-1B and R-1V. These were virtually identical, except that the R-1V replaced some scientific instruments with a parachute for the recovery of the main rocket body. They also carried a variety of equipment to study conditions in the upper atmosphere and included a biological container. An animal would fly in that container, but what type of animal should be used? How could it be trained? What type of container would best protect it during the flight? What would happen to it during a rocket flight? These were all questions waiting to be answered.

PREPARING FOR BIOLOGICAL FLIGHTS

Soon Yazdovskiy headed up a small team investigating those questions. They began by studying translations of published articles about the American animal launches. Although many of these animals had died due to mechanical systems failures, medical data collected from them seemed to indicate that the monkeys suffered no ill effects from acceleration and their brief exposure to weightlessness and cosmic radiation.

Still, the problems that needed to be overcome were daunting: (1) How to prepare an animal for the acceleration forces, vibration, noise and weightlessness encountered during a rocket flight? (2) How to design a payload module that would protect its animal passengers from the vacuum, extreme temperatures, radiation and meteorites of near space? (3) How to solve the problems related to the confinement of animals in a very small space? (4) How to recover the animal safely after the flight?

Which animals would fly on rockets?

Of course, one of the more critical questions was the type of animal that should fly aboard these rockets. Using monkeys, as the Americans did, had one obvious advantage. Physiologically, monkeys and apes were closest to humans. But it was felt that monkeys were too difficult to train under experimental conditions. Small animals – such as rabbits, rats and mice – had proven useful in laboratory studies and on high-altitude balloon and early American rocket flights, but they too had drawbacks. For one, it was difficult to attach to them the various sensors that would be required to read the physiological data during flight. Also, under some experimental conditions, they characteristically had a very high frequency of cardiac contractions and high respiration rates, making them more difficult to monitor [2].

In December 1950 the decision was made to use dogs. A great deal of experimental work with dogs had already been conducted in the Soviet Union. The renowned Russian scientist, Ivan Pavlov (1849–1936) had used dogs in the 1890s and early 20th century for his pioneering work in animal physiology. Dogs were a "known commodity", and were felt to be less excitable than monkeys and closer to humans in their emotional and physical reactions. One disadvantage that raised concern was that dogs were considered to be unique in their personalities, or not "linear". The reaction of one dog can differ dramatically from the reaction of another dog because their life experiences are so different [3]. This presented a problem when trying to determine how a living creature reacts to flight in a rocket. In part because of this, it was decided that two dogs would be used on each flight, so that the reactions of one could be measured against the other.

The kennel that soon appeared at the Institute of Aviation Medicine certainly brought with it a noticeable change in the atmosphere for those who worked there. All of their previous work in biomedical research relating to airplanes had been carried out on human test subjects. Known as "testers", these young men were mainly volunteers from army and engineering schools, helping to investigate the paradigms of flights at high altitudes and reactions under jet propulsion [4]. They were also employed in testing new equipment for the protection and rescue of pilots. Their work was dangerous, but the men were paid well for their efforts and often received coveted government awards. But now these "testers" gave way to a boisterous group of canines that had taken up residence at the Institute.

Selecting the first animal cosmonauts

Preparing "man's best friend" to become an astronaut began with a rigorous screening process. One of the qualities required of the dogs was obvious – they had to be small. To fit in the limited payload space (9.8 cubic feet) of the R-1, along with all of the collateral equipment, the ideal dog had to weigh between 13 and 16 pounds, slightly larger than a house cat. But being small was only the first of many qualities the successful canine candidate had to possess.

Because dogs that were too old or too young did not react as well to harsh conditions, the age cut-off fell between 18 months and 6 years. Since the animals would be filmed during the flight under poor illumination in their cramped container, light-coloured fur was an important factor. Because the sanitary device attached to the dog's clothing fit females better than males, only female dogs were used [5].

Using those criteria, recruitment began. Fortunately, stray dogs were plentiful on the streets of Moscow, and these formed the main pool of candidates for the space dog training programme. Most of the dogs were rounded up from the pound. Purebred dogs had not earned high marks in early lab tests. Mongrels, on the other hand, were made of sterner stuff, especially if they had lived a hard life on the streets, becoming inured to hunger and cold [3]. Such heartiness would serve them well during the rigorous training programme on which they were about to embark.

Each dog got a thorough physical; its height, length, and weight were measured, after which they were categorised by personality. Even-tempered dogs went into one group, restless dogs into another and sluggish dogs into a third. These classifications would help determine the course of their training and their suitability for certain flights. Later in the space dog programme, personality sorting and behavioural profiling would determine whether a dog would be classified as a "rocket dog", suitable for shorter, ballistic flights, or a "satellite dog", more suited for longer, orbital flights.

Among the group of dogs to go into training for the first series of dog launches were those named Bobik, Chizhik, Dezik, Lisa, Mishka, Neputevvy, Ryzhik, Smelaya and Tsygan. Unfortunately, Bobik missed his chance for fame by running off the day before his flight. A stray dog was quickly recruited and given the name ZIB, which is the Russian acronym, derived from the words "substitute for missing dog Bobik". Also in that first class of canine recruits were several dogs that would fly in the second series of launches, including Albina, Kozyavka and Malyshka.

The corps of canine cosmonauts would be in flux over the years. Some dogs did not hold up well under the rigours of the programme and were removed. Others would be in the programme for years and make multiple flights. Still others might be around for years but would only be used to aid in the development of training methods, test equipment or be part of various control groups.

Training dogs to fly in rockets

Preparing dogs to fly in a rocket involved exposing them to the conditions they would experience during a flight. It began with confinement capsules and the patient task of training them to become used to confinement in a small space. For this, they would first be dressed in a restraining suit, composed of a knit vest and short pants of light silk, with rings attached. Chains clipped to these rings held the dog inside a small box. For periods of 1 hour, then 2 hours, then 4, the dogs were required to stay in their "capsule". They had completed this phase of training only when they could submit to confinement for several days [6].

A rocket flight would expose the dogs to both excessive gravity and zero gravity. During the powered phase of the flight, three to five times the force of gravity would press upon the canine passenger. Then, during the gliding phase, the pull of gravity would briefly disappear. These experiences were simulated in a centrifuge and on parabolic airplane flights. Although the dogs would experience a maximum of just over 5 g's during the flight, their centrifuge rides exerted forces on them up to 10 g's. Pressure chambers were also used to introduce the dogs to the feeling of changing atmospheric pressure.

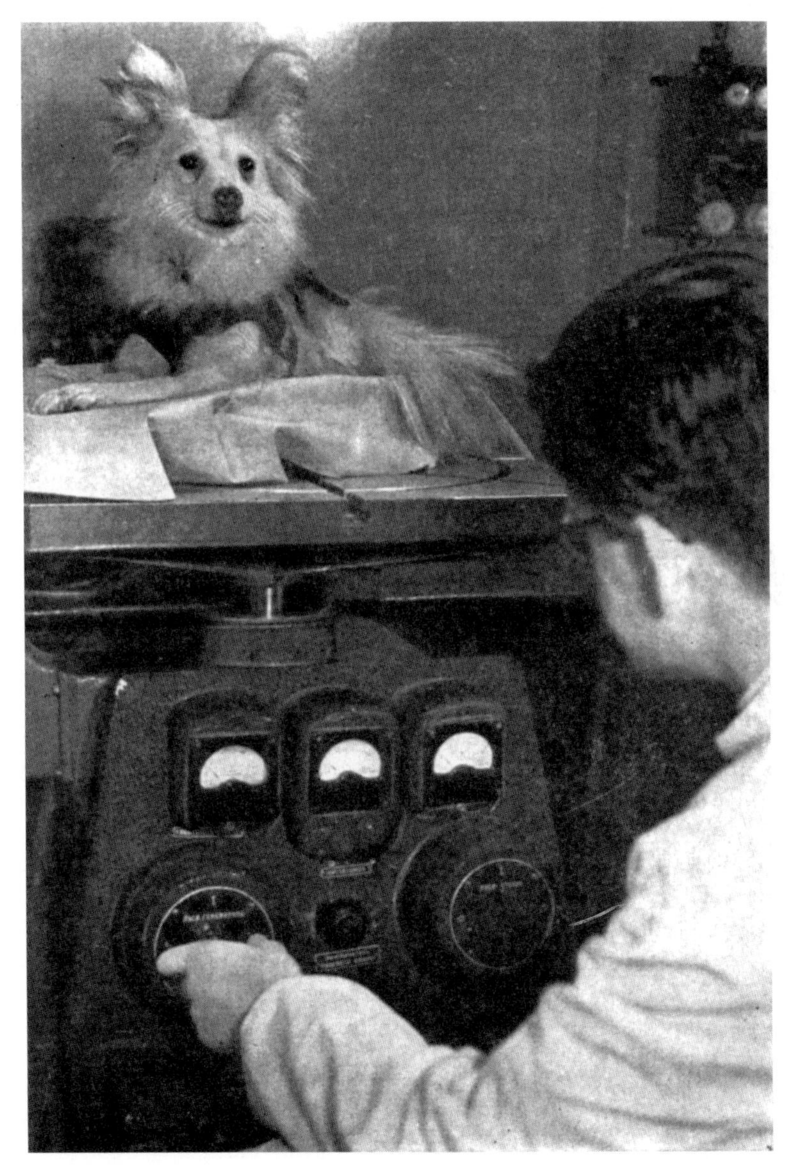

To familiarise them with the vibration they would experience during a rocket launch, dogs were trained on a vibration table such as this one. (Photo: authors' collections)

The use of a vibration table could recreate the noise and shuddering experienced during lift-off. Secured to a cushioned platform mounted on a shaft, with probes attached to monitor its heart rate, breathing and blood pressure, the hapless animal would be shaken about, while the mechanism of the table created a loud and obviously frightening banging. Sensors monitoring the animal's physiological reactions generally indicated a rapid rise in their blood pressure and heart rate during the shaking, but once the animals were removed from the table, these quickly returned to normal. A realistic launch experience was also simulated for the dogs during actual test stand rocket-firings [6].

Physically, the dogs were challenged by the training regimen, but the scientists were equally concerned with their psychological reaction. Every aspect of the dogs' behaviour was monitored closely. Only when their behaviour had been studied over a long period of time, under many different conditions, could an accurate picture be gained of their reactions during the experiments. Not only did these behaviour profiles aid in the selection of the best dog for each flight, they would also help researchers to evaluate the performance of the dogs during a flight.

"RETURN WITH VICTORY": THE FIRST DOG FLIGHT

As Yazdovskiy put the canine charges through their paces in the winter of 1951, he received word that the first series of six biological launches had been scheduled for that summer. The responsibility would fall to him to select the dogs for each of the flights.

Meanwhile, engineers at the Institute were working to design and build a container that would protect the dogs at approximately 62 miles altitude. Both the R-1B and R-1V would be used for the flights. The animal container, hermetically sealed to maintain atmospheric pressure, had to fit inside the tapered upper portion of the rocket, along with a life-support system and the recovery parachute. An injector blew a mixture of compressed air and oxygen into the capsule, while a soda lime cartridge removed carbon dioxide from the air and a silica gel cartridge absorbed moisture resulting from the animal's exhalation. There was no need for a feeding system to sustain the canine occupant, since the combined time from countdown to recovery would be less than a few hours.

Sensors would record the all-important biological data: skin temperature, respiration and pulse rate. A movie camera would also provide a visual record of the dogs' reactions to acceleration and weightlessness.

Tsygan and Dezik lead the way

As 22 July approached, the day of the first launch, Yazdovskiy finally settled on the two dogs to make the historic flight: Tsygan and Dezik, who had both "demonstrated calmness and endurance through all the tests" [7]. Along with Yazdovskiy's team, the dogs were shipped off to the launch site at Kapustin Yar.

The Kapustin Yar site had been selected for its remoteness and certainly not for the amenities it offered the resident workers and VIP guests come to watch a launch.

It was mostly barren steppe, bitterly cold in winter, unremittingly hot in summer and offering the barest accommodations. But it was in the midst of this desolation that an impressive contingent of spectators assembled one July day in 1951 to witness a truly historic rocket launch.

This would be the first time that dogs flew in a rocket. It might also be the first time that animals were ever recovered alive from a rocket flight. The United States had now attempted five monkey launches, and all the animals had died. But perhaps the greatest implication of this flight was how it would affect the continued support and goodwill of the Soviet scientific establishment. The military use of rockets always trumped their scientific use, and if the biological flights could not demonstrate their usefulness, they would be less likely to receive continued support.

Pre-dawn, 22 July 1951, Kapustin Yar. Before Yazdovskiy put the hatch in place to seal the dogs in the rocket, he petted them and said, "Return with Victory," a phrase commonly spoken to soldiers heading to battle [7].

Launch time draws near

In the gathering light of pre-dawn, a crowd of visitors clustered around their cars a safe distance from the launch site. Launches were routinely scheduled 3–5 minutes before sunrise so that the rays of the sun would illuminate the rocket as it climbed, making it easier to track from the ground. Among the VIPs in attendance was a man named Anatoli Blagonravov. He had just left his post as president of the Academy of Artillery Sciences to head the Commission for the Investigation of the Upper Atmosphere, the state commission that was overseeing the biological launches. Other members of the Commission were also on hand.

Tsygan and Dezik had already undergone a thorough pre-flight physical and now sat in the cramped nose cone of the rocket, awaiting launch. Vladimir Yazdovskiy appeared before the spectators, entreating them to remain in place when the dog capsule landed. His recovery team needed time to extract the dogs from the rocket, he explained, and to give the animals post-flight medical checkups.

When the brilliant flash of the rocket engines illuminated the twilight and a mighty roar rolled across the steppes to the crowd, it would have been easy to appreciate Korolev's earlier question to Yazdovskiy about the emotional impact of watching a rocket launch for the first time. Many of these spectators had never before witnessed the launch of a rocket. It was an awesome spectacle and one that would surely stay with them for the rest of their lives.

The dignitaries followed Yazdovskiy's advice during lift-off and throughout the worrying minutes as the rocket ascended to peak altitude of 62 miles. They even managed to contain themselves as they waited for some sign of the rocket's return. Then the spectators saw a long, dark object plunging rapidly Earthward through the grey sky a few miles away. They watched in horror as it neared the horizon and cried out as a sudden flash of light blossomed briefly on the ground, followed soon after by the dull thud of a distant explosion. The rocket scientists assured them that it was merely the main body of the rocket hitting the ground, and the remains of its fuel exploding. Nevertheless, they were still deeply anxious as they scanned the skies, looking for the telltale opening of a parachute. Then they saw it blossom out, and an exultant cry went up as the nose cone containing the dog capsule was slowly lowered to the ground beneath the parachute, now a soft pink with the reflected rays of sunrise. All their previous restraint crumbled. Yazdovskiy jumped in his car and sped off to the landing site several miles away, followed by Korolev and the other dignitaries in their cars, racing across the empty steppe, kicking up great clouds of dust, wanting desperately to be on hand at the recovery site to witness the outcome first hand [7].

The expectant crowd huddled close as the recovery crew removed the hatch, unstrapped the dogs and removed them from the tight space of their capsule. It was hard to tell who was more excited, the dogs or the onlookers. The dogs danced about, accepting the enthusiastic attention of those on hand. A quick examination revealed that Dezik was in fine shape; Tsygan had a scrape on her stomach, but was otherwise fine. They had done it. Tsygan and Dezik had played their role in spaceflight history by being the first animals to survive a ride in a rocket.

Dezik would fly on another rocket the following week, with a dog named Lisa. But disaster struck that flight. A pressure sensor used to trigger the parachute release was damaged by vibration during launch, and the parachute failed to deploy. Both dogs died when their capsule crashed to the ground, although the onboard film camera survived and provided useful data. On learning of the accident, Anatoli Blagonravov announced that Tsygan would not be flying on any more rockets. He took Tsygan home with him to Moscow to be his pet [8].

Shaken but not stirred

Over the course of 2 months in the summer of 1951, nine dogs would fly on six flights. They would travel at a speed of 2,600 mph, in the biggest rocket then in existence, feel the pressure of five times the weight of gravity and the free-floating disorientation of weightlessness for 3 minutes. Four of the dogs would die. The others would return safely, bringing with them invaluable minutes of film, recording their reactions during the flight. Agitated during the brief period of powered flight and during the free-fall back towards Earth, disoriented during the few minutes of weightlessness, the dogs had nonetheless survived. Living beings had flown in a rocket to the edge of space and survived. The implications for manned flights were enormous.

THE LIFE OF A SPACE DOG

It is worth taking a moment here to describe the life of a space dog in training. Although they were submitted to some gruelling training and faced physical challenges during their rocket flights, they otherwise lived a fairly pampered life. Home for them was a kennel, called a vivarium. Square, wooden-floored cages, littered with straw or wood shavings, and raised on yard-high supports served as their cages. The name of each inhabitant was written neatly in chalk on the exterior.

Monitoring the health of the dogs

Always at hand were many dogs that would never see the inside of a rocket, but were simply kept at the Institute for use in related research. They might be employed on the training equipment to establish physiological benchmarks or to test capsules or spacesuits. Their names, like so much Russian music to us now, never made it into the history books; names like Nochka, Bodraya, Zolushka, T'ma, Planeta, Marsianka and many more [6].

Solicitous attendants would walk the dogs at least twice a day, although animals about to be used in an experiment received additional exercise. Twice a day they received a meal from their special diet, which typically included a soup of gristle, bone and grain, as well as bread and meat. Their diet also included vegetables, fish oil and milk. Dogs that were about to undergo a particularly difficult experiment ate a special diet of sausage, bouillon, preserves and sweets.

In some of the long-duration experiments, such as protracted stays in a confinement cage, it was not unusual for the dogs to fret and not eat well. Brought from the

Exercise walks were part of the daily routine for space dogs in training. Even these playful occasions provided opportunities for evaluating the suitability of the dogs for rocket flights. (Photo: Collection of François Viger)

cage weakened and listless, they would require a glucose shot to revive their strength. However, veterinarians closely monitored their health and administered thorough examinations before and after each experiment.

For the behavioural scientists working with the dogs, continuous observation allowed them to create an elaborate profile of a dog's emotional reactions. Even such domestic activities as feeding time, exercise walks and time spent in their cages revealed important information about how the dogs reacted to different circumstances. It created a personality profile of the dog. For instance, it would be determined how much time a caged dog spent in active or passive forms of behaviour. Some of the dogs moved continually, others much less. Some barked loudly when a human appeared in the kennel or food appeared, while others calmly accepted such distractions. The dogs also differed in the nature and intensity of their movements. Sometimes the reactions might be quiet but active, while the reactions of others could be quite sluggish. Other times they could even be quite violent [6].

This study of the nature and extent of the dogs' activities (or lack thereof) in the vivarium and during related experiments provided insights into how the dogs would perform on flights. The dogs Laika, Belka and Strelka, who became famous for their record-setting orbital flights, were selected from the quiet dogs.

THE SECOND SERIES OF DOG FLIGHTS

A great deal had been learned in the first series of canine flights. It had been established beyond question that the occupants of a rocket could survive acceleration forces and weightlessness. There appeared to be no ill effects from a brief exposure to cosmic radiation, and the animal container had kept the dogs alive.

Spacesuits and life-support systems

A second series of flights began in July 1954. The series would comprise nine flights over the course of 2 years, employing first the R-1D rocket and then the R-1Ye rocket. Two radically different design features had now been incorporated, one having to do with the life-support system and the other with the method of recovery. The dogs would now fly in a non-hermetically sealed capsule, necessitating the wearing of a spacesuit to maintain oxygen and pressure. A new, low-tech solution for controlling capsule temperature would also be employed – the use of reindeer fur for insulation because of its innate ability to trap air [9].

The second design feature involved a completely new recovery system. Rather than returning to Earth inside the nose cone of the rocket, each dog would be ejected separately from the rocket, at different altitudes, and be parachuted to the ground in their capsules, or "chassis". An additional innovation for the Soviets was their first use of radio telemetry to transmit biological data during the flight.

The spacesuit used for this series was made of three layers of rubberised fabric, which resembled a canvas sack, with two sealed sleeves for the front paws of the dog. A metal collar allowed for the attachment of a spherical, Plexiglas helmet. The suit

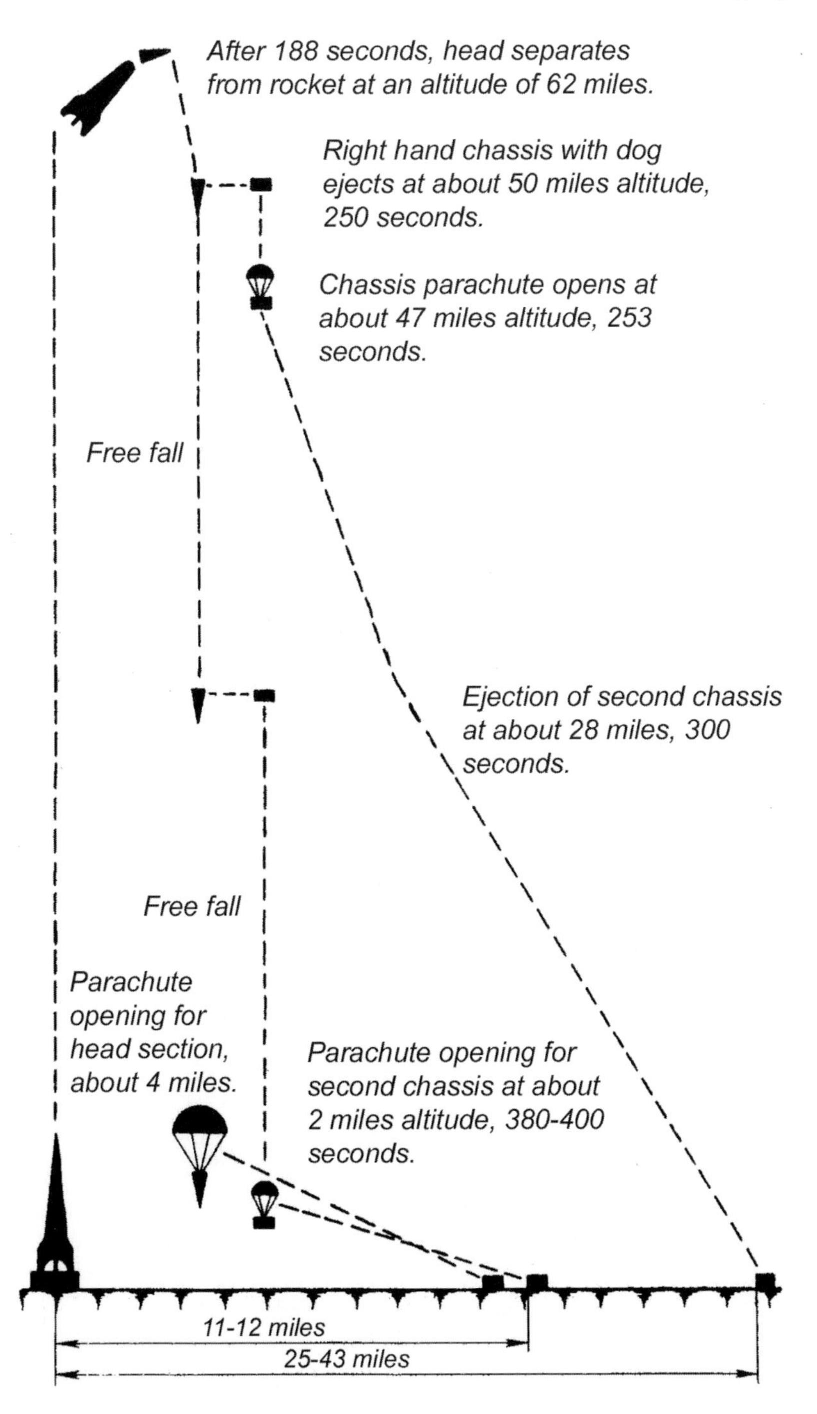

During the second series of launches, the dogs were catapulted from the nose section at different altitudes. (Illustration: authors' collections)

Kozyavka is readied for one of her flights in 1956. This series of flights used pressure suits like this one instead of pressurised capsules. (Photo: authors' collections)

fastened to a retractable tray on top of an ejection trolley. This steel trolley held the oxygen supply, parachute system, all the physiological measurement equipment and ejection timers. A telescopic pyrotechnical ejection mechanism, mounted beneath the trolley, shot the chassis out during ejection.

Physiological equipment was designed to measure and record respiration rate, maximum and minimum blood pressure, pulse rate, and body temperature or temperature inside the spacesuit. This information was relayed via telemetry to the ground, along with "sanitary" measurements of urine production.

The oxygen system in the chassis consisted of three 2-litre bottles that could keep the dog alive in the suit for 2 hours. During the descent, a valve in the animal's helmet opened automatically at around 12,000 feet to allow it to breathe after landing.

One small step at a time

The training regimen for this series of flights was very similar to that for the initial series, except that in addition to becoming acclimated to spending time in a tiny capsule, the dogs also had to be introduced to wearing a spacesuit. Training proceeded in steps. First the dogs were simply strapped down inside small training cabins.

During the second series of dog flights, the dogs were ejected from the descending capsule onboard a chassis such as this. (Photo: authors' collections)

According to Yazdovskiy, at the start the dogs could not stand to be restrained in this way. "They became very aggressive and had to be released from the cabin to rest," he pointed out [1].

Gradually, the time was increased, until they could tolerate 5 hours of such restraint. Then the same levels of tolerance were built up again, first with the dog in the spacesuit without the helmet, then with the helmet attached, then with the full suit and all sensors attached to their bodies. Dogs were selected for flights if they had calmly endured 3 hours in the suit during the week preceding a launch.

Once again flights went to an altitude of approximately 62 miles, at which point the nose section containing the dogs separated from the rocket body. Ejection of the animal in the right-hand chassis occurred at an altitude of about 50 miles, during unstable free-all. Three seconds later the parachute deployed, and the dog had a 1-hour ride to the ground. The nose section continued in free-fall until reaching an altitude of about 28 miles, where the left-hand chassis was ejected. This chassis was allowed to free-fall to an altitude of about 2 miles before parachute deployment.

Findings from the second series of flights confirmed the results from the first series. The blood pressure, respiration and heart rate of the dogs rose during the different phases of the flight, but the changes were not pronounced and always of short duration. The fact that no dramatic, substantial change occurred in the behaviour or physiological functioning of the dogs removed any barrier to longer flights. The stage was thus set for the next series of launches.

THINGS CHANGE IN 1956

In any history of the Soviet animal space programme, 1956 stands out. One could draw a before-and-after line through that year to separate those fledgling years of basic experimentation and the later years of development. It was as though, flush with their successes with these early suborbital flights, a world of possibility had opened for the Soviets.

Already in 1956, several different groups were engaged in preliminary work exploring the possibility of replacing dogs with humans on suborbital flights. Original plans called for manned flights to progress in two phases, utilising the R-5A rocket. They would commence at low altitudes and low speeds, then progress incrementally from there to the full capability of the rocket.

Devising the best recovery system

In his monumental study of the Soviet space programme, *Challenge to Apollo*, space historian, Asif Siddiqi wonderfully details the political machinations that surrounded this work and the directions taken by the research. One group, under the technical leadership of engineer Nikolay Belov, worked on cabin design and methods of recovery. Recovery options considered included the use of air brakes to slow descent and prevent parachute failure. In fact, air brakes would be used on the R-2 rocket for the next series of dog flights. Two radically new designs also given consideration were a rocket-assisted rotor blade that would operate like a helicopter, and a delta-winged capsule for a glider-like descent. By 1958, following the glowing success of additional dog flights and the development of the R-7 rocket, plans for a manned suborbital flight were abandoned [10].

While the media in the U.S. and Europe overflowed with effusive articles about rockets and space travel in the mid-1950s, virtually nothing was reported about the Soviet rocket programme – primarily, because, in the West, virtually nothing was known about it. Ten years of extensive rocket development and 5 years of dog flights had all proceeded in near total secrecy.

The West, and the worst kept secret

There had been a few broad hints in recent years about rocket developments behind the Iron Curtain. One week after the Eisenhower White House announced on 29 July 1955 that the United States would attempt to launch a satellite during the International Geophysical Year (1957–58), Russian scientist Leonid Sedov, the head of a Soviet delegation attending a meeting of the International Astronautical Federation (IAF) in Copenhagen, made an announcement of his own. He proudly proclaimed that the Soviet Union would likely be able to launch a satellite within the next 2 years as well.

As modest as that revelation may have been, it signalled something of a thaw in the obsessive Cold War secrecy that had surrounded Soviet use of rockets for scientific research. It was in December of the following year (1956) that the world first learned about the research being carried out with canine test subjects. Speaking in Paris, at the First International Congress on Rockets and Guided Missiles, Major General Aleksei Pokrovskii, director of the Institute of Aviation Medicine, delivered a report titled "Vital Activity of Animals during Rocket Flights into the Upper Atmosphere" [11]. His presentation summarised the procedures, technology and findings of the first and second series of dog flights, 1951–1956.

According to the report, these experimental flights had clearly established four critical findings: (1) the use of spacesuits protected the animals during flight and recovery, (2) the method of ejecting the animals from the rocket proved effective, (3) the system of parachutes ensured a safe descent and (4) a short flight of 1 hour into the upper atmosphere did not cause any substantial change in the behaviour of the dog or cause any physiological harm.

Based upon those findings, Pokrovskii concluded, "There is no doubt that thanks to the collective efforts of the various branches of science, thanks to the efforts of scientists of all countries, it will be possible to realise manned rocket flight in view of the studies of the upper layers of the atmosphere."

It is interesting to note that in this 1956 report, Pokrovskii is identified as the director of the Institute of Experimental Aviation Medicine. This name was occasionally used for the IAM, primarily as a cover for its space research. Not until 1960 would the name of the Institute be officially changed to the Institute of Aviation and Space Medicine, to better reflect the changing focus of its work.

A further reorganisation of the Institute occurred a few years later, when the space component broke away from the IASM. Actually, the Institute was a military organisation, part of the air force. By 1963, with much of its work being devoted to space research, it no longer seemed appropriate to keep that military connection. In 1963, the Institute of Biomedical Problems (IBMP) was created under the Ministry of Health, to specialise in aerospace medicine. The creation of the new institute involved no major changes. Buildings were transferred from one institute to the other, and many of the same personnel moved over as well.

Oleg Gazenko becomes involved

Beginning in 1956, a different attitude emerged at the Institute, to reflect a new sense of urgency and purpose shared by the scientists who were added to the biomedical research team in that year. Oleg Gazenko, a medical doctor, joined the Institute in August of 1956. He would eventually go on to head the Institute (1967–1987) and play a leading role in U.S.–Soviet cooperation on the Cosmos biological flights.

Like many of the other scientists working on the dog programme, Gazenko came with a background in aviation medicine. After graduating from the Second Moscow Medical School and the Military Medical Academy, he served as an Air Force doctor during the Second World War. After the war, aviation medical research investigated such areas as problems related to high altitude, the protection of pilots during acceleration, and the development of anti-gravity suits, centrifuges and ejection devices. It was referred to as "human factors engineering", and it served as perfect preparation for the biomedical work of flying dogs in rockets.

In a 1989 video interview conducted by the Smithsonian Institution, Gazenko remembered the pleasure and intensity of those early years working on the dog programme.

I'd like to tell you about the general feeling at the time. Everything was developing so rapidly, it was so tightly compressed in time. The nature of our work was so exciting, we needed no prodding, the very notion of day and night was blurred; there was no distinction of weekends and workdays. There was a succession of projects under way, new problems and the necessity to solve them. So that for me it seems like one long day, with all the events compressed into it. I cannot even separate out anything; it was a succession of things. We worked day and night, right into the night, often having no idea what the weather was like, or what was going on in the outside world. All that moved in the background. It was a chain of experiments. A new problem came up and you had to check it out. If you did not find an answer in the literature you went to the lab. And there was no one to consult with. So you had to do experiments yourself... Yes, 1956 and 1957 seemed like one day [9].

Focus at the Institute remained firmly on suborbital flights. The dog programme functioned under the Department of Physiology of Upper Atmospheric Flight. According to Abram Genin, another Institute physician interviewed by the Smithsonian Institution, "At that point space flights were not on the agenda. Anyway, we were afraid to say 'space' out loud" [9].

A new and more powerful rocket

In late 1956 and early 1957 the dogs were being prepared for the next series of launches aboard a new and much more powerful rocket. The Soviet military's desire for a truly intercontinental missile was spurring the development of longer-range rockets. Enter the R-2, an intermediate-range missile. Although it still resembled the V-2 in many respects, the R-2 incorporated several new design features that gave it the power to loft payloads to twice the altitude of the R-1. The new missile was 10 feet longer, lighter in weight because of aluminium alloy construction, and it utilised an improved engine. Rather than travelling to an altitude of 62 miles, dogs would now be lifted to 130 miles aboard the R-2A, the scientific version of the military R-2.

These longer flights, with higher acceleration forces, longer periods inside the capsule and extended minutes of weightlessness, presented new challenges for the dogs and for those who trained them. The focus of research also began to shift. The most basic questions regarding living beings travelling on rockets had been answered by the early flights, namely whether dogs could withstand weightlessness, radiation, and the stress of launch and re-entry. Now research began to focus more closely on the nature of the biological changes that occur during and after a rocket flight.

First, however, the dogs had to be prepared. For that purpose, the scientists sought some rather practical assistance. "I even remember going to the circus with a friend where they had trained dogs and monkeys," Gazenko recalled. "We talked to their trainers to learn something from their experience" [9]. The team also sought help from the legendary circus family of animal trainers, the Durovs. Founded in Moscow in 1911 by Vladimir Durov, the Durov Animals Theater is still in existence today. Durov, who claimed to communicate with his animals through telepathy, was able to get them to perform extraordinary feats for the entertainment of the audience. Maria Aleksandrovna Gertz, an animal psychologist who was a follower of the Durov school of animal training, worked with the dogs at the Institute [9].

As before, the personality of each dog assumed key importance. Their level of aggressiveness or calmness, how well they adapted to their environment, their reaction to stress – these were all telling characteristics. According to Gazenko, the dogs were trained differently than animals in physiology laboratories. "We were more interested in pre-flight training than in biological experiments. Instead of concentrating on the body, we were more interested in the creature itself, the dog's personality. So we observed their behaviour and perhaps learned the principles we used later in the selection and training of the cosmonauts" [9].

An ID card was now created for each dog, complete with photograph and physical description, such as weight, height, temperament, ECG and more. The average weight of the dogs used in these flights was 11–15 pounds. They were now trained as either "rocket dogs" or "satellite dogs", and it was the latter group that would receive the most intense training, if only for the fact that their flights would not last merely a couple of hours, but days and eventually weeks.

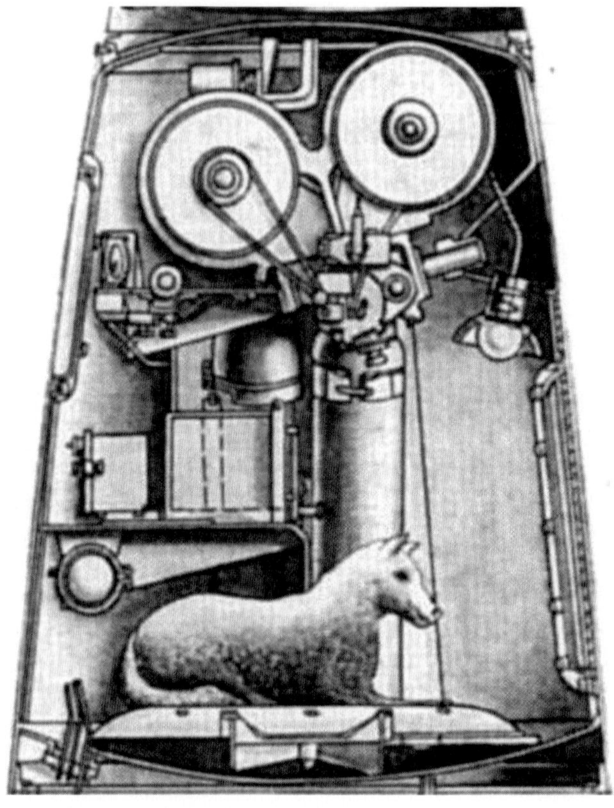

The somewhat more spacious R-2A rocket cabin allowed for direct, overhead filming of the dogs in flight, as opposed to the rear-mounted camera in the R-1 capsule that filmed off of a mirror. (Photo: authors' collections)

The first R-2A dogs

As the year 1957 dawned, the first "satellite dog" flight, of Laika in Sputnik 2, was less than a year away. However, as hugely as that event unexpectedly loomed for the West, and as clear a demarcation it represented in the relentless progress of space flight, it would barely register on the radar of those involved in the dog flight programme. A full schedule of suborbital flights preceded Sputnik 2 in 1957, and an equally robust schedule would follow it for nearly three more years.

For the R-2A flights, the Soviets reverted to the use of hermetically sealed cabins. However, primarily because of the large door in the cabin, there was not a true hermetic seal. Three 7-litre tanks of air–oxygen mix supplied the breathing gases, maintained pressure and vented directly to the outside. The supply was sufficient for 6 hours.

The ejectable capsule design from the previous series had also been discarded. The R-2A allowed for a slightly larger capsule space for the dogs and their equipment, just under $17\frac{1}{2}$ cubic feet. The nose cone of the rocket, containing the dogs, would separate

A dog exits her capsule in the nose section of an R-2A rocket. Note the scorch marks on the air brakes, which were used to slow the capsule before deployment of the parachute. (Photo: authors' collection)

at the peak of trajectory and free-fall. It would then be slowed by drag brakes until a series of three braking parachutes deployed at an altitude of 3 miles. This braking procedure imposed 8 g's on the dogs. At an altitude of about 1 mile the basic parachute deployed, with only a 4-g impact, and lowered the capsule to Earth [12].

Blood pressure, respiration and pulse were recorded continuously during the flight and telemetered to the ground. An ECG was also taken onboard. Pre- and post-flight physicals involved blood analysis, X-ray of the thorax, blood pressure, ECG, pulse and respiration rates, urinalysis, and measurement of body temperature and weight.

The first R-2A dog flight, on 16 May 1957, carried the dogs Ryzhaya and Damka to 130 miles and provided them with 6 minutes of microgravity. In addition to the onboard gathering of biological data, continuous filming monitored the dogs' behaviour.

For the first time in 1957, anaesthetised dogs were used on some of the flights. This had been a common practice on U.S. monkey flights in the late 1940s and early 1950s, but had never before been employed by the Soviets. Using this procedure, they hoped to determine which physiological reactions were a response to weightlessness rather than a response to the general irritants associated with the flight. On the flights on which this was employed, only one of the two dogs would be anaesthetised, using a 10% solution of hexenal injected subcutaneously.

The five R-2A launches in the summer of 1957 were marred by only one tragedy, when the dogs Ryzhaya and Dzhoyna lost their lives during the second flight, on 24 May. Still, there were some worrisome signs. On several of the flights, drops of

Chief Designer Sergei Korolev poses with an unidentified space dog, July 1957. Whenever Korolev came into the training laboratories, he would ask about the dogs and pet them affectionately. (Photo: authors' collections)

blood were found on the walls of the cabin. A number of dogs also had blood on their nose and rectum. A haemorrhage was noted inside Damka's eye after her first flight, presumably from G forces during braking, probably resulting from her being out of proper position [7]. However, the overall conclusion was that these flights to 130 miles did not cause any significant physiological or behavioural problems for the dogs. The final 1957 launch in this series occurred on 31 August, carrying the dogs Belka and Damka.

Although the R-2A would continue to carry dogs and other animals on ballistic flights, a far more powerful rocket had recently made its appearance. Just 10 days prior to the 31 August flight of Belka and Damka, the R-7 rocket had recorded its first successful flight from the new Baikonur launch facility in Kazakhstan. After five failed attempts, it delivered a dummy H-bomb warhead 3,700 miles downrange. This truly intercontinental ballistic missile would soon become the workhorse for all future orbital dog flights.

Creating a biological, orbital satellite

If they were not yet mentioning the word "space" at the Institute of Aviation Medicine, it had long been on the mind of Korolev and some of his design groups. In early 1957, pioneering rocket scientist and engineer Mikhail Tikhonravov was reassigned to Korolev's own design group, OKB-1 (the Russian language abbreviation for Experimental Design Bureau Number 1). One of Tikhonravov's design projects was to begin work on a biological satellite capable of carrying dogs for orbital flights of more than 1 day [8].

By the end of that year, the hastily-fabricated Sputnik 2 capsule would loft a dog named Laika on the first orbital flight of a living creature. Combined with the spectacular launch of Sputnik a month earlier, the Soviet space programme suddenly rated front-page headlines in the world press. Despite all of this publicity, the attendant commotion created barely a ripple in the dog programme, which continued with a full schedule of suborbital flights.

Commenting on his reaction to the launch of Sputnik 2, Oleg Gazenko said, "I am telling you honestly that I did not perceive any gigantic step forward . . . I just thought it was the next step, an interesting one but that's all. I guess the sensation of something new, grandiose, and unexpected came from the reactions of other people, maybe from the highfalutin' press coverage. 'For the first time ever, a man-made apparatus has overcome Earth's gravitation forces,' and so on" [9].

The historic achievement of the Sputnik 2 flight did not mean that dogs had finished their role in the Soviet space programme. No living creature would return to orbit for another 3 years. In the interim, some 9 additional suborbital dog flights would continue to perfect equipment and training procedures and gather critical biological data. The dependable R-2A rocket carried most of these flights to altitudes of 130 miles. However, a scientific version of the R-5 rocket, the R-5A, had been developed in 1956. In 1958 it would once again double the altitude of the suborbital dog flights, lifting the animals to 280 miles and exposing them to 9 minutes of weightlessness.

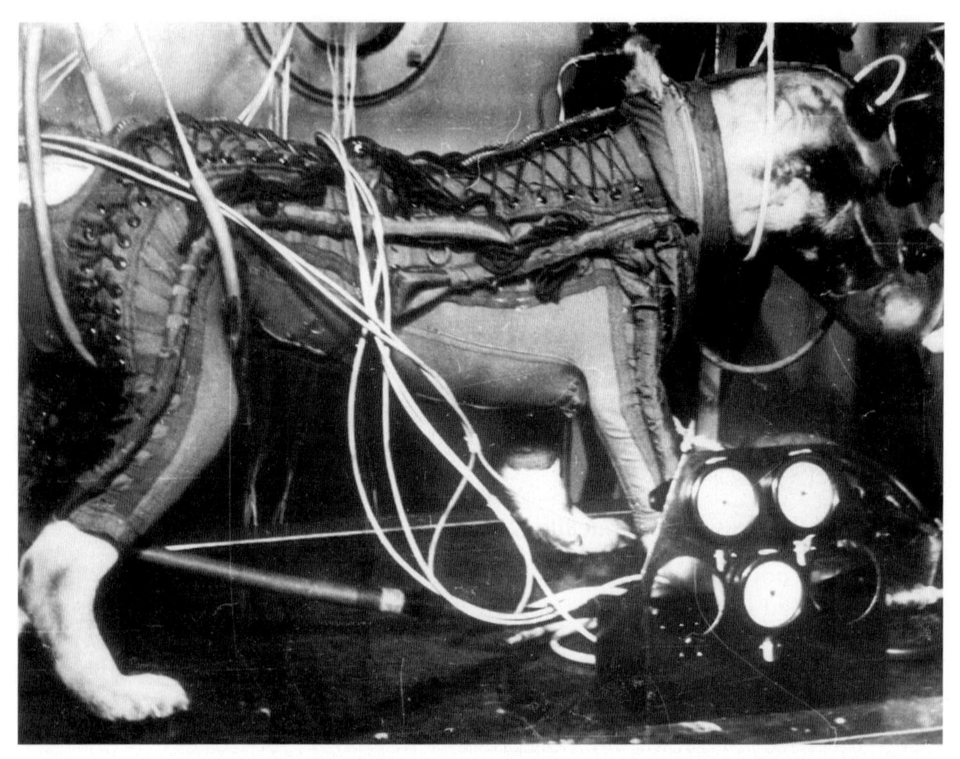

A dog tests the pressure suit developed to protect the dogs at high altitudes. (Photo: authors' collections)

Dogs would ride rockets 5 times in the busy year of 1958. Following the launch of the two Sputnik satellites, the Space Race pushed everything along at a faster pace. The United States launched its first satellite in January 1958. It was also sending monkeys aloft on suborbital flights. In April 1959, the U.S. introduced its first group of space pilots, the seven Mercury astronauts. That same year, in much more secrecy, the Soviet Union began to train its own small group of cosmonauts to fly in space.

A programme ends, another begins

Testing on the first Soviet spacecraft, named Vostok, began in 1960 on an accelerated schedule. NASA had announced plans for a suborbital Mercury flight in early 1961. To Sergei Korolev that meant the Soviet Union had to upstage the Americans by launching a manned Vostok flight before the end of 1960 [10]. The two suborbital dog flights in June 1960 helped to perfect the life-support systems to be used on the manned Vostok capsule. But, they also put a close to 9 years of suborbital dog launches. They had helped to bring the Soviet space programme from the excitement of that first biological launch to the edge of manned space flight. Scarcely 10 months would pass

This rocket launched at 8:06. The clock shows 8:07. The dogs Pestraya and Belyanka are feeling the "G-force" press against them as their rocket climbs to an altitude of 280 miles in 1958. (Photo: authors' collections)

On 15 June 1960, on one of the last of the suborbital flights, the dogs Otvazhnaya and Malek shared their capsule with a rabbit named Marfusha. (Photo: authors' collections)

between the last suborbital dog flight and the historic space voyage of Yuri Gagarin, in April 1961.

The only thing yet required to bridge that 10-month gap would be a series of canine orbital flights that once again called for animals to pioneer the way for humans.

REFERENCES

[1] *Roads to Space, An Oral History of the Soviet Space Program*, compiled by the Russian Scientific Research Center for Space Documentation, translated by Peter Berlin, edited by John Rhea. Aviation Week Group, McGraw-Hill, New York, 1995.

[2] O.G. Gazenko and S. Georgiyevskiy, "Preparation of the Animal Prior to the Experiment," pp. 353–359, in N.M. Sisakyan (ed.), *Problems of Space Biology*, Vol. 1, NASA, Center for Aerospace Information. Russian title: *Problemy Kosmicheskoy Biologii*, USSR Academy Publishing House, Moscow, 1962.

[3] Vladimir Gubarev, "Academic O. Gazenko: Wind of Cosmic Travels," *Nauka I Zhizn*, No. 7, 2001, pp. 30–37 [in Russian]. Russian title: *"Akademik O. Gazenko: Vyeter Kozmicheskikh Stranstviy."*

[4] Valerie Sharov, "For Me, Man's Exit Into Outer Space Was No Miracle," *New Times* (Russia), March 2005.

[5] Evgeny Riabchikov, *Russians in Space*, edited by Colonel General Nikolai P. Kamanin, translated by Guy Daniels. Doubleday, Garden City, NY, 1971.

[6] M.A. Gerd and N.N. Gurovskiy, *The First Astronauts and the First Scouts of Outer Space*, Translation Services Branch, Foreign Technology Division Wright-Patterson-AFB, Ohio. 1963. Russian title: *Pervyye Kosmonavty I Pervyye Razvedchiki Kosmosa*, Akademiya Nauk, SSSR, Izdatel'stvo Akademii Nauk, Moskva, 1962.

[7] V.I. Yazdovskiy, *On the Trail of the Universe: Collection of Cosmic Biology and Medicine in the Exploration of Space*, Firma Slova, Moscow, 1996 [in Russian]. Russian title: *Na Tropakh Vselenniy: Vklad Kosmicheskiy Biologii I Meditsini v Osvoenye Kozmicheskovo Prostranstva.*

[8] Asif Siddiqi, "'There it is!' An Account of the First Dogs-in-Space Program." *Quest, The History of Spaceflight Magazine*, Vol. 5, No. 3, pp. 38–42.

[9] "Soviet Space Medicine, Session Two," transcript of Smithsonian Videohistory Program, interview with Oleg Gazenko, Abraham Genin and Evgenii Shepelev. Cathleen S. Lewis, interviewer, 28 November 1989.

[10] Asif A. Siddiqi, *Challenge to Apollo: The Soviet Union and the Space Race, 1945–1974*, NASA, Washington, D.C., 2000.

[11] A.V. Pokrovskii, "Vital Activity of Animals during Rocket Flights into the Upper Atmosphere," in F.J. Krieger (ed.), *Behind the Sputniks: A Survey of Soviet Space Science*, Public Affairs Press, Washington, D.C., 1958. Originally presented as a report to the International Congress on Guided Missiles and Rockets, Paris, 3–8 December 1956.

[12] George E. Wukelic (ed.), *Handbook of Soviet Space Science Research.* Gordon & Breach, New York, 1968.

4

High-altitude research

A mere 60 years ago, not a single artificial satellite was orbiting the Earth. In fact, the programmes that would eventually put them there had hardly been set in motion. In the post-war years it was known that rockets alone could provide the sustained thrust necessary to loft anything into space, but they were a relatively new and basically unsophisticated technology. In the latter part of the Second World War they had been used as weapons of destruction, and for this reason alone the military was interested in their potential use as ballistic missiles capable of carrying warheads.

Meanwhile, the interest of scientists had been aroused by the possibilities of conducting biological experiments in areas beyond our atmosphere, but the use of captured V-2 rockets would be strictly limited to all but a handful of researchers. However, for those anxious to conduct high-altitude experiments, there was a far more accessible and inexpensive alternative – the balloon. Balloons capable of carrying sizable payloads had been taking to the skies for nearly two centuries, so while rocket technology was rapidly advancing under military administration, scientists once again turned to a more traditional and reliable means of vertical transportation to assist in their investigations. And once again, animals would play a key role.

A CURIOUS PHENOMENON

As children, Joseph-Michel and Jacques-Étienne Montgolfier had often marvelled at a curious phenomenon outside their father's factory in southern France. He was a prosperous paper manufacturer, and they had become fascinated by the occasional sight of paper bags rising in updrafts from the factory's chimney. This led them to experiment over many years with other crude vehicles made of silk and linen, and their innovative work would eventually be recognised as the true genesis of ballooning [1].

First animal passengers

On 19 September 1783, a crowd of nearly 130,000 people had gathered around the magnificent Great Court outside the royal palace in Versailles, France. Surrounded by exquisitely kept, scented gardens, their main interest was focused on less floral matters. They had travelled to the palace to witness the public demonstration (or possible humiliation) of a rather extraordinary vehicle and its two creators. Among the curious onlookers that day were King Louis XVI and his queen Marie Antoinette, who would lose their heads to the guillotine 10 years later.

The mood that autumn day was one of excitement, and the prevailing weather conditions were good. If all went well, and the winds remained calm, Joseph-Michel Montgolfier and his 5-year-younger brother Jacques-Étienne hoped to launch their second full-size hot-air balloon into the skies on its maiden voyage. It was a 40-foot diameter marvel they had christened *Le Martial.*

Standing an imposing 60 feet tall, the elegantly painted balloon was constructed of linen and paper. For this flight a large wicker cage had been suspended beneath the balloon, within which the brothers had placed a cockerel, a duck and a sheep. These three animals were unknowingly destined to become the first hot-air balloon passengers in history.

Le Martial had no transportable heat source of its own, so when ascent time came the gaping hole at the foot of the balloon was suspended over a straw fire smouldering in a large cauldron. The great bag slowly became engorged with hot gases, expanding and rising until it was straining against the guy ropes. Then, to the applause of the crowd, it was unleashed and rose majestically into the sky, eventually reaching an altitude of around 1,500 feet.

The balloon would remain airborne for 8 minutes before it descended to a safe landing over a mile away, on the fringes of the Forest of Vaucresson. The cockerel, duck and sheep had become the first living creatures ever to survive a ride into the skies aboard a man-made ascent vehicle. The only known casualty was the cockerel, which was nursing a broken wing after being kicked by the agitated sheep.

On fire, but safe

Two months later, an adventurous science teacher named Francois Pilatre de Rozier clambered into the open basket of another Montgolfier balloon, accompanied by the Marquis d'Arlandes, François Laurent. This time a portable brazier had been installed below the mouth of the balloon, burning a pungent but effective combination of damp straw, rag and rotting meat that heated and rarefied the air inside the balloon.

Once the guy ropes had been released, the pear-shaped vessel rose to nearly 3,000 feet, drifting at will for the next 22 minutes in the light northwesterly winds. Although contained, flames from the brazier began to lick at the canopy's fabric, which soon caught fire. The intrepid balloonists quickly extinguished the flames with wet sponges and landed safely 6 miles away. They had become the first humans ever to fly in an untethered, lighter-than-air vehicle.

For many decades, balloon flights would remain little more than a novelty, the challenging stuff of sporting achievements and a versatile medium for the establishment of minor records. They also provided amusement for wealthy thrill-seekers.

It wasn't until the very early 1800s that the balloon's potential for scientific observation and atmospheric study was more fully realised. Manned ascents, now as high as 22,900 feet, would be carried out using increasingly sophisticated hot-air balloons, allowing researchers to determine the atmosphere's height, composition and characteristics. As higher altitudes were attained, however, the peculiar hazards of atmospheric exploration would claim many pioneering balloonists, who knew precious little about the insidious effects of hypothermia and oxygen starvation (hypoxia). Some would even perish from high-altitude pulmonary oedema.

The hazards of high-altitude flight

In 1862, two British "aeronauts" undertook a daring series of balloon flights aimed at studying the upper atmosphere, flying as high as possible, but without a breathable oxygen supply. On their first attempt, Sir James Glaisher and Henry Coxwell reached 26,177 feet. Two months later, dressed in nothing more protective than heavy street clothes, they tried again in a balloon called *Mars*. This time they reached nearly 30,000 feet, equivalent to the summit of Mount Everest, but the attempt nearly cost them their lives.

Things had gone well at first. Glaisher, a meteorologist by profession, was busy recording measurements as they ascended. At progressively higher altitudes he would extract one of several pigeons from a cage and fling it overboard in order to record the bird's reaction. He noticed that the higher the balloon ascended, the more sluggish the birds became when released. In fact many simply fell listlessly until they were lost to sight. Curiously enough, this did not seem to alarm the two men.

Things would change as the balloon reached the 5-mile mark. Glaisher now became mildly hypoxic and started to hallucinate. He would later recall the dramatic events in his book, *Travels in the Air*. "Up to this time I had experienced no particular inconvenience," he wrote. "When at the height of 26,000 feet I could not see the fine column of the mercury in the tube; then the fine divisions on the scale of the instrument became invisible. At that time I asked Mr. Coxwell to help me read the instruments, as I experienced a difficulty in seeing them" [2].

Glaisher then found that he had become powerless, and could not even raise his arms off the table. "I tried to shake myself, and succeeded, but I seemed to have no limbs. In looking at the barometer my head fell over my left shoulder ... Getting my head upright for an instant only, it fell on my right shoulder; then I fell backwards, my back resting against the side of the car and my head on its edge" [2].

Then Coxwell, a highly-experienced balloonist, noticed that the rotary movement of the balloon had caused a gas valve line to become tangled in the rigging. It had to be freed or they could not descend. They were now at 28,000 feet, and Glaisher was suffering from oxygen deprivation, with symptoms similar to acute intoxication. He was only vaguely aware of their peril and was unable to help. In a bold but life-saving move, Coxwell climbed onto one side of the open basket and grasped a sturdy metal ring around the base of the balloon. Amazingly, neither man had thought to wear gloves, and even though Coxwell was within reach of the valve line his bare hands had become stuck on the freezing metal ring. In desperation, chilled to the core and beginning to vomit, he grasped the cord in his teeth. Eventually it tore free, together with one of his teeth, and air began to spill from the balloon.

Meanwhile, Glaisher had begun convulsing. "I dimly saw Mr. Coxwell, and endeavoured to speak, but could not. In an instant intense darkness overcame me, so that the optic nerve lost power suddenly, but I was still conscious, with as active a brain as at the present moment whilst writing this. I thought I had been seized with asphyxia, and believed I should experience nothing more, as death would come unless we speedily descended: other thoughts were entering my mind when I suddenly became unconscious as on going to sleep" [2].

Slowly the balloon descended, and despite a hard landing both men would survive their ordeal.

A fatal error

Other balloonists were less fortunate. On 15 April 1875, French journalist Joseph Crocé-Spinelli, naval officer Henri Sivel and a civilian named Gaston Tissandier flew their balloon *Zenith* to 28,000 feet. They were carrying with them a primitive oxygen apparatus manufactured from three small balloons filled with a mixture of air and oxygen, and fitted with India-rubber hose pipes. Unfortunately Crocé-Spinelli and Sivel fainted with the pain and extreme cold, leaving a badly-confused Tissandier to initiate the descent by pulling a valve rope before he also fainted. He somehow survived, but the other two men died before landing. Sadly, none of them had used what little oxygen they carried [3].

Earlier, during the American Civil War, hot-air balloons had provided useful aerial surveillance of enemy positions. They would also be used extensively during the First World War, for observation purposes and as high-flying bombing platforms.

In November 1927, Captain Hawthorne Gray of the U.S. Army Air Service ventured into the stratosphere for a record third time, once again keeping a meticulous log of his reactions to high-altitude flight as he ascended. Unfortunately, he ran out of oxygen on his descent and died before the huge balloon touched down. A barograph, recovered along with Gray's body 2 days later, showed that the balloon had reached a peak altitude of 42,470 feet.

It would soon be determined that – once humans ascended above 30,000 feet – atmospheric pressure had correspondingly dropped to a point where they needed to breathe pure oxygen under pressure to survive, otherwise a loss of consciousness would result in less than a minute. Furthermore, any ascent above 40,000 feet would require the use of a pressure suit or a pressurised cabin. Captain Gray's ill-fated high-altitude flight would be the last conducted in an open basket until 1955, when the development of pressure suits became an imperative.

Research balloons and rockets

As aircraft became involved in high-altitude flights for research purposes, balloons once again became the poor flight cousins of the skies. But that would not last long. With military jet aircraft flying ever faster and higher, and eager test pilots scything relentless trails into the very thresholds of space, the U.S. Army Air Force (USAAF) – later the U.S. Air Force (USAF) – became an active participant in biodynamics and space biology research, principally carried out at Holloman Air Force Base in Alamogordo, New Mexico.

While rockets would one day carry animals aloft from nearby White Sands Proving Ground, military high-altitude balloons would likewise contribute to outstanding accomplishments in two broad fields of space biology research – cosmic radiation and controlled artificial environments. For many years, physicists and flight physiologists had been especially concerned about the effect of heavy nuclei on the human body in the upper atmosphere and were keen to test their theories.

This would eventually lead to the inclusion of many living specimens on high-altitude balloon and rocket flights. Rats, cats, dogs, mice, rabbits, hamsters, guinea pigs, goldfish, frogs, chickens and monkeys – all would participate in high-altitude balloon flights that benefited researchers in the development of protective pressure suits and breathing apparatus, and help scientists investigating potentially detrimental effects of cosmic rays on the living tissues of both human and animal subjects.

These animals would innocently assess many of the dangers that lurk beyond our protective atmosphere and prove that humans, given proper life-support systems and protection, can function quite successfully on high-altitude missions.

Conducting experiments with cosmic radiation

Just as the Montgolfiers had sent three animal "guinea pigs" aloft in 1783 to see if it could be done, so scientists began including living organisms on balloon flights to test their theories. Fruit flies, seeds, fungi and small animals became payloads in scientific experiments aimed at determining whether their exposure to high-altitude cosmic radiation, or other unknown phenomena, might impact significantly on a subject's short- and long-term health, behaviour, growth, or reproduction.

As aerospace writer Lloyd Mallan once noted of prevailing thoughts on such research, specifically mentioning the year 1949, "it was a subject reserved for the lunatic fringe of science and the lurid fiction fan." He added, however, that many eminent specialists in aviation medicine, particularly those from Germany, England and the United States, had been giving the subject serious thought. "Pioneering experiments in aviation medicine had stimulated an awareness of the practicality of exploring the reactions of man's body to the conditions of space" [4].

The USAF had certainly become interested in space biology research, particularly that related to sub-gravity and cosmic radiation. The first, preliminary balloon flight undertaken on behalf of the Air Force's Missile Development Center took place from Holloman AFB on 8 September 1950, conducted by the Aeromedical Field Laboratory (AMFL). An indeterminate number of white mice (recorded as "14 or 16") were

A variety of animals, such as this St. Bernard, were used in tests of protective pressure suits. (Photo: USAF)

hoisted to 47,000 feet, but their *Albert* capsule – coincidentally the same programme name as the one used for their primate rocket launches in New Mexico – sustained a leak and depressurised. Recovery took place 7 hours after lift-off, by which time all of the mice were dead [5].

The next AMFL balloon flight took place 20 days later. On this occasion only eight mice were carried in the *Albert* capsule. The balloon reached 97,000 feet and was successfully recovered after a flight of 3 hours and 40 minutes. The eight mice had survived their ascent, becoming the first animals to reach that height in a balloon.

The third flight on 18 January 1951 did not end successfully. The balloon burst at 45,000 feet and the mice on board perished on impact with the ground. Next, the AMFL instituted a series of 39 numbered flights, during which a variety of study animals, small and large, would be exposed to heavy primary cosmic rays. This series of biological balloon flights would last through to December 1953.

The first of these flights, numbered AMFL 2, was launched on 23 August 1951, but it ended in a balloon failure during ascent at 59,000 feet, and the hamsters on board did not survive.

Results of other flights varied. Balloon or equipment failures were the norm, resulting in the loss or death of biological specimens, ranging from fruit flies, black mice, white mice and hamsters through to a number of cats and dogs. On subsequent histological examinations of surviving animals, no evidence of tissue damage was found. A number of albino mice were flown to determine if cosmic radiation might cause cataracts to develop on the rodents' super-sensitive eyes. No evidence of this resulted.

The return of David Simons

Slowly, painstakingly, knowledge was being accrued, and surviving specimens provided valuable data on the effects of cosmic-ray penetration, life-support systems and capsule construction. Several helmet variations were also monitored for cosmic-ray particle hits [5]. In the latter part of 1953, USAF Flight Surgeon Major David Simons returned from a 30-month assignment to Korea, ready to return to the world of high-altitude research. He was reassigned to Wright Field in Ohio, from where he would transition to Holloman AFB in Alamogordo, New Mexico.

By January 1953, the cosmic-ray programme had become a function of the USAF Missile Development Center at Holloman, which would ensure it received better support than when it was just one of numerous programme activities being carried out at Wright Field. At Holloman, Simons recalled, he would serve as chief of the Space Biology Laboratory, exploring problems associated with cosmic radiation exposure to animals at high altitudes, using balloons.

Dr. Herman Schaeffer, a physicist at Pensacola in the Navy, had done a number of calculations in terms of exposure to cosmic rays in space that would indicate that there were very significant biological hazards. The concern was heavy atomic nuclei, like carbon and nitrogen, impacting tissues at speeds approaching that of light. There was enough question about this that somebody had to do some experiments to find out whether it was that bad, not that bad, or worse.

Exposing animals to these particles at high altitudes was the only way known to do it. There were no accelerators that could accelerate atoms of that size. The cosmic rays are deflected by the magnetic field that produces the Van Allen belts at the equator [6].

The first flight for which Simons became a principal investigator was AMFL flight 39, achieving 88,000 feet on 16 December 1953. The biological payload carried aloft was little more than skin excised from 10 mice, and some barley seeds. On the subsequent flight, however (AMFL 41), two dogs were lofted to 60,000 feet. Both animals would perish shortly before recovery due to overheating in their containers, which lacked cooling systems.

The dogs were part of an experiment to develop prototype containers specifically designed to carry animals weighing up to 20 pounds to altitudes between 80,000 and 100,000 feet and "to maintain internal conditions at a level to permit flight durations of 36 hours" [7]. Sponsored by the Air Research and Development Command (ARDC), the New York University's College of Engineering Research Division had developed and manufactured the container in collaboration with the Aeromedical Laboratory and the Wright Air Development Center. Prior to the first canine flight, chamber and ground tests had verified the container's integrity, but the lack of any cooling system was a lamentable oversight.

Problems continue

AMFL flights 43 and 44 fared little better. On 12 March 1954, another two dogs lifted off from Holloman on flight 43 and flew to 75,000 feet. This time a cooling system had been incorporated. Unfortunately, the animals perished in their capsule after a tracking aircraft lost contact during the descent phase, delaying its recovery. Three months later, on 24 June, a number of mice were on board when a balloon was destroyed by heavy winds during the launch phase [7].

While biological research balloon flights in the early 1950s gave no evidence of radiation damage in the mice, hamsters, fruit flies, cats and dogs, scientists had begun to realise that the flights were being conducted too far south to obtain a significant exposure to cosmic rays. This led to them undertaking many more northerly flights.

The next AMFL flight (45) was launched from Fleming Field in Minnesota, carrying some mice, *Neurospora* and cats to 79,000 feet. The flight summary states that "all biological specimens were recovered in good condition" [5].

Philippine macaques (*Macaca fascicularis*) would take to the skies on the following flights. AMFL 46 was an ambitious mission, the first of eight to be launched from Sault Sainte Marie, Michigan. It was carrying "dry corn and barley seeds; 3 monkeys; 11 white mice; 31 black mice; radish seeds; *Neurospora*; 1 rat; 6 pieces of human skin; 19 fertilized chicken eggs." The balloon attained a height of 96,750 feet, and the later flight summary simply records that "most animals survived the flight satisfactorily" [5].

The next test took place 3 days later carrying a similar biological payload, which included the same pair of macaque monkeys. The flight peaked at 94,300 feet, but on this occasion the results summary recorded that "most animals did not survive the flight" [5]. AMFL 48 met with even less success. The balloon failed at around 50,000 feet and the biological capsule free-fell to the ground: "61 white mice; 42 black mice; radish seeds; *Neurospora*; 3 rats; 8 fertilized chicken eggs" [5] were all lost.

A rabbit was one of numerous biological passengers carried on AMFL 49 on 25 July 1954, but all of the animals succumbed following a capsule depressurisation.

Monkeys on instalment plans

In their report for the April 1956 edition of the *Journal of Aviation Medicine*, titled "The 1954 Aeromedical Field Laboratory Balloon Flights, Physiological and Radio-biological Aspects" [8], co-authors David Simons and Charles Steinmetz state that one successful innovation they employed was an "instalment plan" for lengthening the exposure of specimens to cosmic rays. This meant sending the same organisms on consecutive flights. In this way a batch of radish seeds was exposed on several

In preparing for AMFL flight 56, containers of mice are secured in place for their ascent on 3 February 1955. In a touch of serendipity, their containers resembled wedges of cheese. The carriage balloon reached 40,000 feet but unhappily the mouse capsule was lost and never recovered. (Photo: USAF)

consecutive flights for a total of 251 hours above 82,000 feet, while some animal subjects were exposed for more than 74 hours at the same altitude.

In observance of this "instalment plan", each of the two following flights carried the same pair of monkeys in their respective capsules, with both balloons flying to around 97,000 feet. In both cases all specimens were said to have been "recovered satisfactorily" [8].

According to a 1956 report prepared by Harry Harlow, Allan Schrier and David Simons [9], the purpose of this particular investigation was to determine the effects of primary cosmic radiation on the behaviour of primates. In all, four macaque monkeys were given a series of behavioural tests before and after two of the subjects made successive flights to altitudes above 90,000 feet for a cumulative total of 62 hours. Preliminary post-flight examinations conducted on the two flown and two control monkeys included "re-tests of discontinuous response pursuit and colour-discrimination performance" [9]. All four animals underwent further examination on oddity and delayed-response problems, and their appetite for peanuts and raisins. These tests were repeated some 4 months after the high-altitude flights.

The weight, general behaviour and neurological conditions of the flown animals were found to be quite normal. Post-flight test performances showed continued improvement, equalling or bettering those of the unflown control animals. It was tentatively concluded that prolonged exposure to primary cosmic radiation at high altitudes did not produce any general behavioural loss in the animals [9].

Further studies of the possible biological effects of day-long exposure to primary cosmic radiation were conducted on 85 mice carried aboard AMFL 67, launched from International Falls, Minnesota. Although the specimen payload was only planned to fly above 80,000 feet, the balloon ultimately reached closer to 109,000 feet, remaining there for nearly 23 hours. For comparative purposes, the flown and some identical ground-based control mice would be allowed to live out their lives while undergoing periodic scrutiny and testing. "Taking into consideration the minor differences exhibited by the experimental and control animals in longevity, incidence of neoplasms, and in reproductivity and aging," the later report concluded, "there was no definite evidence that a day's exposure to light- and medium-weight primary cosmic particles in the stratosphere had any adverse long-term effect" [10].

Summarising the flights

In his book *Biomedical Aspects of Space Flight*, James Henry later stated that the effect of radiation depended on the particular group of cells involved in the track of the impinging radiation:

For instance, if a hair is struck at the root – in the follicle where a few cells determine the growth and colour of the whole hair – then it is possible to obtain direct evidence of cosmic ray hits. Indeed, black mice exposed to cosmic rays have shown white hairs. But the tremendous duplication of the body's cells makes serious damage as a result of these cosmic "needle pricks" unlikely [11].

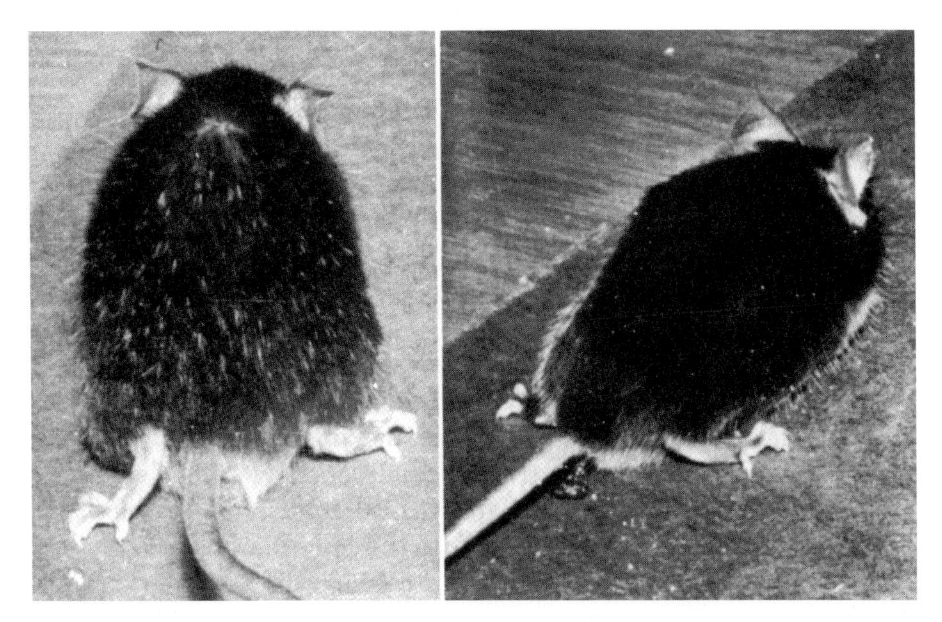

These two black mice were flown in research balloons for the same length of time at the same altitude, but the mouse on the left received exposure to the more intense cosmic radiation found in northern latitudes. The mouse on the right flew much farther south and shows no sign of the same grey speckling. (Photo: Brown University)

In summarising the animal test flights on balloons, David Simons said there was relief expressed at many of the conclusions.

Over a period of four or five years, we gradually learned how to build life support systems that would maintain 200 mouse units' worth of animals in a capsule for thirty-six hours at altitudes up around 130,000 feet. We used "mouse units" because sometimes we flew guinea pigs, sometimes we flew rats; lots of times, of course, mice, and even occasionally monkeys. And, life support systems are quite critical in terms of being sure you have provided enough oxygen and carbon dioxide absorbing capability and temperature control to cover the load that you're carrying. I remember that a guinea pig was two mouse units, a monkey nearly seven mouse units. This way, we could add it up and know we had not exceeded our 200-mouse units for the capsule and that it would do well. After about four or five years of this, we'd done enough experiments that it was clear there could be some effects detectable, but they were certainly not as bad as was first prognosticated, and very clearly not worse than was theoretically envisioned.

So, the problem looked like it was one of those things you ought to know about and not forget, but wasn't really a major concern. Except for unusual, and violent, solar flares, it would not be a problem in near space for things like moon shots. Now, if you're going to go to Mars and beyond, with transit times of two years or so and round trip times of four years or more, it involves a much different order of exposure, which may require some reconsideration [6].

Project Man High is born

With animal experiments on high-altitude balloons coming to an end, Simons was approached by his chief at the Aeromedical Laboratory, Lt. Colonel John Paul Stapp, who wanted to know if he felt that the time was right to do balloon flights with manned pressurised gondolas suspended beneath them. This would allow advanced studies into the effects of cosmic rays and the determination of the physical and psychological capabilities of human subjects during extended flights into near-space conditions. "I hadn't thought much about it, but it's true," Simons later recalled saying to Stapp. "Our data at this point looked like it shouldn't be dangerous" [6]. Stapp then asked if he would be willing to undertake such a mission.

"Well, basically, with what I know, I surely have no compunctions," Simons told his chief. "I think that it would be very safe from a radiation point of view and it could very well give us a lot of information that we can't get from animals. You can make observations that the animals can't tell you about" [6].

One of the potential effects a human passenger might observe would be the penetration of the retina by a heavy-particle cosmic ray. Simons reasoned that a person at high altitude should be able to experience a flash of light behind the eyelids and, unlike an animal, report on the phenomenon. He agreed to become a human "guinea pig" on an upcoming flight, but he had lingering concerns. Although most of the animal research flights on balloons had indicated that the effects of cosmic radiation on tissue were negligible, Simons knew that a pioneer cosmic-ray researcher in Europe had been reaching different conclusions.

Dr. Jakob A.G. Eugster of Berne, Switzerland, had sent batches of oat seeds aloft in our capsules and after planting the oats he reported major mutations through three generations.

Even more exotic and somewhat alarming, Eugster excised and dried samples of his own skin and sent them to us to fly in the upper atmosphere. The dried skin was keyed to photographic plates which mapped the exact points on the specimens that were penetrated by cosmic particles. When we returned the exposed skin samples, Eugster re-implanted them on his body. Later, the skin developed dark granules suggesting cancerous growth at the points penetrated by cosmic particles [12].

This effect had not been noticed in returning live animals, and Simons was keen to prove that Eugster's dried skin had reacted differently from his own live skin. "But I had to acknowledge that my eagerness might earn me a batch of minute skin cancers instead of confirmation of my own theory," he added [12].

They were going to call it Project Daedalus, but when the name was suggested to their higher-ups in the Air Force they were told that the service already had a highly-classified programme by that name in operation. Simons therefore gave their project a simple but descriptive name – Man High.

At first, Simons was going to be the first Man High pilot, but it was wisely decided that he should serve as reserve pilot for the initial test flight, leaving him free to conduct the second, full-scale research flight. Instead, the experienced test pilot

The Man High II balloon carrying Dr. David Simons on his high-altitude flight ascends into the sky from a mine pit on 19 August 1957. The balloon would reach a record 102,000 feet. (Photo: USAF)

Captain Joe Kittinger was given the first flight, and was carried to 95,000 feet on 2 June 1957.

Two months later, on 19 August, Simons would ascend to 102,000 feet – 19 miles – in his sealed Man High II gondola, leaving around 99% of the Earth's atmosphere behind him. He remained at that altitude for 32 hours.

A third and final Man High flight would take place on 8 October the following year, this time with Lieutenant Clifton McClure aboard.

A man and a monkey?

According to USAF historians Drs. James Hanrahan and David Bushnell in their report (later released as a book), "Space Biology: The Human Factors in Space Flight," the Chief of the Neuropathology Section of the Armed Forces Institute of Pathology, Webb Haymaker, had been keen to send a monkey aloft with Simons on his flight – not just to keep him company, but to allow another cosmic-ray experiment. Haymaker, the report says, "had actually selected the monkey and had it shipped from Washington to Wright Field in Ohio to take part in a pre-flight test of the capsule. But at this point the animal was firmly grounded by order of Colonel Stapp, Chief of the Aeromedical Field Laboratory" [13]. The only biological specimens that would accompany Simons on his world altitude and endurance record balloon flight were some samples of *Neurospora*.

Later, in the summer of 1958, Haymaker would be given the opportunity to expose several small primates to cosmic radiation in six balloon flights orchestrated by the AMFL, as discussed by Hanrahan and Bushnell:

These flights were to use the newly designed balloons of five-million-cubic-foot capacity. The plans even called for frogs and goldfish to take part – the goldfish to travel in plastic-bag containers for a special study on the effects of radiation on their pigmentation. But only one batch of specimens – a monkey, a goldfish, Drosophila *[fruit fly] larvae, and some* Neurospora *– was actually assembled, and in three attempts from International Falls and Crosby [both in Minnesota] this cargo never managed to get above 63,000 feet. The summer's flight program was cancelled, on the ground that more work was needed to perfect the balloon vehicle* [13].

Following the completion of the three Man High flights, it was decided to terminate all further animal balloon flights at Holloman, allowing the AMFL to concentrate on offering operational support for other biosatellite activities.

For his part, David Simons would find himself reassigned to the U.S. Air Force School of Aviation Medicine in January 1959.

GERMAN ROCKETS FOR SCIENCE

Meanwhile, as the United States and Russia experimented with captured V-2 rockets and biological balloon flights, a badly defeated nation had begun to re-emerge from

Preparing to launch a Kumulus rocket. (Photo: Deutschen Museum, Munich)

the ashes of war. There had been a slow but determined revival of rocket science in Germany, mostly based on wartime technology breakthroughs, but now with peaceful and scientific purposes in mind.

From 1957 to 1964, the German Rocket Society (later renamed the Hermann Oberth Society, after the pioneering rocket designer) would once again utilise the test-firing site at Cuxhaven to launch nearly 500 sea rescue and research rockets. This was a co-operative effort, carried out in partnership with the Seliger Research and Development Society. Under the auspices of rocket designer Ernst Mohr, the Hermann Oberth Society would launch Kumulus and Cirrus meteorological rockets, while the Seliger group would eventually send three-stage rockets to altitudes exceeding 70 miles.

Kumulus and Cirrus rockets take to the skies

The first launch of a single-stage Kumulus rocket took place on 1 November 1959, carrying a simple radio transmitter designed by Professor Max Ehmert of the Max Planck Institute. The Kumulus rocket reached an altitude of just under 10 miles, but it could not be tracked during its flight as the transmitter's batteries had frozen during launch preparations. The Kumulus launch programme recommenced in February 1961 when two of the 10-foot rockets, with a mass of around 62 pounds, successfully reached a height of almost 10 miles [14].

On 16 September that year, the German Rocket Society lofted a total of four Cirrus and Kumulus rockets into the air. The first Cirrus rocket achieved an altitude of 22 miles, and the second soared to a height of 31 miles.

A Cirrus rocket takes to the skies. (Photo: Deutschen Museum, Munich)

Later in the day, both Kumulus two-stage sounding rockets would carry biological specimens on short ballistic flights. The first living passenger was a 9-inch salamander of the *Ambystoma mexicanum* species, a completely aquatic amphibian known as the Mexican axolotl or Mexican walking fish, which had been given the name Lotte. The launch went ahead as planned, and the Kumulus rocket streaked to an altitude of nearly 8 miles before parachuting safely back to ground.

The second Kumulus launched that day carried a goldfish named Max, enclosed in a small Plexiglas tank. Like the salamander on the previous flight, Max was filmed during his brief journey into the sky, reaching an altitude of 9 miles. Unfortunately, there was a problem with the full deployment of the parachute system, and the Kumulus rocket slammed into the ground much harder than anticipated. Max did not survive the heavy landing.

Gerhard Zucker and his problem rockets

Cuxhaven would continue as a launch site until June 1964, when the German government issued a provisional order prohibiting all further civilian rocket launches. This followed an unrelated but deadly accident involving experimental mail rockets in Braunlage the previous month, in which up to three people were reported killed. The grossly under-performing rockets involved in this incident used a powder fuel very similar to that packed into common fireworks, and were the brainchild of inventor Gerhard Zucker. He had hoped to make a fortune out of flying unofficial but profitable postal covers on his rocket "mail delivery" flights, but his questionable activities only met with limited success. On 7 May 1964 he was demonstrating his latest series of rocket mail flights on the Hasselkopf hill near Braunslage in Lower Saxony when a fatal explosion took place. Much to the outrage of the Cuxhaven rocket societies, this incident led to the total prohibition of all non-military rocket flights in Germany.

As a postscript to this story, Gerhard Zucker (who died in 1985) would resume launching his fraudulent postal covers in the 1970s [15]. Even though his antics had effectively curtailed serious research rocketry in Germany, the postal covers that did survive his self-serving "rocket flights" are today seriously regarded as highly-prized collectables.

"THE FASTEST MAN ALIVE"

In March 1947, 36-year-old USAAF Major John Paul Stapp, M.D., Ph.D., the Brazilian-born son of Baptist missionaries and a graduate of the School of Aviation Medicine, had been handed a challenging assignment. As a project officer and medical consultant at Wright Air Development Center's Aeromedical Laboratory in Ohio, he had already come to the notice of his superiors for his work on testing oxygen systems in unpressurised aircraft above 40,000 feet, and he would quickly prove he was more than equal to a new and demanding task.

Understanding forces that can kill

Post-war, the U.S. Army Air Force (which separated to become the U.S. Air Force on 18 September 1947) was well aware of a pressing need for a comprehensive series of studies on ways to better protect occupants of military aircraft involved in serious accidents. At the same time, urgent work needed to be done on creating a powered ejection seat to handle bailouts at near-supersonic speeds, amid fears that ejecting pilots could be maimed or killed as a result of the forces involved. There were sudden deceleration and windblast effects pilots could suffer as they ejected into a virtual wall of air before their parachute opened.

The Wright Air Development Center had been tasked with developing equipment and instrumentation that would not only simulate an aircraft crash, but investigate human tolerances to high g-forces. Stapp, working under the auspices of the Aeromedical Laboratory, was charged with finding out what stresses, particularly deceleration forces, the human body could be subjected to and survive in a series of simulated ejections and crashes. From these experiments, it was envisaged that new procedures and safety equipment would evolve to better protect service personnel in peril.

On the path to medical research

As a youth, Stapp had modestly intended pursuing a literary career, but two incidents would have a profound effect on his life and eventually propel him into aviation medicine. He was 17 years old, enjoying a Christmas break at an uncle's house, when his infant cousin ventured too close to an open fireplace. The boy's bedclothes caught fire, and he suffered horrible burns. Stapp was at his little cousin's hospital bedside for more than 2 days offering support and nursing the critically ill infant until his tiny body gave up the fight and he died. Then a few years later, his fiancée was badly hurt in a highway accident. She too died of shock as a result of her terrible injuries. Stapp consequently decided to take on a medical career, principally in the field of high-shock research. With this in mind he took courses in zoology, chemistry and biophysics at Baylor University and the University of Texas. He was awarded a Ph.D. in biophysics after he had already enrolled in the University of Minnesota Medical School. He received his M.D. there in 1943, and the following year entered active duty with the Army Air Corps. After attending the Medical Field Service School, the School of Aviation Medicine and the Industrial Medical Seminar, he was assigned post-war to the USAAF Aeromedical Laboratory and placed on detached duty as a research project officer at Muroc (later Edwards) Air Base in California's Mojave Desert.

USAF Flight Surgeon Major John Paul Stapp secured onto the rocket-propelled decelerator sled at Edwards Air Force Base. (Photo: USAF)

John Paul Stapp's "Gee-Whizz" machine

Stapp's first deceleration experiments had begun on 30 April 1947 using a 2,000-foot B-1 launching track at the base. The track had originally been laid to allow testing of the American version of Germany's V-1, the destructive, so-called "buzz bomb". He would convert this facility for his own work and oversee construction of a 1,500-pound, rocket-powered sled, specially fabricated from aluminium tubing by Northrop Aviation Inc. This sled, dubbed the "Gee-Whizz", would rapidly accelerate along the track and come to a shuddering dead stop in around 1 second with the aid of 45 sets of mechanical friction brakes installed between the rails.

As a precaution, Stapp required the use of 185-pound mannequins (known as anthropomorphic dummies) on the first 35 trial runs to check the equipment. It would prove to be a prudent decision; on the first run the sled ran off the tracks, while on another the restraint harness broke as a result of the massive deceleration forces, and the unfettered dummy sailed through the air, smashing down 700 feet beyond the end of the track.

It was intended that anaesthetised chimpanzees would be used in some initial test runs, but when their arrival at Edwards was delayed, Stapp decided to carry out the first sled tests using himself as the test subject, employing only one of the sled's rockets. During later test runs, in which power and speed was raised incrementally, Stapp and other volunteers momentarily reached acceleration forces of around 45 g's – three times the gravitational force that informed speculation had given as the maximum limit a human body could tolerate. While no lasting injuries or ill effects were recorded, there were some instances of retinal haemorrhage, broken ribs, abrasions, lost teeth fillings, concussion and unconsciousness resulting from the sudden acceleration and deceleration forces.

Chimpanzees would later ride the rocket sled a total of 88 times in a variety of positions; some sitting up and others lying down. According to Stapp, "The maximum deceleration sustained by a chimpanzee was from 169 miles per hour to a stop in 18 feet in the supine head-first position. It is many times what would be encountered in any automobile collision or plane crash short of complete demolition of the vehicle" [4]. The animal in question had been anaesthetised and strapped onto the sled in a head-first configuration, blasted down the rails, and brought to an abrupt stop measured at a crushing 270 g's. One of the researchers would later describe the luckless chimpanzee's remains as "a mess."

The sad saga of the Holloman hogs

Some anaesthetised hogs would also be involved in impact tests, although these would be carried out using carriages that moved along a monorail on a horizontal ejection seat catapult. As aerospace writer Lloyd Mallan wrote of these tests:

An explosive charge sent the carriage flying forward to be stopped abruptly by impact with one of several heavy lead cones. Size and weight of a particular cone determined the braking power. The newly developed harness systems were used to tie the hogs upright in

Looking the picture of misery, this chimpanzee has been strapped onto a deceleration sled, with sensors taped to its body to record physiological data. (Photo: USAF)

the carriage. "Uninjured survival of anesthetised hogs," says Stapp, "occurred in all experiments up to 80 g's in the backward-facing seated position and up to 125 g's in the forward-facing seated position."

Then the animals were tested using only a partial harness or none whatever. "The same gravity forces," continues the Colonel, "that could easily be sustained without injury while the subject was restrained with webbing, now produced fatality when the subject was impinged (or thrown) against solid test objects" [4].

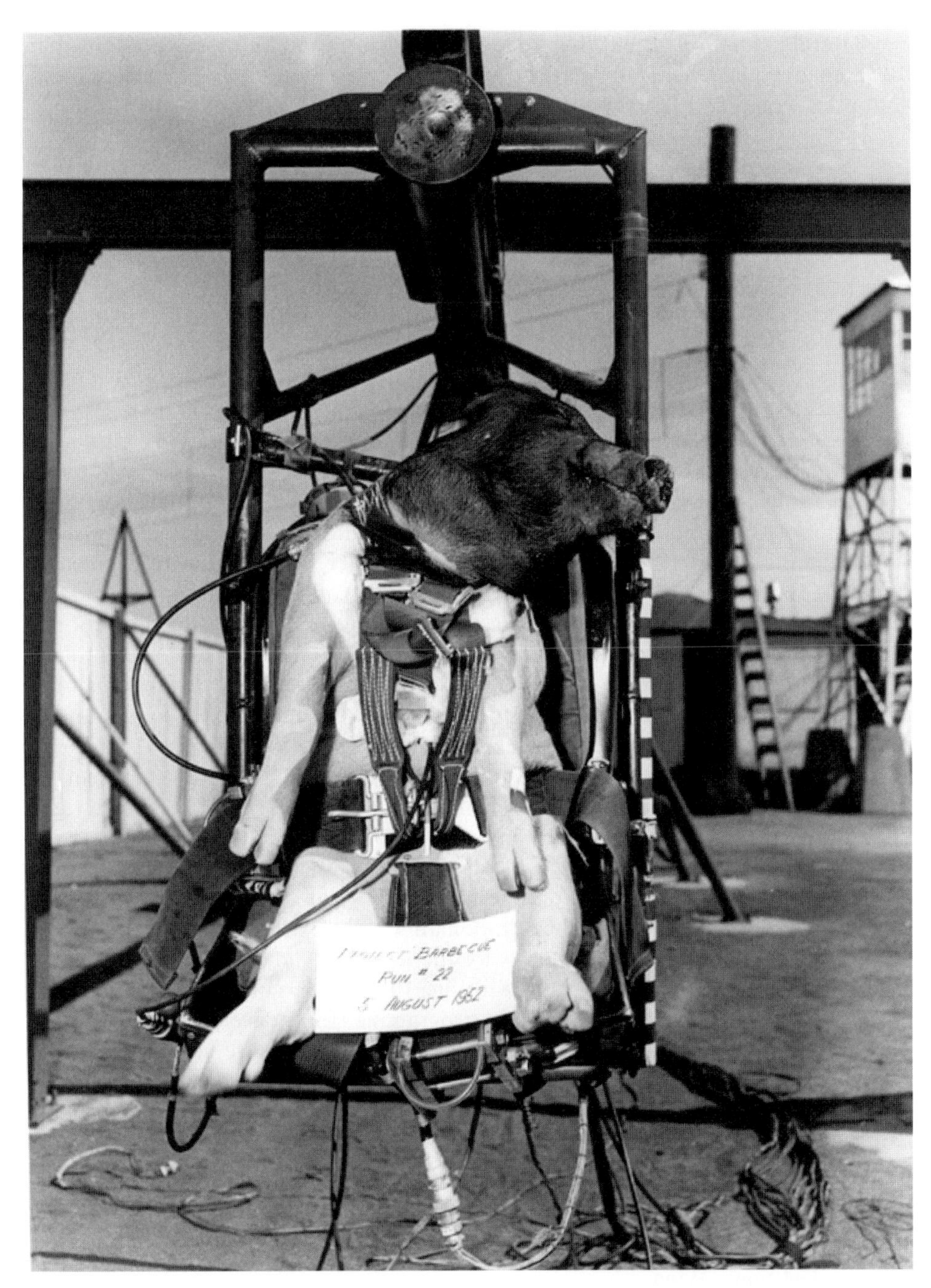

Hogs were used in crash simulations that had applications in both the aviation and automobile industries. The sign "Project Barbecue, Run #22, 5 August 1952" referred to the fact that – following investigative autopsies – the unfortunate animals were cooked and eaten. (Photo: USAF, Courtesy New Mexico Museum of Space History)

Several animals died, but they rode the sleds faster and on more occasions than the human volunteers, and were purposefully subjected to far greater g-forces. Autopsies would be carried out and all of the results carefully collated.

Fast rides and fractures

By 8 June 1951, human volunteers had been involved in 73 deceleration runs on the sled, which had reached the limit of its propulsion and braking system. Nineteen of these had been conducted with the subjects facing backwards, and 54 with them facing forwards. Always the first to volunteer, Stapp had twice sustained fractures to his right wrist during separate test runs.

Within 6 months, Stapp had personally completed 16 runs, and he was sustaining braking forces of around 35 g's – twice what experts had predicted human flesh and bones could tolerate. But he needed better and more powerful facilities to continue his work, and he would soon be reassigned to Holloman AFB. Newly promoted to the rank of lieutenant colonel, he would assume command of the Aeromedical Field Laboratory (AMFL) at Holloman's Air Development Center in April 1953.

Beginning life in 1950 as a small operation under the direction of Lieutenant James D. Telfer, the AMFL was nothing more than an annex to much larger research centres under the banner of the Air Force. Prior to this, Holloman's main function in aeromedical and related activities had been to act as a support service to such projects as the V-2 rocket programme. By 1951, the work being carried out at Holloman had reached a level sufficient to warrant its designation as an Aeromedical Field Laboratory, and construction had begun to expand and upgrade its facilities.

On 10 October 1952 the Holloman Air Development Center also became an Air Force Air Research and Development Command designated centre in its own right. Over the next few years the important aerospace biology and biodynamics work carried out by the AMFL resulted in rapid growth, and a corresponding renown for the many achievements credited to the laboratory's staff. The immediate attraction for Stapp in resuming his research at Holloman was the existence of a 3,550-foot rail track set on concrete strips – almost twice the length of the one at Edwards.

Holloman's track had been created in June 1950, 3 years prior to Stapp's arrival, and in the infancy of America's missile programme. First developed as a captive-missile launch facility, it had proved a far less costly and more efficient means of testing and uncovering faults in prototype Snark missiles, then identified as Project MX-775. Over the next 2 years, rather than send missiles skywards on expensive launch tests, several prototype Snark missiles were bolted down onto rocket sleds and blasted along the steel tracks in what engineers would call "pre-flight testing and recovery of the test item" [16].

A sled called Sonic Wind

Working in full co-operation with the Northrop Corporation, Stapp would organise the construction of an advanced and more powerful rocket-powered sled known as Sonic Wind No. 1, which went into service with an initial test run on 24 November

1953. The sled was equipped with a water-brake system, in which a scoop attached to the sled would plunge into a series of water dams between the rails at the end of each run. Two months later, on the third test run, a chimpanzee was strapped to Sonic Wind 1 for the first time, and Stapp's work had begun in earnest.

As Gregory Kennedy, the executive director of the Alamogordo Space Center in New Mexico, observed in his 1988 history of the AMFL:

Colonel Stapp began his research using the Holloman rocket sled track under the aegis of the Air Research and Development Command Test Directive 5200-HI, "Biophysics of Abrupt Deceleration". This directive, dated April 15, 1953, called for a program of experiments to study tolerance and survival limits for deceleration, windblast, tumbling and combinations of these factors. The major thrust of these investigations was to study escape from high-speed, high altitude aircraft [17].

During Stapp's earlier deceleration experiments at Edwards, extremely useful information had been obtained on the amount of wind drag a crewmember would encounter in the ejection process from a subsonic aircraft. Now, more needed to be known about the forces involved in ejecting from a supersonic aircraft, and Holloman's chimpanzees would take an active part in these experiments to further that knowledge.

The chimps and Project Whoosh

Related work was continuing elsewhere. According to the NASA History Office, one Air Force research programme involving Holloman chimpanzees was known as Project Whoosh, which aimed to "evaluate escape from a high speed aircraft at approximately Mach 2" [18]. This involved strapping an anaesthetised chimpanzee into an open ejection seat, which was then secured inside a specially-modified Cherokee missile. The missile would then be carried aloft beneath the belly of a modified Boeing B-29 Superfortress. Once released, a solid-fuel rocket motor would blast the missile to supersonic speed, and the monkey woud be ejected.

The first two supersonic test ejections took place at Edwards AFB on 26 January and 8 June 1954, and the following year all project activity was moved to Holloman. In July 1955 two low-speed ejections took place, in which chimpanzees strapped into ejection seats were dropped directly from a Douglas C-47 Dakota aircraft. The B-29 was then flown down to Holloman from Edwards in order to take part in advanced supersonic tests, which were eventually carried out on 21 October at Mach 1.5 and then on 3 April 1956 at Mach 1.4. The tests, as the NASA History Office confirms, were not altogether successful.

Problems of coordination were multiplied several times over for the last test by confusion and misunderstandings at command headquarters, Wright Field, Holloman, and Edwards as to whether the entire project was or was not being cancelled. It was cancelled beyond any doubt soon after the final Holloman test. Not one of the animals ejected at supersonic speeds had managed to survive, for in each case there were equipment difficulties (with parachute system or ejection seat) that led to the death of the subject and overshadowed any possible evidence of injury through supersonic windblast, tumbling and deceleration. Nevertheless, the project was not a total loss. Even the failures were instructive, and the work performed on Whoosh led directly to further ejection experiments at the Supersonic Military Air Research Track, Hurricane Mesa, Utah [18].

The track in Utah was administered by the Wright Air Development Center, and chimpanzees would continue to be used as test subjects in what was basically a ground-based extension of Project Whoosh. At this facility the track actually ended right at the edge of a cliff, with a 1,500-foot parachute drop to the canyon floor below. Five chimpanzees would take part in these frightening test runs in the fall of 1956, ejected outwards at the end of the track in a special ejection seat. Three were reported to have been recovered uninjured, with successful ejections taking place between Mach 0.95 and Mach 1.1.

In March 1957 a "chimpomorphic dummy" was used in a successful ejection at Hurricane Mesa, registering Mach 1.1, but on the successive run a chimpanzee was killed when the sled malfunctioned on the track and crashed.

Animals, humans and g-forces

Meanwhile, back at Holloman's Aeromedical Field Laboratory, Stapp's own deceleration work had continued, once again using chimpanzees as the test subjects. In one series of sled runs some chimpanzees, facing backwards, were subjected to a combined total of 88 runs in order to help perfect new safety harness designs for military pilots. On one of these runs the carrier sled, propelled by flaming rockets, reached a speed of 169 miles per hour before it hit the braking water-barrier trap at the far end, causing it to come to a complete stop in just 18 feet. The chimpanzee was badly shaken, but recovered quickly from his unexpected ordeal.

In September 1954 a chimpanzee was strapped into what was effectively a "tumbling seat" attached to the sled. This experiment had been designed to evaluate the results of tumbling when combined with windblast and deceleration. It had been determined that bailouts in many instances, especially at high altitude where the air is thinner, often resulted in violent tumbling after ejecting. Stapp was keen to examine this problem in association with his deceleration and windblast studies. He would also trial a windshield device that could be jettisoned once the sled reached a certain speed, but several test animals were injured by the jettison device, and it was soon discarded from further evaluation.

Once the chimpanzee runs had been concluded, a further series of test runs was conducted using hogs. While most people would probably not regard a docile hog as a particularly clever animal, it actually rates very closely to the monkey in intelligence. The internal tissue of the animal is also very similar to that of human beings.

In the first tests the anaesthetised hogs were patiently strapped into their backward-facing seats in experimental harness rigs. At the end of their wild rocket-propelled ride down the tracks, they would be subjected to a force of around 80 g's when Sonic Wind 1 came to a violent stop in the water trap. The water braking system was described by authors Clyde Bergwin and William Coleman as being "a series of precisely placed beaverboard dikes in the concrete trough between the two thick longitudinal concrete dikes which support the rails of the track." They added further details:

Sections of this center trough between the rails are filled with water and maintained at various levels by dams placed at 10- and 20-foot intervals. A scoop, mounted beneath the sled, picks up the water from this trough and throws it out from all sides of the sled, thus slowing it to a stop. The rate of deceleration is determined by the depth of the water and the size, shape and draft of the sled scoops. These factors are predetermined precisely, believe it or not, by mathematicians before each sled run [19].

All of the hogs are said to have survived these runs without serious injury; and then came the second series of runs. This time the animals were strapped into their couches facing the direction of travel, and one hog endured a crushing deceleration force of 125 g's at the abrupt end of a sled run. Once again, it was reported that all of the test animals survived these rocket-powered rides down the monorail track. The safety harnesses would undergo further development and testing for the life-saving use of jet pilots.

FASTER THAN A SPEEDING BULLET

Animals would not be the only living things to ride the sleds. John Paul Stapp, now 44 years old, had decided that someone had to find out first hand if a human subject could also take these extreme decelerative forces, and he volunteered to be that person. It would take a lot of courage, and this was something Stapp had in abundance.

First to volunteer

Eventually, the Headquarters of the Air Research and Development Command gave approval for human test runs to proceed at Holloman, and Stapp made all the necessary preparations. As he stated at the outset of his sled run experiments, he was desperate to find out "What causes some men to die and others to survive crashes that occur under similar circumstances and with similar gravity impacts," and to uncover reasons for this apparent contradiction. It would also help in the development of more efficient safety devices [4].

Ed Dittmer recalls that when he came to work at the laboratory's Space Biology Branch, assigned directly to David Simons, he was an Air Force crewmember on flying status. As such, he was able to participate as a volunteer subject on any of the sleds or other harrowing devices being tested at Holloman, but Stapp would only allow anyone to participate under one condition.

On any of these tests, Dr. Stapp was Number One to ride them. He'd take the test first, he'd ride the sled first ... He never let anybody ride it unless he tried it out himself and

Lt. Colonel John Paul Stapp aboard Sonic Wind No. 1 prepares for a rocket-powered sled ride to test the effects of acceleration and deceleration research. (Photo: USAF)

made sure that it was comparatively safe for everybody. Of course there was always a risk in volunteering for any of these things, but Dr. Stapp took the first one, so everybody else just fell in easy [20].

The sled known as Sonic Wind No. 1 was not the only deceleration device in operation at Holloman. There was the Swing Drop, basically a rigid swing attached to an extendable cable that men and animals rode at varying speeds and cable lengths as it fell to the bottom of its arc, where the cable would suddenly become taut. The swing and its restrained occupant would come to a correspondingly abrupt halt. For other restraint-testing there was also a device known as the Bopper, which was essentially a small sled on a short track. The Bopper was propelled by huge bungee cords that would be wound back by the test team, like a slingshot, and released. The sled, racing along at around 35 miles an hour, would end its run by slamming into a huge air bag.

Faster and still faster

Stapp would ride three times aboard the 1,500-pound rocket-powered Sonic Wind 1 while tightly strapped into an exact replica of a jet pilot's seat at the front of the vehicle. The massive red-and-white sled, which carried instrumentation and high-speed cameras aimed at the test subject, was fabricated from chrome–molybdenum tubing by Northrop Aviation.

On 19 March 1954, Stapp made his first run along the Holloman high-speed track, his 27th rocket-powered sled test overall. On this occasion Sonic Wind 1 reached 421 miles an hour – the fastest any person had ever travelled across land – and he was subjected to a force of 22 g's when the sled decelerated. Five months later, on 20 August, the test run focused on the physiological effects of windblast on a human passenger, and the speed was consequently ramped up to 502 mph. Stapp would suffer no lasting ill effects from this test.

Several high-speed runs involving chimpanzees then took place, in order to verify the sled's performance and stability at speeds in excess of 600 miles per hour.

For his final run on 10 December 1954, Stapp was tightly strapped into a tubular steel seat mounted on the sled. The sled was actually two vehicles: a propulsion sled weighing 3,500 pounds and the test sled itself, weighing another 2,000 pounds. He was facing forwards and wearing a helmet and protective clothing. During his first run in March, six rockets offered 27,000 pounds of thrust, but for this run three more rockets had been added. It was now a formidable speedster, powered by nine 4,500-pound thrust jet-assisted take-off (JATO) solid-fuel rockets that developed close to 70,000 horsepower and provided a massive 40,000 pounds of thrust. They would be ignited by remote control from a nearby blockhouse.

No protective windscreen had been fitted, and Stapp's arms and legs had been tightly secured to prevent them from flailing about in the high winds, which could cause serious injuries. As he sat completely immobilised, his team was anxiously scanning the skies over Alamogordo, waiting for some threatening, overcast weather to clear before the run could be attempted. The high-speed motion picture and still cameras mounted on the prow of the sled facing Stapp required good, strong light, and they had to wait for conditions to clear. He would later describe the wait to begin as "that firing squad sensation" [21].

I had been sitting there during an hour and a half of lashing my helmet to the head rest and putting on double thickness of 6,300-pound-test nylon shoulder straps three inches wide, attached to my seat belt buckle, from which similar straps passed downward around my thighs to the rear corners of the seat. The chest belt had been drawn too tight for comfort, but I could not speak while gripping the protective rubber bite block between my teeth. I could only look down the 3,500 feet of track through the Plexiglas face plate of the fibreglass helmet completely enclosing my head. The cotton sweat-shirt under my regulation wool flying coveralls gave little protection from the chill breeze. Intermittent clouds that had delayed the countdown were clearing, and I could hear a T-33 jet camera plane make a low practice pass overhead, as the pilot prepared to cover the experiment from the air. Below him, a row of high-speed cameras prepared to film the sled run in overlapping profile coverage from a hundred yards, and one brave cameraman at the north end of the track would focus on the approaching sled to the last minute, then run to a foxhole [22].

The weather cleared sufficiently soon after lunch, and the support team checked the restraint straps yet again. When the countdown hit zero, a T-33 chase aircraft flown by Joe Kittinger was hurtling low above the track as the sled began its run. A cameraman was in the back seat of the aircraft in order to record the run, but the jet was rapidly overtaken as the sled's acceleration piled on at an incredible rate. "I will never forget, as long as I live," Kittinger later stated, "the incredulous awe I felt at that moment as Colonel Stapp accelerated like a bullet away from my own speeding airplane" [23].

Forty times the pull of gravity

Sonic Wind No. 1 reached a record Mach 0.9, or 632 miles an hour, in just 5 seconds. Then, at the far end of the run, the rocket sled's scoops plunged into the water pool barrier, absorbing the energy and bringing the vehicle to an abrupt halt in a misty spray of water, 50 feet from the end of the track, in just 1.4 seconds. Stapp's body weight, momentarily, was around 6,800 pounds – almost three-and-half tons. Impact-wise, it was the equivalent of hitting a brick wall in a car travelling at 120 miles an hour. Bound tightly to his seat by heavy nylon webbing, Stapp was unable to move so much as a finger after the sled had come to a halt, and he could only wait as safety and ambulance personnel rushed to his assistance. He was suffering from extreme shock and severe upset to his respiratory and circulatory systems. For several desperate minutes he didn't know if he was permanently blinded, as aerospace author Lloyd Mallan recalls:

It was the 632 mph run and he was struck with an impact, momentarily, of over 40 times the pull of gravity. It was like being crushed back suddenly from forward motion by a lead trip hammer weighing four tons – while his eyeballs continued to speed ahead. For eight minutes he felt sure that the retinas had been torn away from his eyeballs by the distortion. But they were not and this gave added proof to his conviction that the human body is much tougher than any machinery man can invent or build [4].

By extrapolation, Stapp had just proved that ejection from an aircraft at 1,800 miles an hour and an altitude of 35,000 feet was entirely survivable. In making a total of 29 wild powered rides down various sled tracks, he had suffered several retinal haemor-

During Stapp's heroic, wild sled ride of 632 miles an hour on 10 December 1954, Sonic Wind No. 1 was actually travelling faster than the jet flying overhead to film the event. (Photo: USAF)

rhages, cracked ribs and two broken wrists. In its 12 September 1955 issue, *Time* magazine accorded Colonel Stapp the title of "the fastest man on earth and No. 1 hero of the Air Force." In fact the renowned tag of "fastest man on Earth" would remain with Stapp for the rest of his life.

Even as he was recovering from his injuries, Stapp was openly contemplating an even faster sled run, this time in excess of supersonic speed at around 1,000 miles an hour. It was therefore devastating (but probably life-saving) when he read in newspapers in June 1956 that the Air Force had decided to "ground" him from any future high-speed runs, as his life and his work were far too valuable to risk on even more hazardous exploits, however well intentioned and beneficial they might be.

Days of the Daisy Track

In later years at Holloman, in the frantic run-up to the first manned space flights, training for some chimpanzees would involve not just acceleration and launch simulation tests in newer, more compact rocket-powered sleds on the high-speed test track. Milder splashdown impact tests and deceleration studies were carried out on another specially-designed facility also located at Holloman, formerly inaugurated in the summer of 1955 and known to the test team as the Daisy Track.

Instead of using rocket power, a huge air compression unit would hurl these sleds down a 120-foot long, 5-foot wide track. It was a propulsion principle used in Daisy air rifles, from which the sled track quickly derived its unusual name. A controlled decelerative force was supplied through a metal piston mounted on the front of the sleds, which would slam into a water-filled brake cylinder at the far end of track. According to Dr. David Bushnell, the Daisy Track was extremely useful, as "the g-loading at deceleration could be tailored by varying the configuration of the water brake" [24].

Test runs using live subjects could now begin.

Animal research continues

While a chimpanzee would be the first to ride the Daisy Track (and Ed Dittmer stressed there would always be a veterinarian present on the animal runs), the first human passenger was task scientist Lieutenant Wilbur Blount, who completed a run on 17 February 1956. By September the following year over 200 high-speed journeys had been made along the track, carrying humans, animals and mannequins. Thirteen months later, in October 1958, that number was nearing 400.

Captain Eli Beeding would unintentionally hold the human record on the Daisy Track. On 16 May 1958, facing backwards to the run, he recorded a peak decelerative force on his chest of 83 g's – the highest any person has ever experienced on a test track – but only because a mechanical failure had caused a more abrupt halt than planned. Beeding would later state that he felt this was probably the most a human body could stand in the backwards-mounted test configuration that was used on the day, and he would probably not have survived had he been facing forward.

There were also a couple of smaller survivors on Beeding's sled that day – he had taken along two anaesthetised albino rats as part of the experiment. One was mounted on a special rotating anti-g platform and was found to be unharmed at the end of the run, while the other had been secured head first on the sled. Although rodents generally have a relatively high g-force tolerance, this second rat was reported to have had a bad time, although it would fully recover. This particular experiment was administered by German-born sub-gravity scientist, Dr. Harald von Beckh, who explained the significance of the anti-g device to Holloman biographer George Meeter.

It's known that maximum g tolerance is at right angles to the long axis of the body ... when this platform pivots about the lateral axis of the aircraft or space vehicle it's through the action of acceleration itself that it assumes a position perpendicular to the heart–head line of the body and therefore changes longitudinal g's into more easily tolerated transverse g's. In other words it avoids the action of significant parallel-to-spine loads ... The unprotected or control animal received noticeable internal effects by taking the g load in the vulnerable longitudinal axis. The animal on the anti-g platform took the load in the less critical transverse position – in which four or five times as much deceleration can be endured [16].

In his report on the use of the two Holloman test tracks, Dr. David Bushnell wrote that animal aerospace experiments had figured less prominently in Daisy tests than on the long, high-speed track.

Most test configurations to date have not been of an order to cause serious injury, and therefore it has normally been possible to use human subjects. Nevertheless, chimpanzees did take part in some of the early tests and helped check out the facility for human use. On two later occasions, hogs, which have never been privileged to ride the long track, took part in preliminary experiments with a new test configuration and received spinal fractures from an impact force measured at less than thirty g's. This unfortunate result was due to the particular combination selected of g-forces and body orientation (forces parallel to spine), and to the nature of the hogs themselves, including the virtual impossibility of properly restraining these animals on the sled [24].

The rocket sled bears

Even black bears, four of which first appeared at Holloman in late 1957, were used in Daisy Track test runs, as their pelvic region is very similar to that of humans. But – like the hog tests – these were more related to automobile crash restraint systems, according to Bushnell, "seeking correlation between spinal injury in bears and humans" [25].

The first use of a bear in a 20-g Daisy Track run would attract some unfavourable publicity when, following the run in which the animal showed no outward ill effect of the test ride, it was euthanised and dissected in order to look for any possible internal injuries. Nevertheless, the same fate would befall the other hapless animals.

Water-immersion tests were also carried out at Holloman using chimpanzees and other subjects strapped into water-filled capsules to determine if fluid protection could annul the forces acting on them during deceleration. While it seemed to help, there were many complications that caused these plans to be shelved as a possible tool for space flight, including vehicle design and weight.

Some more high-g sled runs at Holloman involving chimpanzees took place in March 1957, after which the historic 3,550 foot-track was extended until it was almost 10 times longer. It would now be used for a continuation of windblast studies and devising means of protection against strong g-forces.

By the time it was completed, however, Colonel Stapp had left Holloman, having been reassigned to the Aerospace Medical Laboratory at Ohio's Wright Field. Here he would continue his work on windblast studies using chimpanzees and anthropomorphic dummies on the 10,000-foot track at Wright Field – later extended to 20,000 feet – as well as the Supersonic Naval Ordnance Research Track (known as SNORT) at China Lake in California. Here he collaborated on tests with another medical doctor, Captain John D. Mosely, who would later head the chimpanzee training programme at Holloman in association with NASA's man-in-space venture, Project Mercury.

An application to automobile safety

In later years Stapp would serve as chief of the Armed Forces Institute of Pathology, following which the Air Force would assign him indefinitely as chief scientist to the

Department of Transportation's National Highway Safety Bureau, to assist them in their research into motor vehicle restraint systems. Today his work on restraint and crash systems and their application to the automobile industry is recognised as having saved countless thousands of lives. John Paul Stapp died peacefully at his Alamogordo home on 13 November 1999, aged 89, but he will always be renowned and recognised for his courageous, lifelong work.

Stapp's work continues at Holloman

In his report on track-test programmes applicable to space flight at Holloman from 1958 to 1960, Bushnell reports that the best facility for studies of the rapid onset of acceleration and simulation of an actual rocket launch was the now-lengthened high-speed test track at Holloman. The first such run took place on 6 August 1960, with the rocket sled carrying three primate capsules. In this way, three chimpanzees could be exposed to exactly the same acceleration profile at the same time, allowing comparison studies to be conducted, which could never be accurately duplicated on three separate runs.

For the first track run, only one capsule was occupied, while the other two contained ballast of an equal weight. The test proved less than fruitful, only reaching around 390 feet per second, which was far less than predicted. At the end of the run the deceleration from the water brakes was proportionally low, and it was further found that the water brakes had sent up a plume of water that obscured the camera view of the test chimpanzee. The chimp was then strapped onto the Daisy Track sled, where, according to Bushnell, the animal "received a good 20-g simulation of a water impact" [25].

No primates were carried on the next high-speed test on 25 August, on which a new propulsion system configuration was tested, which entailed the addition of a fifth booster to the sled's fourth stage. On this occasion a higher speed was achieved – around 540 feet per second – but there were still problems with the water brakes. By 29 September the technicians figured they had solved the water brake problem. On this next run, again with no passengers on board, the deceleration worked well, even though the velocity was once again less than desired.

Gregory Kennedy said that for the next run, on 26 October, a chimpanzee rode the high-speed sled.

Again, the performance was less than predicted, but a deceleration of 8 g's was obtained, which was high enough to satisfy test conductors. As with the first chimpanzee test, a Daisy Track run followed the rocket sled. This test also included a psychomotor panel in the capsule. Results were encouraging; the subject showed no performance decrement during the high speed track or Daisy Track runs. After this test, further high speed track runs were suspended until the time came to test chimpanzees for orbital flight [26].

Stapp's research had helped a great deal in helmet improvements, leg and arm restraints, and provided criteria for much stronger safety harnesses – work that would later prove beneficial and certainly life-saving to the automobile industry. He had also devised techniques for best positioning the body to absorb crushing g forces associated with windblast and rapid deceleration. Animals would also prove beneficial in programmes implemented to develop better protective clothing and pressure helmets for pilots. These reluctant subjects would face the unknown elements of extremely high altitudes and low temperatures in special altitude chambers designed to reproduce conditions at the top of the stratosphere, while wearing carefully fitted miniature suits, masks and helmets.

Colonel Stapp's extensive work at Holloman, and his renowned rocket-sled ride, became the subject of a well-received 1956 Hollywood bio-pic called *On the Threshold of Space*. Directed by Robert Webb and starring Guy Madison, Virginia Leath and John Hodiak (who died suddenly of a heart attack just a week before filming was finished), it placed special emphasis on historical accuracy and was partially shot at Holloman from July to September 1955, with input and assistance given by U.S. Air Force personnel. Some high-speed sled runs were specially staged for the film's producers. While the story centres on an Air Force doctor who conducts tests to discover how the human body will stand up to the rigours of space travel, the central character is actually a composite of pioneering researchers such as Stapp, David Simons and Joe Kittinger.

Chimpanzees begin training for space flight

As Holloman biographer George Meeter explains, the AMFL's doctors and their talented assistants continued Stapp's work while America moved ever closer to sending the first human into space, and the Holloman chimpanzees would contribute greatly to that effort by helping to solve many vexing problems.

Thus, as the Laboratory's human population had grown so had its animal population, at times to as many as 400 clinically observed creatures representing as many as nine species. (The largest numbers being mice, a conveniently small biological unit.) Most of them seemed to thrive with the unusual care they got, and were both useful and continuing subjects [16].

In the early days of the AMFL, a colony of around 40 chimpanzees had been living in three buildings in a part of the laboratory known as the Vivarium. Prior to the importation of the chimpanzees, the Vivarium had been home to an eclectic group of research animals – a few bears, a number of hogs, cats, dogs, fish, frogs, rats and mice.

Air Force Captain Jim Cook was a veterinarian and the AMFL's chief pathologist, and he said that most of his chimpanzee charges had come from the Cameroons in equatorial Africa. "In fact they're classified as an anthropoid ape of Africa," he told Meeter. "In the Latin known as *Pan satyrus*, sometimes as *Pan troglodyte* – a primitive little man not too unlike ourselves. Usually we try to get them as babies, one to three years of age.

"After they reach Holloman we segregate them for about six weeks and build them up to gain confidence. Also to give them full physical checkups and eliminate parasites and disease carriers. It's sort of quarantine – as soon as it's over they're allowed to join our regular colony and in about six months we can use them with other chimps in the space biology programs" [16].

In the following years, standout Holloman students such as those named Enos, Duane, Paleface, Bobbie Joe, Chang (later renamed Ham), Elvis, Tiger, Roscoe, Little Jim and Minnie would take part in NASA space biology programmes.

REFERENCES

[1] "Montgolfier Brothers", Wikipedia on-line encyclopaedia. Website: *http://en.wikipedia.org/wiki/Montgolfier*

[2] James Glaisher, Camille Flammarion, W. de Fonveille and Gaston Tissandier, *Travels in the Air*, Richard Bentley & Son, London, 1871 (translated from 1867 French edition).

[3] Reverend John M. Bacon, *Dominion of the Air: The Story of Aerial Navigation* (Chapter 19), Cassell, London, 1902.

[4] Lloyd Mallan, *Men, Rockets and Space Rats*, Julian Messner, New York, 1958

[5] Dieter E. Beischer and Alfred R. Fregly, *Animals and Man in Space: A Chronology and Annotated Bibliography through the Year 1960*, U.S. Naval School of Aviation Medicine, Pensacola, FL, for the Office of Naval Research, Department of the Navy, Washington, D.C., 1964.

[6] Interview with David G. Simons conducted by Center Director Gregory P. Kennedy for the Alamogordo Space Center Oral History Program. Interview conducted 30 September 1987 at the International Space Hall of Fame.

[7] W.D. Murray, "A Gondola for Physiological Research in the Atmosphere," *Journal of Aviation Medicine*, issue 25, 1954, pp. 354–360.

[8] David Simons and Charles Steinmetz, "The 1954 Aeromedical Field Laboratory Balloon Flights: Physiological and Radiobiological Aspects," *Journal of Aviation Medicine*, issue 27, 1956, pp. 100–110.

[9] H.F. Harlow, A.M. Schrier and D.G. Simons, "Exposure of Primates to Cosmic Radiation above 90,000 Feet," *Journal of Comparative Physiology*, issue 49, 1956, pp. 195–200.

[10] I.J. Lebish, D.G. Simons, H. Yagoda, P. Jannsen and W. Haymaker, "Observations on Mice Exposed to Cosmic Radiation in the Stratosphere: A Longevity and Pathological Study of 85 Mice," *Military Medicine*, issue 124, 1959, pp. 835–847.

[11] James P. Henry, *Biomedical Aspects of Space Flight*, Holt, Rinehart & Winston, New York, 1966.

[12] David Simons and Don A. Schanche, *Man High: 24 Hours on the Edge of Space*, Sidgwick & Jackson, London, 1960.

[13] James S. Hanrahan and David Bushnell, *Space Biology: The Human Factors in Space Flight*, Basic Books, New York, 1960.

[14] "Cumulus (Kumulus)", Encyclopedia Astronautica. Website: *http://www.astronautix.com/lvs/kumulus.htm*

[15] "Zucker Rocket", Encyclopedia Astronautica. Website: *http://www.astronautix.com/lvs/zucocket.htm*

[16] George Meeter, *The Holloman Story: Eyewitness Accounts of Space Age Research*, The University of New Mexico Press, Albuquerque, NM, 1967.

[17] Gregory P. Kennedy, Report IAA-87-651, *The Aeromedical Field Laboratory: Space Medicine at Holloman Air Force Base*, The Space Center, Alamogordo, NM, 1988, p. 4.

[18] David Bushnell (ed.), Report, *History of Research in Space Biology and Biodynamics, Part IV: Other Work on the Escape Problem*, U.S. Air Force Missile Development Center, Wright Field, OH, December 1958, pp. 2–4.

[19] Clyde Bergwin and William Coleman, *Animal Astronauts: They Opened the Way to the Stars*, Prentice Hall, Englewood Cliffs, NJ, 1963, pp. 49–50.

[20] Edward Dittmer, Oral History Interview conducted by George M. House, Curator, Alamogordo Space Center, NM, 29 April 1987.

[21] Space Log magazine, article *Stapp Honored at ISHF*, vol. 4, no. 3, July 1987, p. 2.

[22] John Paul Stapp, "Rocket Sled," *Above and Beyond: The Encyclopedia of Aviation and Space Sciences*, p. 1986, New Horizons, Chicago, 1968.

[23] Joseph W. Kittinger, Jr. and Martin Caidin, *The Long, Lonely Leap*, E.P. Dutton & Co., New York, 1961, p. 18.

[24] David Bushnell, *History of Research in Space Biology and Biodynamics, Part I, The Beginnings of Research in Space Biology at the Air Force Missile Development Center, 1946–1952*, Air Force Missile Development Center, Holloman Air Force Base, Alamogordo, NM, 1955.

[25] David Bushnell, *The Aeromedical Field Laboratory: Mission, Organization and Track-Test Programs, 1958–1960*. Air Force Missile Development Center, Holloman Air Force Base, Alamogordo, NM, 1961.

[26] Gregory P. Kennedy, Report IAA-89-741, *Mercury Primates*, The Space Center, Alamogordo, NM, 1989, p. 5.

5

Able and Baker lead the way

1958 was a truly momentous year in the history of space exploration. Within the confines of those 12 months, America's first satellite, Explorer 1, was launched into orbit, while Russia retaliated by lobbing up a massive space laboratory known as Sputnik 3. On 2 April that year President Eisenhower had placed a proposal before Congress, calling for the creation of a civilian space agency to be known as NASA – the National Aeronautics and Space Administration. A compromise bill was approved and passed by voice votes of both houses of Congress on 16 July and signed by the President just 13 days later.

Within weeks, NASA had boldly announced plans for an ambitious manned spaceflight programme that would be known as Project Mercury.

NASA AND THE ARPA

The compromise bill signed by President Eisenhower meant that NASA would be in overall control of space development in the United States, while the Department of Defense would remain in charge of "activities peculiar to or primarily associated with development of weapons systems, military operations or defense ..." [1]. The President was charged with resolving any conflicts between the civilian space agency and the defence department's space weapons unit, the ARPA (Advanced Research Projects Agency).

Project Mouse-In-Able

In 1958 a series of three launches using Thor IRBMs took place from Cape Canaveral – each of them carrying a single mouse in an undertaking known as Project MIA, or Mouse-In-Able.

The programme was directed by the Space Technology Laboratories of Los Angeles in conjunction with the Air Force Ballistic Missile Division. The scientist in charge was a tenacious woman named Franki Van der Wal, who had managed to convince the Air Force to place female mice into each of three Thor–Able missile re-entry test flights [2].

Project MIA was planned as a non-interference experiment in conjunction with the Project Able re-entry test programme, which involved launching a two-stage missile consisting of a Douglas Thor IRBM topped by a modified Aerojet-General 1040 liquid propellant second stage.

In all, three Thor–Able rockets were sent aloft on 23 April, 9 July and 23 July, each with a nose cone built by General Electric and Avco, which was designed to absorb the intense heat of atmospheric re-entry by shedding thin layers of surface material. These nose cones also carried, respectively, mice named Minnie, Laska and Benji in a 20-pound bio-capsule. The third mouse was also known as Wickie, popularly said to have been named after the daughter of well-known female Cape journalist, Mercer Livermore. However, according to Air Force handlers, the white mouse was given the nickname simply because it "drank water from a wick" [3].

The principal purpose of these tests was to obtain re-entry data for nose cone development, while the mice were included as a secondary experiment to gauge the reaction of living specimens to spaceflight conditions.

A very small unit with limited space

Ed Dittmer was one of the principals involved in building the environmental capsule that would contain the mice. Together with project officer Captain Grover Schock, he travelled to the Ramo-Woolridge plant in California, where the capsule would be manufactured, to explain exactly what was needed. As it would only be a very small unit with limited space, Dittmer went armed with a basic design for a capsule capable of carrying a mouse as well as a life environment system.

The result, as Dittmer explained, was "a very small capsule that incorporated an air generation system. The air would continually pass through the mouse's chamber, all the impurities would be collected, and then the air was regenerated back through the chamber containing the mouse. In order to keep the mouse at the same attitude as the rocket after take-off, we designed a small carrier which had a swing apparatus. This meant that whichever way the rocket turned, the mouse in his little carrier would swing the same way" [4].

Urine collection for later analysis was another problem, but it was found that a small wad of cotton wool placed beneath the mouse was an ideal and simple solution. Another small challenge Dittmer faced was actually securing the mouse to a small platform for the rocket flight. As he told interviewer George M. House in 1987, "I had worked with guinea pigs in attaching track plates and so forth, and we decided to use acetone-sensitive paper. You just run it in the acetone and it becomes real pliable, then when it dries it stays in that form. So, just run a few strips of that over him and run under [the platform] and when it dried it held him there so he couldn't run off the couch" [4].

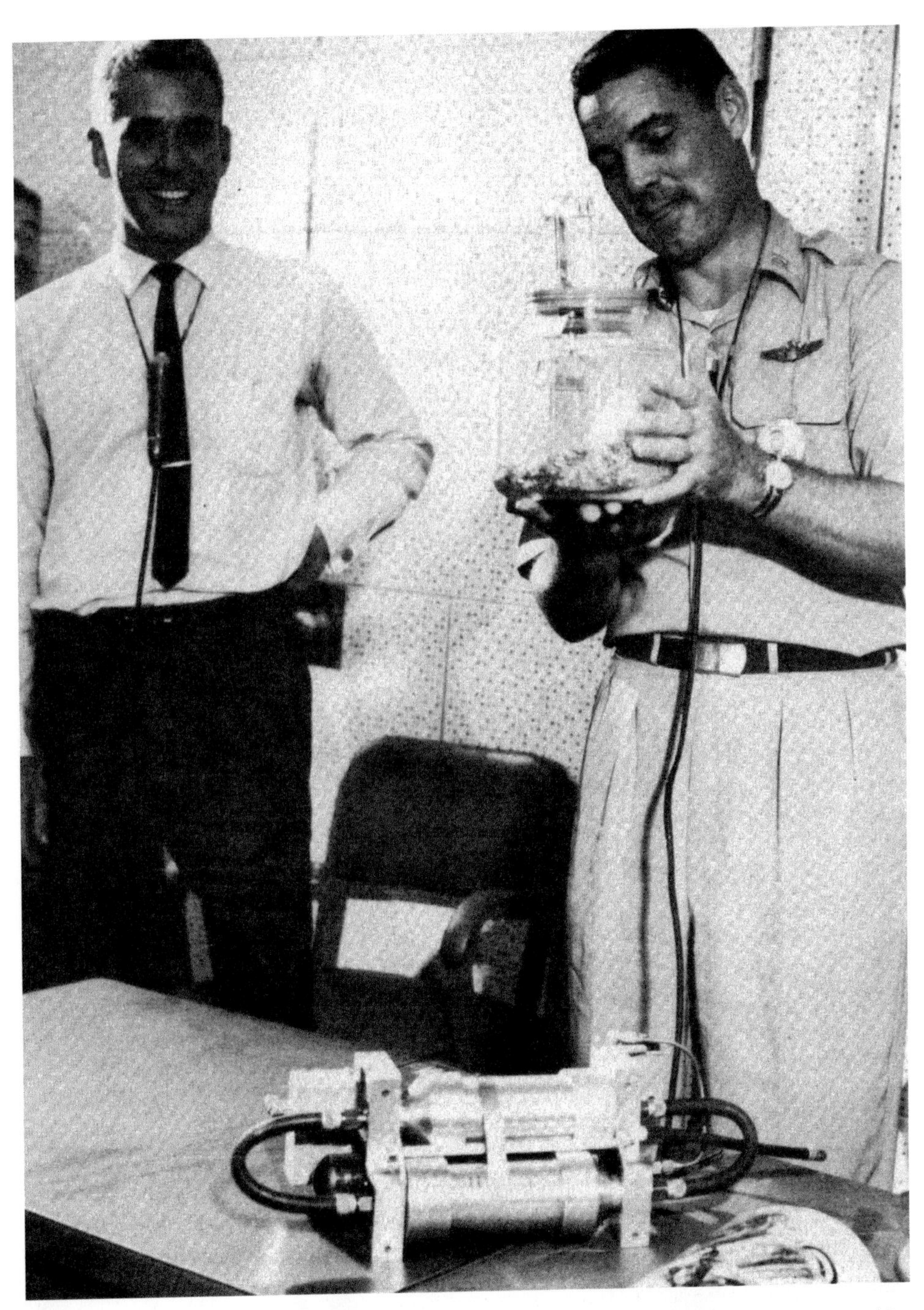

Captain M.E. Griffith (right), a Thor–Able project officer, holds up a glass jar containing white mouse Wickie. The life-support apparatus designed to hold and sustain the mouse is on the table at front. (Photo: USAF)

Dittmer then travelled to Trinidad where Navy recovery boats would arrive after retrieving the nose cones post-flight. He briefed the crews on how to open the capsules, provided them with mouse-care manuals and instructed them on what they had to do to preserve the data so it wouldn't be lost. Once everything was ready the mice and their capsules were delivered to Cape Canaveral.

The Thor–Able III rocket carrying Wickie the mouse thunders off the Cape Canaveral launch pad on 23 July 1958. (Photo: USAF, courtesy Joel Powell and Art LeBrun)

A victim of the space age

On 23 April the first Thor–Able lift-off from Launch Complex 17A carrying the mouse Minnie began well, but 146 seconds later there was a major failure in the first-stage turbo-pump, which caused the rocket to explode, and the nose cone was lost in the Atlantic. The second and third Thor–Able launches carrying Laska and Wickie were quite successful, each shooting up to an altitude of around 600 miles and travelling nearly 6,000 miles at around 15,000 mph before the nose cones splashed down into the South Atlantic half an hour later, at the far end of the United States' missile testing range. On these two flights the mice were instrumented for registering the electrical activity of the mouse's heart through a single-commutated telemetry channel. Both nose cones emitted radio signals, confirming they had survived the fierce heat of re-entry, but an intense search by aeroplanes and ships failed to locate and retrieve them.

Telemetered physiological records for both of the successful test firings indicated that both Laska and Wickie had survived their flights, and would have been recovered alive had their nose cones been located. In fact, Wickie, sealed in a waterproof, air-conditioned chamber marked "mouse house" had enough food, water and oxygen to last for about 2 weeks. However, the search was called off after 2 days, and little Wickie became yet another tragic victim of the space age. The amount and nature of the data drawn from both space flights was extremely limited, which meant that no real conclusions could be drawn regarding the behaviour of the mice during their journeys.

On the positive side, information gathered from the flight indicated that conditions during lift-off had not been sufficiently severe to cause any adverse response from the mice. However, the two mice did respond differently to their flight experiences. The acceleration or g-loads during the ascent, measured at a peak of 17 g's, were essentially paralleled by Laska's climbing heart rate, although this characteristic was not evident in Wickie's responses after lift-off. Laska's heart rate after first-stage burn-out decreased gradually, but rose again sharply following the second-stage burn-out. By comparison, Wickie's heart rate barely changed until a noticeable increase with the onset of weightlessness.

Despite their unfortunate loss, and that of Mia (Minnie) months earlier, both mice were truly record-breakers. They each returned to Earth from a higher altitude than any previously flown by other living organisms in an American spacecraft, and they were both weightless for longer periods than any animals other than Russian space dog Laika.

Training the satellite mice

Several American publications were starting to realise that space flight and associated research were popular subjects, and would regularly include feature articles on these – especially after Russia sent up the first Sputnik satellites. This public interest would soon culminate in *Life* magazine being granted exclusive access to the astronauts and their families in the first years of the manned spaceflight programme. In the October 1959 issue of *National Geographic* (vol. CXII, no. 6), one article said that orbital space flights with mice on board were under consideration by the Air Force's Aeromedical Field Laboratory at Holloman AFB in New Mexico, where Major David Simons was chief of the Space Biology Branch. As the magazine reported, this project involved placing mice inside a "bio-satellite" to observe their reactions as they sped around the Earth.

"Satellites are in effect weightless, because once they are in an orbit their velocity balances the pull of gravity. Knowing the physical and mental reactions of organisms at zero gravity is of great interest to scientists planning manned flights at great altitudes. Mice and monkeys have already been sent up 37 miles in rockets, and thus put into the weightless state for about 2 minutes.

"A bio-satellite might keep a mouse supplied with food and oxygen for 30 days; television apparatus could let us observe the mouse at regular intervals, say 1 minute out of 5.

"The satellite mouse would be trained – to hit a switch, for example, in order to avoid an annoying stimulus. Once at zero gravity, the mouse would presumably be upset and therefore unable to hit the switch as often as it might want to. But after being up for some time, it might again remember what it had learned and hit the switch at a rate more in keeping with its previous performance" [5].

PROJECT "DOWN TO EARTH"

Meanwhile, the U.S. Army's Office of the Surgeon General had made a rather extraordinary request to the Army Ballistic Missile Agency. They wanted the ABMA to make space available in its missile nose cones for spaceflight experiments involving animals and other life forms as an essential part of a military programme to be known as "Project Down To Earth".

The Army looks to space

The proposed biological flights would serve many purposes. They would not only help the Army develop missile countdown and launching procedures, but would also determine the physiological response and well-being of primate passengers during ballistic rocket flights into space. Additionally, these flights would demonstrate that living creatures could survive unharmed if adequate life-support systems were provided, while ensuring they could also be quickly located and recovered after an ocean splashdown.

Earlier, because difficulties were perceived with the concept of enclosing a monkey in such a confined space, much consideration had been given to first sending up a "safe" cargo of plants and much smaller animal subjects. It was felt by some researchers that these specimens had a far better chance of survival for later study. At this time the project also acquired the whimsical codename of "Noah's Ark", but – once it had been established that a small monkey could indeed be used – these tentative plans (and the biblical nickname) were scrapped.

Gordo, a.k.a. Old Reliable

On the first of the flights, launched on Friday, 13 December 1958, a male South American *Saimiri sciureus* squirrel monkey was catapulted aloft from Cape Canaveral aboard a U.S. Army Jupiter IRBM (Intermediate Range Ballistic Missile), on behalf of both the U.S. Army and Navy Medical Corps. During his training at the Navy's Aviation Medical School in Pensacola the cute, bushy-tailed monkey had been known to everyone as "Old Reliable", due to his ability to learn quickly and well. It also happened that the highly dependable Jupiter rockets, the same type that would carry him, were commonly endowed with that name.

Squirrel monkeys by nature are highly excitable and can even be mildly aggressive in captivity, which makes them difficult to tame and train. Conversely, this sensitivity to their handling environment also causes them to be excellent biological indicators in stressful situations.

Four days prior to the first flight a few laboratory facilities, two flight-ready capsules and six monkeys were transported by Navy personnel from Pensacola to Cape Canaveral. Then, 11 hours before the scheduled lift-off, two teams of doctors selected a pair of monkeys with the best all-round performance during training and prepared them for placement in the space capsule and a secondary, back-up unit.

Creating a Bioflight capsule

The heated Bioflight 1 capsule into which the prime candidate Old Reliable would be inserted and launched into space had been developed and tested at the School of Aviation Medicine, together with design assistance provided by the Army's Ballistic Missile Agency. The tapered, $29\frac{1}{2}$-pound fully-instrumented capsule, custom-built to contain Old Reliable's cylindrical container, was precisely manufactured to fit into the bottom section of the nose cone to allow for easy access. Covered with insulating foil and fibreglass, it provided around 750 cubic inches of space for holding the couch-restrained monkey as well as his life-support systems and monitoring equipment, all of which would be inserted through an access port 3.75 inches in diameter. The capsule was totally self-contained apart from a single electrical connection, which provided electrical power and allowed for the transmission of data.

As with earlier animal flights, an environmental control unit had been fitted to eliminate any excess moisture from the capsule, while pellets of baralyme, a mixture of calcium and barium hydroxides, would help absorb the monkey's exhalations and minimise any possible hazards associated with re-inhaling potentially hazardous levels of carbon dioxide.

GORDO BLAZES A TRAIL

Up to this time the prime candidate for the flight was still called Old Reliable, but as the time for launch drew nearer he was assigned another pet name, phonetically much better suited to communications – Gordo.

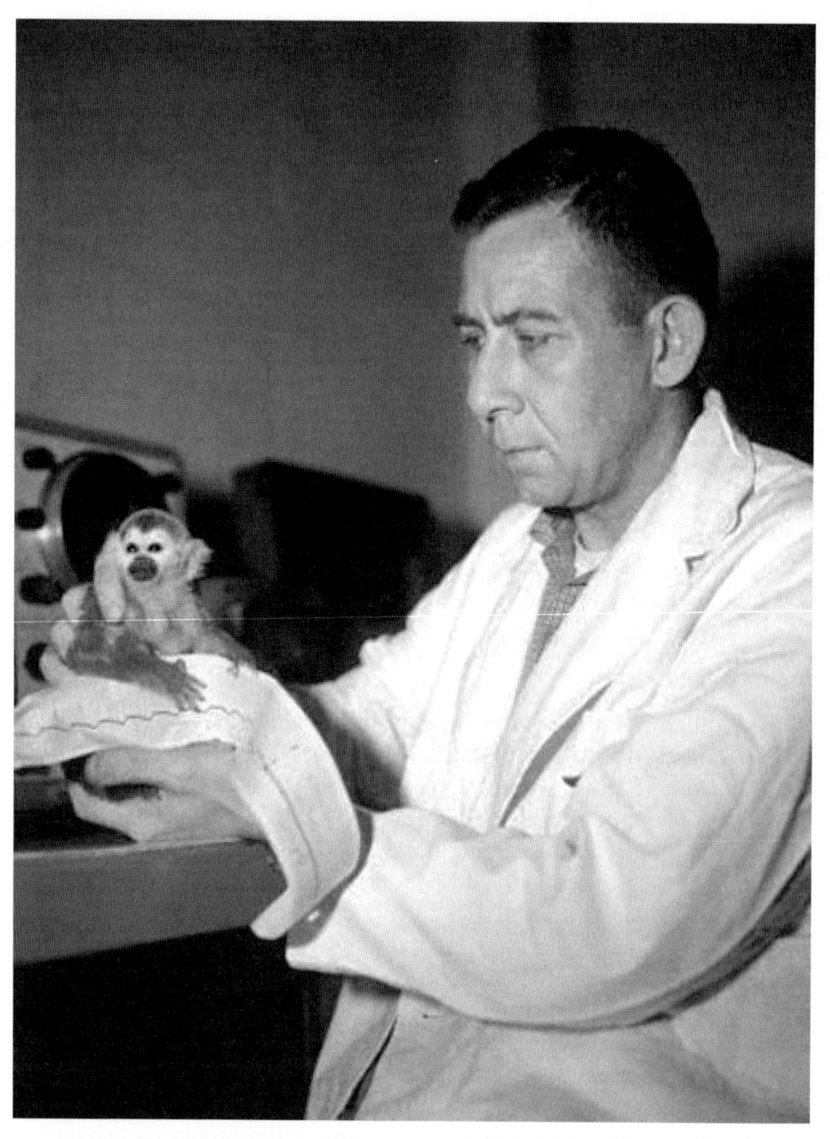

A handler with Gordo, also known as Old Reliable. (Photo: USAF)

Ready for lift-off

In the late evening hours of Thursday the 12th of December a start was made on preparing the petite, 1-pound monkey for his journey early the next morning. He was suited out with a chamois leather-lined plastic helmet made of moulded plastic compound and then strapped onto an individually-fitted silicon rubber couch. Next came a thin overlaying sheet of foam rubber with an inbuilt microphone to monitor his heartbeat. A thermistor – a semiconductor resistor device used in the measurement

Jupiter 21 with Gordo in the nose cone prior to launch. (Photo: USAF)

of body temperatures – was placed under one of his armpits and a respiratory sensing device was secured with model glue just above the monkey's nostrils.

Small straps extending from Gordo's helmet were secured onto rubber posts that formed part of his moulded bed in order to immobilise his head, after which further foam rubber overlays were secured across his body. Gordo's knees had been drawn up in a supine position, which would allow him to cope better when undergoing the severe stresses associated with acceleration. Once this had been accomplished and all the monitoring instruments were in place, Gordo's container was carefully slipped into the waiting capsule for insertion in the base of the Jupiter's nose cone [6].

Gordo takes flight

Lift-off of the 50-ton Jupiter AM-13 flight took place at 3:38 a.m. EST from the Cape's Launch Complex 26. Gordo apparently handled this phase of his flight well, although telemetry indicated that the tremendous surge of acceleration during lift-off, measured at around 10 g's, slowed the monkey's respiratory rate and caused his pulse rate to fluctuate and then speed up. During an estimated 9 minutes of weightlessness both of these readings returned to normal, and he made an otherwise uneventful quarter-hour flight through space.

Gordo is thought to have survived the 10,000 mph re-entry, following which his nose cone splashed down in the South Atlantic, just over 1,500 miles downrange from Cape Canaveral. Then success turned to calamity, when plans for his retrieval from the ocean went horribly awry. A recovery ship that had steamed out to the splashdown area was unable to pick up any signals from beacons within the nose cone, and this made locating the spacecraft virtually impossible. Despite their best efforts, the search was abandoned after 6 hours. It was later thought that the most probable explanations for the loss were a parachute malfunction or a mechanical failure in the flotation system attached to the nose cone.

There was, however, *some* good news; after examining data from the flight, Navy doctors reported that respiratory and heartbeat telemetry signals from Gordo's capsule had provided strong evidence that a human being could have survived a similar journey.

Some naysayers did suggest after the loss of Gordo that the launch should never have taken place on Friday the 13th, as the U.S. Navy has always deferred to widely-held superstition by not launching any new ships on this date.

ABLE AND BAKER

With the results from Gordo's flight in hand, and more information to be harvested, a second Jupiter IRBM biological mission was manifested for the following year. The two female passengers chosen for this particular flight were a rhesus monkey and a much smaller squirrel monkey, each of whom would be individually trained by two separate divisions of America's armed forces. The two primates would eventually become widely known by the first two letters of the American military's phonetic alphabet – Able and Baker.

The U.S. Navy joins in

The Navy was once again participating in this flight, having been assigned the task of recovering the nose cone off the coast of Antigua. They would also supply and train the squirrel monkey destined to ride the rocket.

Meanwhile, Alton Freeman from the Miami Rare Bird Farms in Florida had contacted his friend Ralph Mitchell, the chairman of the Riverside Park and Zoo in Kansas. He had received a rush order for two dozen healthy, acclimated and tuberculosis-tested rhesus monkeys, and asked if Mitchell's zoo could help. Coincidentally, for a variety of reasons, Mitchell had wanted for some time to rid the zoo of its rhesus monkeys, and he quickly made a deal to exchange 26 rhesus monkeys for a like number of spider monkeys. The deal was done, and the 26 monkeys were eventually shipped through Freeman to Wisconsin University's research department [7].

The Ralph Mitchell Zoo in Independence, Kansas, is proud to be recognised as the original home of the space monkey that came to be called Able. (Photo: courtesy of Mike Myer)

From these, a pool of eight suitable candidates was selected by specialists from the Army Medical Research Laboratory (AMRL) in Fort Knox, Kentucky, and the Walter Reed Army Institute of Research in Washington, D.C. In a humorous revelation, a spokesman at Walter Reed admitted that there appeared to be some "executive monkeys" amongst the candidates, whose reaction to the stress and decision-making led to them developing ulcers, just like their human counterparts! [8].

A monkey is chosen

After exhaustive tests a 7-pound rhesus monkey born at the Riverside Zoo Park in December 1957 was selected for the flight, although she was not the original choice as prime candidate. In an interesting diplomatic move a fully trained, Indian-born rhesus monkey chosen earlier to fly the mission was replaced 2 weeks before the launch on specific orders emanating from the White House. It seemed that President Eisenhower had been briefed on the forthcoming mission, but when he noticed the birthplace of the monkey he decided this might offend the people of India, who considered their rhesus monkeys sacred animals. The one-time prime candidate was quickly replaced by another born in a pet shop in Independence, Missouri. This alternative monkey

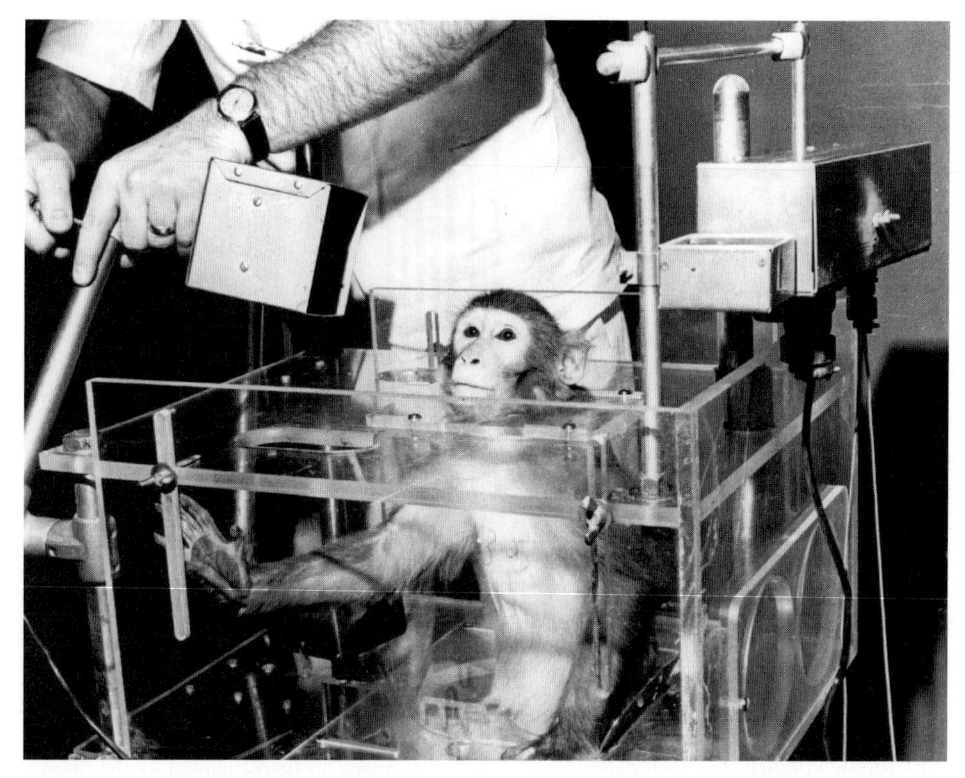

The monkey that would become known as Able undergoes flight-testing. (Photo: NASA)

would only be assigned the rather unimaginative code name Able shortly before the flight.

A team of physicians and enlisted men charged with readying the monkey for her flight would begin the extraordinarily lengthy process by strapping Able into her contoured couch some 64 hours prior to launch.

For her protection, Able wore a specially-tailored spacesuit and fibreglass helmet with inbuilt sensors, which would also allow data on her condition to be collected throughout the flight and transmitted back to the ground. Altogether, telemetry from 16 channels of biomedical information would be received for later examination and dissemination. This involved instruments such as an electrocardiogram to record her heartbeats, and an electromyogram to monitor muscular reaction. Information would also be collected from Able's pulse and respiration rates, body temperature, behavioural response and respiration rate. Other instruments would measure the temperature, pressure and relative humidity inside the monkey's capsule.

Her individual support couch had been constructed from a life-sized model of the monkey made of concrete-like moulding compound. The couch was then formed over this model using a vacuum technique of fibreglass fabrication. In order to restrain Able in the custom-built couch, her legs and abdomen were held tightly in place by nylon mesh straps, while a fibreglass chest plate would immobilise her upper body. The strapped fibreglass helmet, also custom-built to suit each candidate, served to hold her head firmly to the top of the couch.

To allow for a performance study of the primate, a 16-mm motion film camera would also be loaded into the capsule to photograph her upper chest and head area. The camera shutter would operate at 16 frames per second over 400 feet of film, which would allow around 17 minutes of operation.

Miss Baker

Able's flight companion, Baker, was a tiny South American squirrel monkey, originally born in Iquitos in the Peruvian jungle sometime during 1958, but captured by hunters and brought to the United States. She weighed in at just 11 ounces. Together with 25 young spider monkeys purchased from a pet shop in Miami, she had been transported to the U.S. Navy Aviation Medical School in Pensacola, Florida.

As the monkeys would have to fly in very confined capsules, they underwent extensive training in similar, small enclosures for periods lasting up to 24 hours. Each

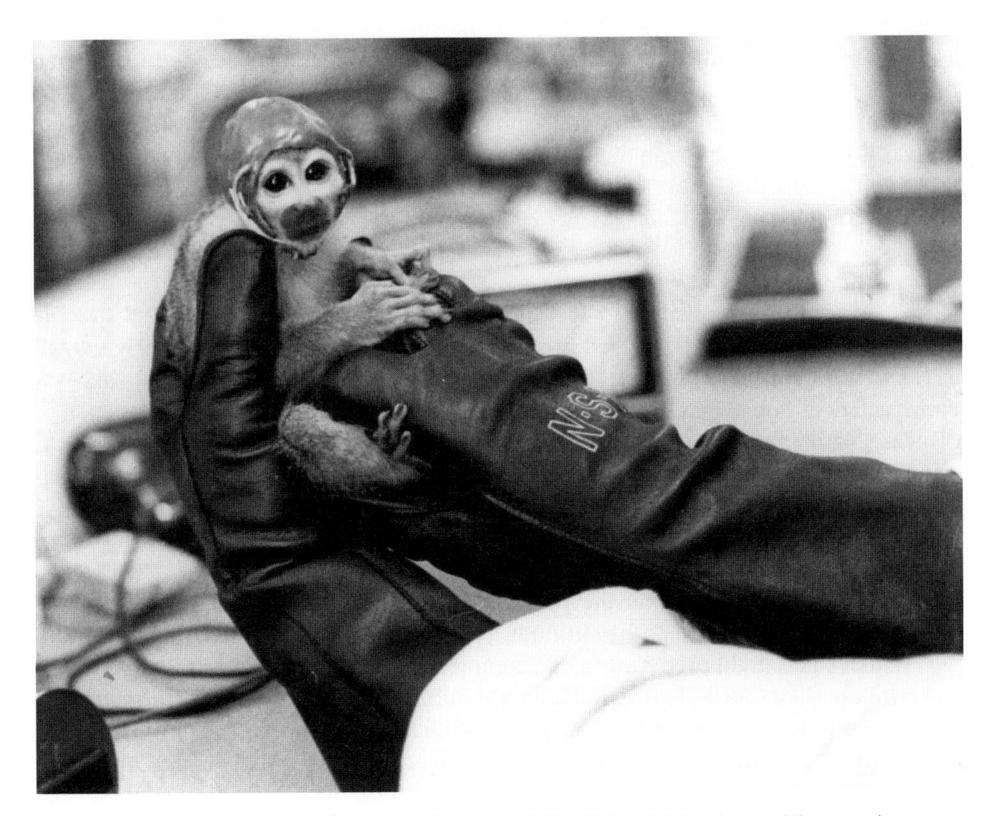

The squirrel monkey that would become known as Miss Baker is kitted out with some tiny space gear. (Photo: NASA)

of them also took part in simulated flights and laboratory tests, and gradually became inured to having tiny electrodes plastered all over their bodies and their physiological impulses monitored day and night. Some, of course, just could not cope: they preferred their cosy zoo homes and a standard diet of grapes and bananas to the rigid discipline.

Eventually, 14 of these tiny candidates would prove suitable for the flight, but one in particular stood out from the rest because of her intelligence and loving, docile manner. She had been nicknamed TLC, as she seemed to enjoy being handled with tender loving care by the doctors and trainers.

When asked post-flight by a news reporter about the attitude of the monkeys during their training, Navy Captain Ashton Graybiel responded, "These monkeys are almost volunteers. During the pre-flight testing, we didn't force a monkey to take a test if it objected to it" [6].

A RIDE ABOARD A JUPITER

In the days leading up to the Jupiter launch, the outstanding candidate was still known as TLC, but this would change when Army and Navy specialists decided to christen the two monkeys with more appropriate names on their dual-service space flight. The Army settled on the phonetic name Alpha for their subject, and the Naval School dryly followed suit by naming theirs Bravo. The identifying names quickly transmuted into identifying words more commonly used in traditional military phonetics – Able and Baker.

Preparations continue

Having been selected and declared fit for the flight, the tiny monkey formerly known as TLC was housed in a separate, smaller capsule than Able's. This capsule was constructed of somewhat heavier aluminium sheeting than Bioflight 1 in order to dispense with reinforcing ribbing that consumed a lot of space. The capsule was about the size of a large shoebox, measuring $9.75 \times 12.5 \times 6.75$ inches and was insulated with rubber and fibreglass. Bottled oxygen with a pressure-activated valve would be supplied to the monkey's capsule during the flight, while an absorber removed excess moisture from the surrounding atmosphere. Instead of the chemical pellets of baralyme carried on the previous flight for this purpose, the life-support system had been modified to use lithium hydroxide.

While Able's flight preparation had commenced nearly 3 days earlier, getting little Baker ready (like Gordo before her), only required some 8 hours. Telemetry sensors were taped onto Baker's body, or surgically implanted, to record her body temperature and heart action, while other instruments would monitor the air pressure inside her capsule. In order to record the monkey's respiration during the flight a glass bead thermistor was soldered to fine wire leads, which were used to position the bead in the main stream of Baker's exhaled air. The end of the thermistor lead was then fixed to

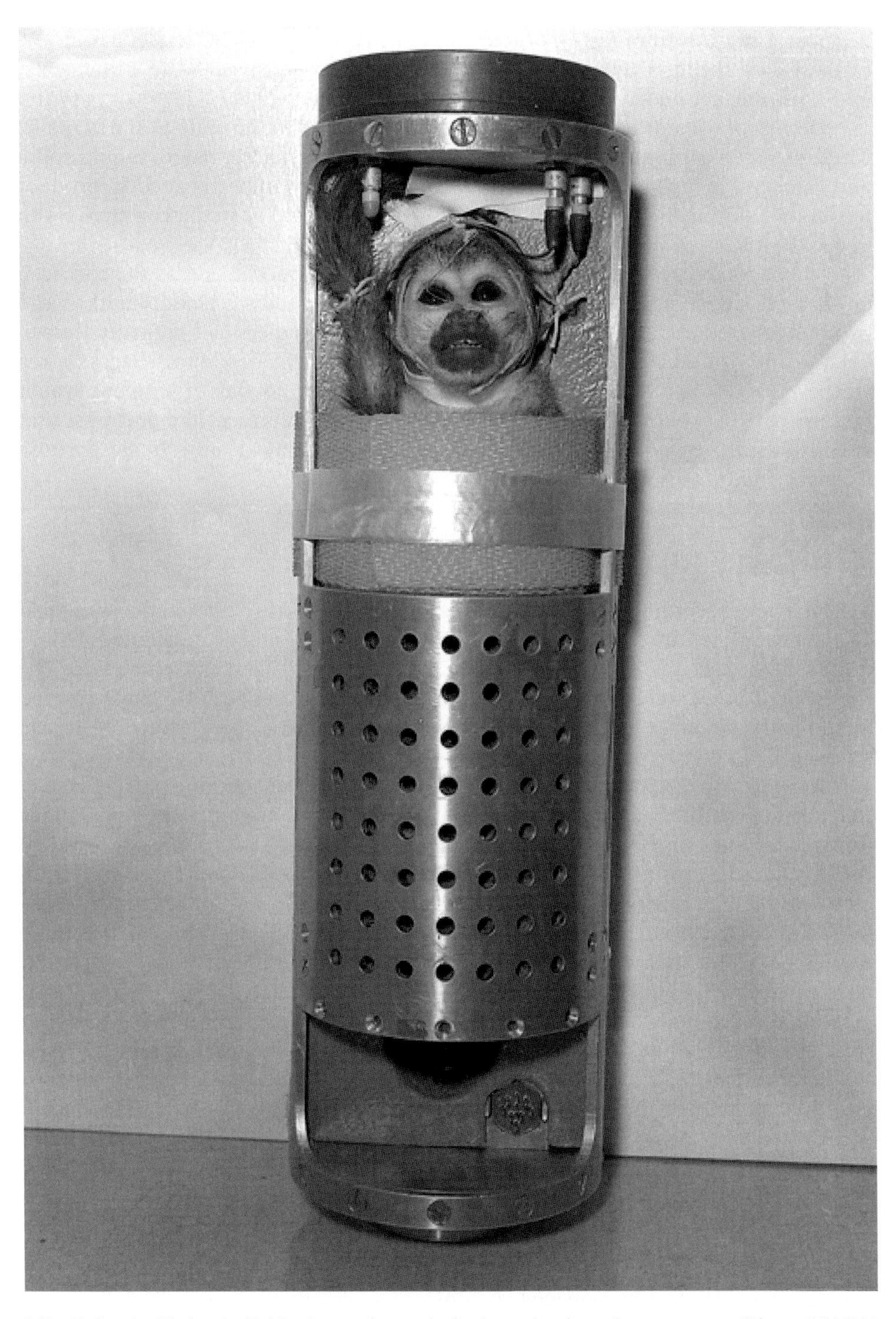

Miss Baker inside her individual capsule, ready for insertion into the nose cone. (Photo: NASA)

her nose using a drop of fast-drying model cement. The other end was attached to a connector on the monkey's cylinder. And, just like Gordo, Baker wore a tiny jacket and a soft helmet lined with chamois leather and foam rubber.

Dozens of other test subjects were also being carried in the biological nose cone. These included corn and mustard seeds, fruit fly larvae, human blood, mould spore and fish eggs, as well as sea-urchin shells and sperm, carefully triggered to produce fertilisation during the flight. Each of these specimens would be studied on recovery to gauge the effects of phenomena such as cosmic radiation.

Unlike Baker, Able was permitted a small degree of movement, as she had been given a relatively simple task to perform in order to study her coordination and psychological reactions to space flight. Just as Army handlers had patiently trained her Indian-born predecessor to press down with her right paw on a telegraph key whenever a red light blinked inside the compartment, so this experiment would become part of Able's schedule. If successful, it would represent the first time that scientists had been able to receive actual performance data from a living creature during a flight into space.

Set for launch

The launch was set for the early hours of the morning of 28 May 1959, and all went just as planned in the early stages. Lift-off of the giant Jupiter missile designated AM-18 came right on schedule at 2:35 a.m. from Cape Canaveral's Launch Complex 26. The rocket fired beautifully, propelling the biological nose cone on a precisely charted flight nearly 360 miles into space, during which it reached a speed exceeding 10,000 miles an hour.

The nose cone separated flawlessly right on schedule and continued to fly through weightlessness a further 9 minutes before it re-entered the atmosphere. Tremendous forces associated with deceleration now began to act on the two monkeys. As these forces reached a crushing 38 g's, Baker developed some cardiac inhibition with sino-atrial block, but otherwise exhibited little change in her respiratory rate. Able's heartbeat soared from 140 beats a minute to a high of 222, and she was breathing three times faster than normal. As these forces rapidly diminished, so the flurry of physiological activity in both monkeys correspondingly fell to regular levels.

Forty-five minutes after lift-off the nose cone finally parachuted down to an ocean splashdown 1,700 miles downrange from the Cape, hitting the water nearly 250 miles southeast of San Juan, Puerto Rico. Members of the Navy's underwater demolition team were quickly on the scene. They hooked up lines to the nose cone, after which it was hoisted onto the deck of the ocean salvage tug, USS *Kiowa*, arriving on board at 5:08 a.m.

"Are the monkeys safe?"

Once Army personnel had opened the nose cone and carefully removed the larger capsule containing Able, they retreated to the ship's hospital where they proceeded to

Able after recovery from the nose cone. (Photo: NASA)

remove the monkey from her capsule. Likewise, Navy doctors and scientists extracted Baker's capsule and carried her into a separate section of the small hospital where she was also released from her cylinder. The restraint straps and covering blanket were removed and Baker was gently lifted from her couch, following which the tiny jacket, helmet and external sensors were removed. She then received a small cookie as a reward for her patience during the extraction process, after which the preliminary post-flight medical tests began.

Meanwhile, scientists back at the Cape were anxious to know if the two monkeys had survived the trip, but radio communications were poor with the *Kiowa*. Finally the ship picked up a faint question through the static: "Are the monkeys safe?" Their equally weak response, "Yes, yes," caused great excitement back at the launch site. It was enough to tell the waiting scientists that the flight, and Project Down to Earth, had been a success.

Army doctor Major Gerald Champlin gave both monkeys a physical examination on board the *Kiowa*, and further reassured his medical colleagues back at the Cape by transmitting the message: "Able/Baker perfect. No injuries or other difficulties" [6].

Flight results come in

The sea-urchin fertilisation experiment, designed to shed light on problems that could be associated with human reproduction on multi-generation space flights, did not go quite according to plan. The sea-urchin eggs somehow disintegrated during the flight, although those that had been fertilised prior to the flight continued to develop normally.

Much to the consternation of project scientists, the experiment in which Able would press a key in response to a blinking light proved fruitless. Officially, it was blamed on a transmission failure just prior to the launch which rendered the experiment useless, but it was also rumoured that the late replacement of the original candidate had resulted in the lesser-trained Able being too easily distracted from completing even this simple task.

Within hours of plucking the nose cone from the sea, the recovery ship had sailed into port at San Juan. The two monkeys, who appeared to be in fine spirits, were carried ashore and quickly flown to Washington, D.C., for a brief welcome-back "press conference".

Losing Able

Later, Able was given a thorough post-flight physical examination at the Walter Reed facility, where a total body radiation count was conducted. Once again the monkey was found to be in excellent condition apart from several slight abrasions where her helmet had rubbed. She was then flown back to the AMRL in Fort Knox, where the implanted ECG electrode would be surgically removed in order to prevent possible infection. All that was required in the operation was a shallow, half-inch incision.

On 1 June, just 4 days after her recovery at sea, physicians at the laboratory began squirting a cloud of the mild anaesthetic trichloroethylene into Able's cage to spare her any undue stress during the minor operation. The intention was to harmlessly put the monkey to sleep for a few minutes.

Then, without warning, Able's heart suddenly began convulsing rapidly and she stopped breathing. Despite a mighty 2-hour effort to revive her, which included a civilian doctor giving the hapless monkey mouth-to-mouth resuscitation, she never regained consciousness. Two *Life* magazine photographers, who had been on hand to take shots of what was ostensibly a minor operation, came away instead with dozens of photos of the dramatic but ultimately futile efforts to revive Able. Several would appear spread over two pages in the following week's issue [8].

A deputy chief at the centre, Colonel Robert Hullinghorst, later attributed Able's death to cardiac fibrillation resulting from the anaesthetic. "We have used it a thousand times on about seven hundred monkeys at Fort Knox alone, and this has never happened before," he emphasised. "This is the type of anaesthetic death that every surgeon fears. We don't know exactly why it occurs." He did report, however, that following Able's sudden death, an autopsy had indicated "absolutely no evidence of any injury" that might have contributed to the monkey's demise [9].

The body of Able was carefully preserved through taxidermy after her death, and the pioneering monkey can now be viewed at the Smithsonian's National Air and Space Museum in Washington, where she is displayed wearing her helmet and strapped into her fitted couch, just as she was when she made spaceflight history in May of 1959.

Miss Baker makes friends

Little button-eyed Baker had also undergone a similar operation to remove two tiny electrodes from under her skin (albeit without any anaesthetic), but fortunately suffered no ill effects. There had been some dark grumblings from Army medical personnel at Fort Knox following the death of Able, when they discovered that Navy and Air Force space biology teams routinely steered away from using general anaesthetics on animals that had recently undergone prolonged stress. Far too late for Able's operation they had mentioned that administering a general anaesthetic could prove fatal, as the animal might be suffering from stress-related haemorrhages. However, no such haemorrhage was found in Able's lungs.

Miss Baker (as she was now popularly known) was eventually returned to the Navy's Aviation Medical School at Pensacola for some well-earned rest and recreation, where she proved an incredibly popular attraction for the staff and visitors. She had lost 1 ounce of her normal 11-ounce weight during the flight, but this was soon regained with a daily diet of two peanuts, two biscuits, one slice of fruit and half a cup of milk. For a time she shared a cage with a pair of squirrel monkey space candidates named Sugar and Spice, but this was only a temporary arrangement. Soon she had her own 7-foot square cage with Formica-covered walls and ceiling, a tiled floor, lighting system and air-conditioning. The cage was also fitted with a large one-way mirror that allowed observers to discreetly monitor her behaviour and well-being.

Dr. Dietrich Beischer, a space biology expert, headed the Navy team that continued to monitor Baker's progress with daily weight and temperature checks, but the little monkey continued to display no adverse effects from her flight into space. Dr. Beischer's team simply could not take to the name Baker, however – to them she was (and would always be) known as TLC.

In 1962 the popular space monkey was even "married" to a slightly larger male named Big George in a small ceremony at the school, although there would be no offspring. The Navy continued to care for Miss Baker and Big George in the 10 years following her space flight, but it was expensive to do so, and with looming budget cuts the Navy was looking for someone else to provide them with a suitable home.

Moving on

Finally, in 1971, the two monkeys were moved to the newly-opened Alabama Space and Rocket Center in Huntsville, following a request made by centre director Ed Buckbee, with the eminent support of Dr. Wernher von Braun. Here the two monkeys lived quite contentedly in a comfortably large, plastic cage, complete with

Miss Baker with her capsule and a Certificate of Merit for Distinguished Service presented by the ASPCA. (Photo: NASA)

air-conditioning, running water and exercise bars, where they ate together and either sat or played happily on the bars.

By now, Miss Baker had achieved enduring celebrity status for her space mission, and would even appear on television shows hosted by Dinah Shore and Mike Douglas. On her 21st birthday in 1978 she was honoured by a reception for nearly a thousand invited guests at the centre, during which she was presented with a birthday cake made from red jelly topped with strawberries and bananas. One of the centre's staff remarked, "Of course no one is sure of the day she was born, but we like to give her a birthday party every year to mark her getting older and the advances we've made in space" [10].

There was also some regular fan mail from children. One young girl wrote, "I like you because you are cute. I haven't seen you before, but I have seen you in a book." For her efforts she was sent a photo of Miss Baker, which bore her tiny paw print [11].

A much-loved monkey

Aged 17, Big George passed away on 7 January 1979, and 3 months later Miss Baker was paired off with another male named Norman, from the Yerkes Primate Center, also in Alabama.

Baker outlived her spaceflight companion Able by more than a quarter of a century. On 29 November 1984, at the grand old age (for a spider monkey) of 27, she died of acute kidney failure at a small-animal clinic at Auburn University's School of Veterinary Medicine. She was documented as the oldest known squirrel monkey in captivity.

Baker was buried at the entrance to the space centre in Alabama (which by then had been renamed the U.S. Space and Rocket Center), where her polished stone marker proudly reads: "Miss Baker, Squirrel Monkey. Born 1957. Died November 29, 1984. First U.S. Animal to Fly in Space and Return Alive. May 28, 1959." In 2005, Miss Baker was inducted into the Alabama Veterinary Medical Association's Hall of Fame [12].

Today, with the permission of the staff at the centre, a banana is often quietly placed atop the marker by visiting children, in fond remembrance of a much-loved spaceflight pioneer.

REFERENCES

[1] Excerpt from National Aeronautics Space Act of 1958, subsection *Declaration of Policy and Purpose*.

[2] James S. Hanrahan and David Bushnell, *Space Biology: The Human Factors in Space Flight*, Basic Books New York, 1960.

[3] E-mail correspondence with Ken Havekotte, Florida, 12 September 2005.

[4] Edward C. Dittmer, Space Center Oral History interview conducted by George C. House, Director, Alamogordo Space Center, 29 April 1987.

[5] "How man-made satellites can affect our lives," Joseph Kaplan, Ph.D., D.Sc., *National Geographic*, vol. CXII, no. 6, December 1957, pp. 791–810.

[6] Clyde R. Bergwin and William T. Coleman, *Animal Astronauts: They Opened the Way to the Stars*, Prentice Hall, Englewood Cliffs, NJ, 1963.

[7] "An Island Native", Kansas *Independence Reporter* newspaper, article, author not named, issue 1 June 1959.

[8] "New U.S. Advances in March to Space," *Life*, vol. 46, no. 24, June 15, 1959, pp. 20–31.

[9] Ashton Graybiel, Robert Holmes *et al.*, "An Account of Experiments in which Two Monkeys Were Recovered Unharmed after Ballistic Space Flight," *Aerospace Medicine*, vol. 30, no. 12, December 1959, pp. 871–931.

[10] "A U.S. Space Pioneer Marks 21st Birthday," D.J. Herda, *New York Times*, issue 24 June 1978.

[11] "Miss Baker, Big George Await Merry Christmas," Edd Davis, *The Huntsville Times*, issue 19 December 1971.

[12] "Miss Baker: America's First Lady of Space," Alabama Veterinary Medical Association online article. Website: *http://www.alvma.com/displaycommon.cfm?an=1&subarticlenbr=52*

6

The most famous dog in history

June 1957. Flying more than 70,000 feet above Soviet Kazakhstan, an altitude which exceeded the reach of any Soviet interceptor aircraft, an alert pilot of an American Lockheed U-2 spy plane spotted something interesting in the distance and departed from his prescribed course to get some photographs. What he found would astound his intelligence chiefs in Washington. He had inadvertently stumbled upon the Baikonur launch facility. This was the "crown jewel of Soviet space technology, whose existence had not even been suspected", according to the memoirs of Richard M. Bissell Jr., director of the American U-2 spy plane programme and of photo-reconnaissance at the Central Intelligence Agency [1].

A "SIMPLE" SATELLITE

Within a week of that U-2 flight, intelligence analysts and photo interpreters had constructed a scale model of Baikonur. "Photo intelligence also allowed analysts to determine the size and power of Soviet rockets," Bissell wrote, "based upon burn marks and the configuration of the pads for exhaust gases" [1]. From this, the CIA alerted President Eisenhower that the Soviets would likely soon send a series of small satellites into space.

Actually, sending satellites into space was not the top priority at Baikonur that spring and summer; instead, it was simply getting their new intercontinental ballistic missile, the R-7, to fly. The R-7, designed to deliver an atomic warhead to another continent, had been modified to allow it to lift a satellite into orbit. Chief Designer Sergei Korolev's original plan had been to launch a massive satellite, known as Object D, in the spring of 1957 prior to the start of the International Geophysical Year (IGY – July 1957 to December 1958), but production of Object D progressed too slowly and test-firings of the R-7 engines had not produced sufficient thrust for it to place Object D into orbit.

In January 1957, prompted by concerns that the United States might launch its satellite into space first, Korolev set aside plans for Object D in favour of a lighter spacecraft, officially designated "Simple Satellite No. 1" (PS-1). In fact, revised plans called for two new satellites, PS-1 and PS-2, each weighing about 220 pounds, to be launched in April or May of 1957. These dates fell before the start of IGY and prior to when the U.S. planned to launch a satellite during IGY [2].

GETTING THE R-7 TO FLY

The first R-7 rockets had arrived by rail at Baikonur that spring. (The launch facility was actually located at Tyuratam, but in an effort to disguise its location, it was given the name of Baikonur, a small town some 300 miles away.) Technicians laboured around the clock to perform all the necessary tests to prepare the rocket for launch. The R-7, also known as the *Semyorka* (The Seven), was an enormous rocket for this period, consisting of a central core rocket surrounded by four strap-on boosters. Its 876,000 pounds of thrust would be capable of throwing a 5.3-ton warhead some 5,000 miles.

According to Alexander Maximov, who oversaw the R-7 missile development for the Ministry of Defence, a festive mood prevailed at Baikonur that spring. The entire state commission had arrived in April, along with senior representatives from the design bureaux, research institutes and manufacturing plants that had participated in the development of the R-7. After all, work had first been authorized on the R-7 in early 1953, and now they had come to the culmination of that work. After two successful launches, the R-7 would be ready to launch the Soviet Union's – the world's – first satellite. All of the principals wanted to be on hand to finish and celebrate the R-7's success [3].

But a more sobering tone set in over the next few months. The first R-7 launch attempt, on 15 May, ended in failure when one of the strap-on rocket boosters tore away. Then a succession of R-7 launch failures that summer completely soured the mood at Baikonur, threw off the satellite launch schedule and brought much criticism onto the R-7 programme.

On 16 May, the day after that first R-7 failure, the reliable R-2A rocket lofted two dogs (Ryzhaya and Damka) on a suborbital flight from the launch complex at Kapustin Yar, about 500 miles to the west, astride the Kazakhstan–Russian border.

Throughout the spring and summer, the two endeavours – the suborbital dog flights and the programme to perfect the R-7 rocket – ran parallel to each other, rushing headlong towards a nexus that would be played out later that year. These efforts, which had transpired in the highest secrecy for so many years, would suddenly be thrust into the very bright light of world publicity and international politics. As with many developments in this momentous year, the launch of the Soviet Union's two "simple" satellites in the autumn of 1957 would have as much to do with politics as with space science.

Space dogs move centre stage

If one was looking for a hint during that summer that a dog might shortly fly into space, the signs were all there. Even as the CIA was constructing its model of the Baikonur launch facility that June and the R-7 suffered its second failed launch attempt, the Soviet Union forsook its usual secrecy. It began holding press conferences and sending news releases to the world press, ramping up publicity for the anticipated launch of the first Soviet satellite, Sputnik, as part of the IGY.

Three of the dogs that had made suborbital flights – Linda, Malyshka and Kozyavka – were trotted out for foreign reporters at a June press conference held by the State Committee for Cultural Relations with Foreign Countries. The animals were in good health, it was reported. Films made during their flights showed that they had behaved normally. "I would like the British correspondents to inform the British Society of Happy Dogs about this," stated Aleksei Pokrovskii, director of the Institute of Aviation Medicine and a member of the Soviet IGY committee, "because the Society has protested to the Soviet Union against such experiments." The chorus of complaints would escalate considerably later in the year with the launch of Sputnik 2 [4].

At the same time in the United States, the press sensed a growing excitement amongst notable Russian space scientists, whose carefully-worded articles in *Pravda*

Kozyavka, Linda and Malyshka, who had just completed suborbital flights, on display at a press conference in June 1957. (Photo: authors' collections)

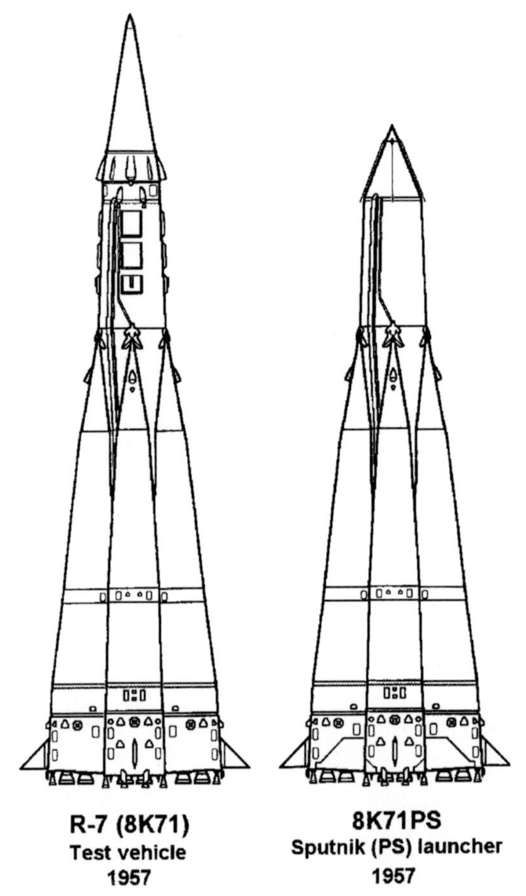

With the first successful launch of the R-7 rocket in August 1957, the stage was set for the launch of the Sputnik satellites. Sputnik 2 carried the dog Laika, the first living creature to orbit the Earth. (Illustration: NASA)

and elsewhere gave credence to mounting speculation about the imminence of a Soviet satellite launch. Seven times during the month of June the *New York Times* carried articles about Soviet plans to launch satellites, including one that claimed dogs would be passengers on one of these spacecraft [5].

On 21 August, when the R-7 recorded its first successful flight, travelling 3,700 miles from Baikonur to Kamchatka, all obstacles to the launch of the first satellite seemed to have been removed.

Meanwhile, the concurrent series of suborbital launches concluded in late August and early September with three R-2A flights to an altitude of 130 miles. On each of these flights, one of the two dogs was anaesthetised in order to test physiological reactions without the complicating factors of fear and stress. Nearly a year would pass before suborbital flights resumed in August of 1958, with a series of flights to an altitude of 280 miles. In that hiatus, the Soviet space programme would make history –

and shock the Americans – by launching the world's first artificial Earth satellite, Sputnik, on 4 October 1957. That accomplishment drew such an enormous amount of favourable publicity to the Soviet space programme, and to the Soviet Union in general, that it contributed to the hasty decision to launch a second satellite, Sputnik 2.

One very busy month

The story has come down in several variations, but the basic facts are these. Hungry for more of the flattering attention that the Soviet Union garnered from the launch of the Sputnik spacecraft, Soviet Premier Nikita Khrushchev asked his chief designer, Sergei Korolev, if he could possibly launch a second satellite for the anniversary of the Russian Revolution on 7 November – one month away. After checking with the engineers in his design bureau, Korolev not only agreed to do so but added that this second satellite would carry a living creature, a dog. It was an astonishing promise for him to make [6].

Although much of the preliminary work had been done on many of the components that would be necessary for such a satellite, it would still be a monumental task to make it happen within one month.

Korolev's first chore, however, would simply be to gather his workers. Following the successful launch of Sputnik, they had been given a well-deserved vacation, their first in years. Senior staff had been sent off to the Black Sea resort of Sochi. In short order, they were summoned back to the job and plunged once again into a gruelling work schedule. They had re-gathered by the second week of October, between the 10th and 12th. Sputnik 2 did not exist, not even in the planning stage, and it had to be launched by 7 November. This would obviously call for certain compromises.

Cutting corners on Sputnik 2

Korolev explained to his workers the expedited procedure they would follow. Engineers would make drawings and pass them along to the workers. In fact, the designers would move into the factory workshops so that designs could be handed over directly. There would be no special drawings, no organised quality control. Instead, everyone would be guided by his own conscience. It was a risky approach that called for shortcuts and corner-cutting.

Since the technology did not exist for returning a payload from orbit, this would be a one-way trip for Sputnik 2's canine passenger, and that was a considerable advantage for the hasty design and construction of an orbital capsule. The dog could travel in a relatively simple, pressurised cabin, similar to those used on ballistic flights.

They would begin with an R-7 stripped down to make room for a satellite. The "stripping down" would be accomplished mainly by eliminating the satellite separation mechanism, permitting the spacecraft to go into orbit while remaining attached to the upper stage of the rocket [3].

In *Roads to Space*, a collection of oral history interviews with those involved in the Soviet space effort, engineer Arkadiy Ostashov recalled the scramble to pull things together, saying that "life support cabins had already been built for the R-5A and

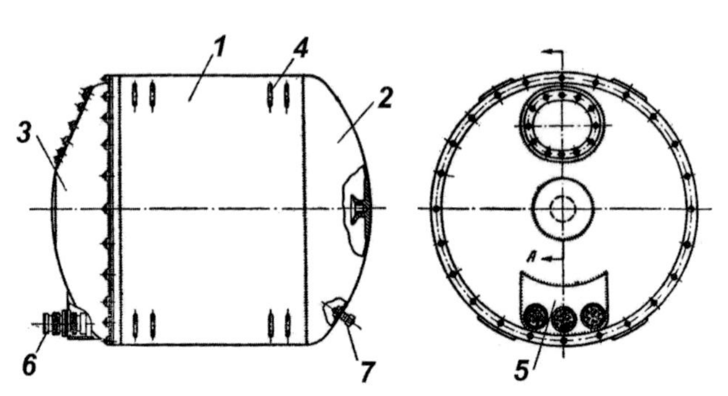

Side and front views of the pressurised capsule used in Sputnik 2: (1) cylindrical shell, (2) base, (3) lid, (4) ridges for mounting, (5) pocket for airtight connectors, (6) airtight connectors, (7) connecting pipe. (Illustration: authors' collections)

R-5B rockets and that one of them could be re-used after installing an automatic feeder to nourish the dog for a prolonged period of time. We suggested that he [Korolev] save a few kilograms by designing the feeder for one meal only, since we were mainly interested in knowing whether the dog would be able to eat at all'' [3].

This food-dispensing shortcut appears to have been adopted. During extensive testing on the nutritional and water needs of the dogs, the Soviets had determined that a dog could survive, without loss of weight for up to 8 days, if presented with a single portion of the entire amount of food–water mix it would need for that period.

''Since the flight of the dog Laika on the second artificial Earth satellite was due to last for only seven days, it was possible to dispense with the use of any automatic food dispensing equipment and it was sufficient to open up, prior to the launching, the access to the entire quantity of food to insure maintenance of its life for this period'' [7].

This single-serving approach had the advantage of simpler and lighter construction and it eliminated the electrical system necessary to activate the system during flight. A simple half-gallon tin box held all of Laika's food. Prior to launch, the lid was opened electrically allowing her access to the food.

Many decisions were made without sufficient research, according to physician Oleg Gazenko, trainer of the space dogs, who also worked on the Sputnik 2 capsule. ''We did not have time to produce blueprints and then look for someone who could do the job. They'd say, 'Maybe we can do it this way, alright, let's do it.' '' He recalled problems with the fan that forced air through the ventilation system of the animal cabin. Rather than measuring and calculating the air stream, they simply turned the fan at different angles until it provided the best airflow [8].

The capsule that began to take shape was no longer quite so ''simple'' as the 220-pound design proposed in January. Sputnik 2 would have a final weight of 1,120 pounds, six times heavier than its predecessor. It would stand about 13 feet high, 7 feet wide at the base and contain three components stacked atop each other. At the bottom sat the airtight cabin for the dog passenger. Constructed of aluminium alloy, it measured 25 inches in diameter by 31.5 inches in length. Above that perched a sphere

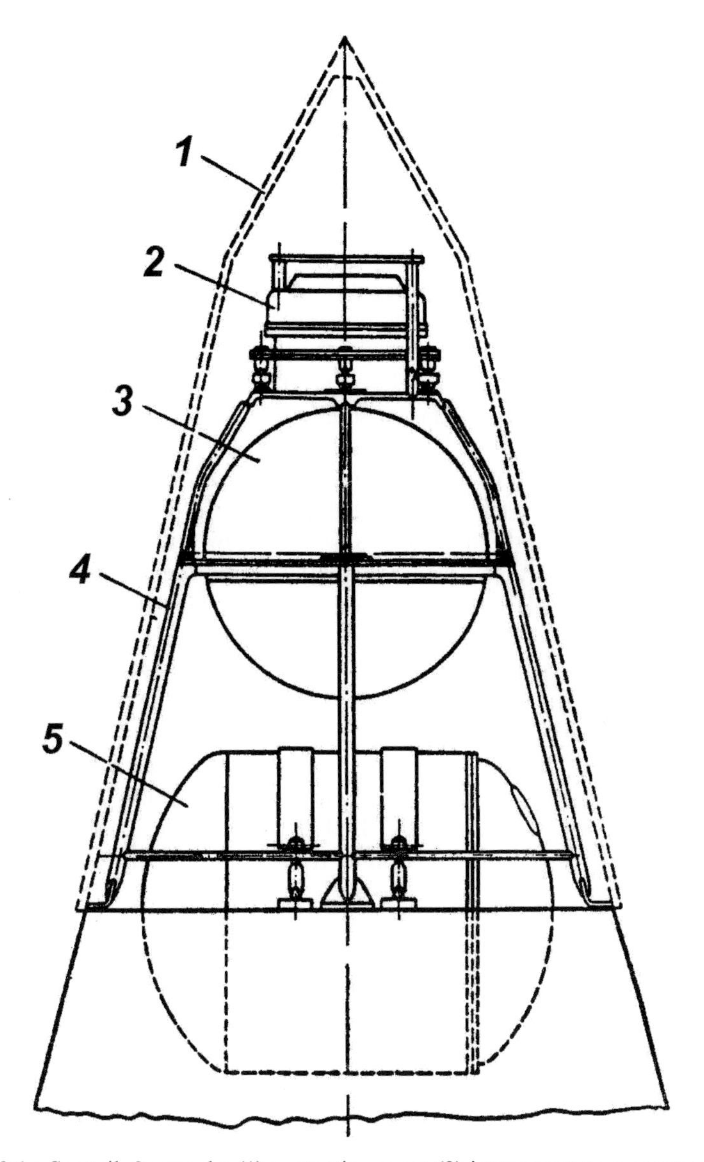

Cutaway of the Sputnik 2 capsule: (1) protective cone; (2) instrument to measure cosmic rays plus X-ray and ultraviolet emissions from the sun; (3) a duplicate of Sputnik 1, containing radio transmitters; (4) reinforced frame; (5) airtight cabin for the experimental animal. (Illustration: authors' collections)

resembling the original Sputnik, which contained the radio transmitters, batteries and instruments to register cabin temperature. Topmost sat a container holding two spectrophotometers for measuring solar radiation (ultraviolet and X-ray emissions) and cosmic rays.

Other equipment included a device for air regeneration, which absorbed carbon dioxide and water vapour and released the necessary amounts of oxygen. A fan provided ventilation for the regenerative equipment and regulated temperature in conjunction with a heat-conducting screen. The heating element automatically shut off when the cabin temperature rose above 59°F. Air regeneration and food were planned for a flight duration of 7 days.

Rather than having the capsule separate from the rocket once in orbit, as with the earlier satellite, Sputnik 2 was designed to remain attached to its rocket booster. Although it has been reported in some publications that Sputnik 2 failed to separate from the rocket by accident, that was not the case. To Korolev and his scientists, keeping the satellite attached seemed easier, faster and presented fewer opportunities for problems to arise. Not having the separation device on board also saved an enormous amount of weight. The larger satellite–rocket combination would be easier to track visually as well [3].

One additional advantage of keeping Sputnik 2 and the rocket core attached to each other was that it enabled the satellite to use the booster's telemetry system. The R-7 rocket utilized a "Tral-D" telemetry system that broadcast two signals, at 66 and 70 MHz. The capsule itself contained its own telemetry signals, the same used by the first Sputnik. One, at 40 MHz, broadcast a steady tone, which made for easier Doppler study. The second, at 20 MHz, was the familiar "beep-beep" signal made famous by the first Sputnik. The Tral-D system would be utilized to transmit physiological data collected on the dog. The flight of Sputnik 2 would be the first time that physiological data were transmitted by telemetry from space [9].

Although the 20- and 40-MHz signals would broadcast continuously, to the delight of satellite trackers around the world, Tral-D would not. Because of the limited strength of onboard batteries, Tral-D data transmission would be limited to one 15-minute burst of data per orbit as the satellite passed over Soviet territory.

One vital aspect of keeping the rocket and satellite attached remained poorly understood – how it might affect the temperature in the animal capsule. Overheating had been a major concern of the engineers involved in building it. The sun would heat the capsule from the outside, while the equipment and the dog's body temperature heated it from the inside. Korolev believed that the additional metal of the rocket would help to dissipate this heat into space and thus aid in keeping the dog cabin from overheating. However, in his 1996 memoir, Vladimir Yazdovskiy, director of the Institute of Aviation and Space Medicine, suggested just the opposite. "It seemed to us that that would be easier and more reliable. But, we didn't take into account that the metal of the construction could bring more heat to the animal" [10]. Given the critical role that heat would play during the flight of Sputnik 2, this was a serious consequence of the accelerated launch schedule.

SELECTING THE DOG TO MAKE HISTORY

For nearly a year, as the ballistic dog flights continued, Oleg Gazenko and his team at the Institute of Aviation Medicine had been busily training the next generation of canine cosmonauts – the "satellite dogs" – preparing them for all they might experience in space. Gazenko had joined the Institute in 1956 as this training was getting under way.

The impetus to move from suborbital to orbital dog flights grew out of Korolev's plans for a series of satellites, including one designed to carry animals. This scheme had been advanced publicly as early as August 1955, at a meeting of the Military–Industrial Commission [2]. It is worth noting that this meeting occurred just 1 month after the Eisenhower Administration announced that the United States would launch small satellites as part of the International Geophysical Year. The possibility of being beaten into space by the Americans motivated Korolev and the Commission to expedite planning.

When Korolev first won official approval of plans for his ambitious first satellite (Object D) in January 1956, he stipulated eight scientific explorations to be carried out on board. The final item on that list: "Studies of possibility of survival and life of animals during long-term residence on board a space craft" [11]. Not long after this, the Institute took on additional staff and created a separate training programme for satellite dogs.

Gazenko and the other physicians on the staff recognised the pioneering nature of their work. Everything they were learning from the dogs – in training, in suborbital rocket flights – was defining a new field of medicine: space medicine. Like their American counterparts they were, in essence, writing the book as they went along – making things up, trying things out, pushing the boundaries of their understanding.

According to Gazenko, this new field of space medicine research took an approach opposite to that of the aviation medicine research that had preceded it. Gazenko and most of his colleagues at the Institute had come to the dog programme with backgrounds in aviation medicine, where they had worked to protect pilots from the greater acceleration and atmospheric pressures of high-performance aircraft. He tried to articulate the distinction in a 1989 interview conducted by the Smithsonian Institution.

"Space medicine scientists had time on their hands. They could sit down and try to anticipate the surprises that a space flight might have in store for its crew and what should be done to insure their safety. Aviation, on the other hand, took off without giving much thought to the consequences on humans. Only later did medicine try to catch up and improve the needs of the crew inside the aircraft" [8].

Preparing dogs for space travel

The Institute of Aviation Medicine team would have a year to anticipate any surprises that orbital flight might pose for a living creature, and to train the dogs to overcome them. The laborious process of training satellite dogs proceeded in stages, initially focusing on acclimating them to confinement. The first steps proceeded through a progression of smaller and smaller capsules. The "Cyclops" chamber had a single observation window that trained the dogs to be under poor illumination. Aluminium "tight cells" had many large openings but only enough space for the dogs to turn around. The dogs typically protested with barks and whining when placed in this restrictive space for periods of 2–3 days [12].

Next, they moved on to tighter confinement that allowed only a few inches of movement. The dogs wore a garment, consisting of a light vest with metal restraining chains, which permitted them to stand, sit, recline, and move forward and backwards a little; just the sort of restrictions on their movements that they would endure in an orbital capsule. The duration of their stay in these compartments was gradually increased to 20 days. Only those dogs that could tolerate confinement for this amount of time would be allowed to continue in training.

Other stages of the training focused on their reactions to the stimuli they would encounter – acceleration, vibration and noise. First, the animals had to be conditioned to eliminate their first defensive reactions to the equipment. Then, their general physiological reaction to a stimulus could be studied. Finally, the individual reaction of each dog was carefully noted.

Out of the 10 dogs that completed the training in confined cages, 6 were selected for tests in a pressurised capsule. Conditions inside the capsule could be varied by changes in pressure, temperature and the composition of gases. According to Gazenko, the dogs did well during these experiments, except when the levels of carbon dioxide or temperature went beyond preset limits. "For example, during the training of Laika (experiment 119, of 19.5 days duration), the behaviour of the animal was calm with the exception of the first two days and an increase of motor disturbance during the 14th day when the concentration of carbon dioxide rapidly increased as a result of a stoppage of the ventilator motors in the regenerating plant . . . This required separate tests to determine the upper limits allowable in the concentration of oxygen and the upper limit on temperature in the small capsule" [13].

During the later stages of their training, the dogs were introduced to the rubber sanitation device that consisted of a urine and faeces receiver attached to the pelvic area and fastened with a shoulder harness. Once on board the actual satellite, the suit would attach to an airtight reservoir containing activated charcoal and dried moss. It was a cumbersome device, and it took the dogs a long time to simply get used to having it attached.

The far more difficult challenge with the sanitation device was getting the dogs to actually use it. Nature had accustomed them to adopt a certain position to urinate and defecate, and now they had to be reprogrammed to do it differently. Initially, the dogs retained all wastes rather than use the unfamiliar device. Even the use of laxatives did not encourage them. Only slow and consistent training finally acclimatised them to the novelty of this new approach [14].

Careful study of the dogs' energy requirements led to the creation of a unique food, containing 40% breadcrumbs, 40% powdered meat and 20% beef fat. One hundred grams a day supplied a dog's needs. Water and agar were then added to form a gelatinous consistency that supplied both food and water requirements [15].

Now that the first orbital flight was imminent, one dog had to be selected from the final group of six candidates to undertake this historic mission. The selection process involved a meticulous review of the training records of each dog. Scientific workers from the training programme each gave their report on the physiological reactions of the dogs to the various tests conducted during this time. They reviewed the records of how each dog had comported itself during prolonged stays in the confinement capsule, and studied extensive diaries that recorded the behaviour of each dog. They also reviewed the data from veterinary examinations.

Which dog would fly?

The candidate finally selected was a 13-pound, 2-year-old female by the name of Kudryavka (Little Curly), who like many of the dogs serving in the canine cosmonaut corps had been a stray plucked from the streets of Moscow. Two other dogs also joined the launch team. Albina (Whitie) would serve as backup, should anything happen to Kudryavka, while Muhka (Little Fly) would be the "technical dog", used for the testing of all of the satellite equipment and life-support systems. Each of the dog flights had a control dog like Muhka, one that travelled around with the flight dogs, ate what they ate and was kept under the same conditions, but did not fly.

One interesting side note on Mukha's behaviour at this time was reported by Oleg Ivanovskiy in *The First Stages* [16]. Ivanovskiy said that the little dog suddenly began acting strangely and quite out of character. On one occasion, when she had spent a few

Laika, the most famous dog in space history, was known for her easy-going disposition. (Photo: authors' collections)

Albina and her pups. The pups may have prevented her from being selected to fly on Sputnik 2. (Photo: authors' collections)

days alone in a training cabin, a scientist checked in on her only to find her eyes filled with tears and her food untouched. The research team never determined what was wrong, but they joked that she probably became depressed because she had not been the one selected to go into space.

It was actually surprising that Kudryavka was chosen for this important flight, which would ultimately make her the most famous dog in history, as she was not considered the most qualified of the satellite dogs. Kudryavka was known for her calm disposition, and she had also scored high marks in training by quickly adjusting to extreme conditions. But, according to Vladimir Yazdovskiy, Albina was the best candidate for the flight [10]. However, Albina missed her chance at fame for several reasons. For starters, she had already flown twice before, and so it was reasoned that she had done her part for science. Second, she had recently given birth to a litter of puppies and it would not be right to take her from them. And, perhaps the reason that carried the most weight – she was everyone's favourite, and they did not want to sacrifice her on this one-way mission.

Kudryavka, the dog that would fly in Sputnik 2, had not yet officially acquired the name Laika. She apparently was known by several names while in training. Aside from being the name for a breed of dog similar to a husky, "laika" also means "barker" in Russian. Like all of the space dogs-in-waiting, Kudryavka was a mixed breed, and like others she had some husky (Samoyed) in the mix. She may have been referred to as a laika. Kudryavka also had a rather loud, resonant bark. The term "lai" means "bark" in Russian. Adding the feminine diminutive "ka" to the end could loosely be interpreted to mean "barker". So, it is likely that – at least among the workers – Kudryavka was referred to as Laika. She also had a nickname of Zhuchka (Little Bug). Immediately following the launch of Sputnik 2, some Soviet sources claimed the dog was named Limonchik (Little Lemon). The name "Laika" seems to have attached itself permanently to Kudryavka during the flood of press coverage that followed the launch.

This casualness with the names of the dogs was not uncommon, especially during the latter years of the dog programme. Some dogs actually flew under different names for different flights. Take the case of Zhulka. Even though Oleg Gazenko adopted Zhulka as his pet, and she lived with him for 12 years, she was also destined to make several orbital flights, but not under that name. Perhaps the name Zhulka ("Mutt" in Russian) was not glamorous enough for one who travelled in a rocket. So Zhulka flew under the name Zhemchuzhnaya (Pearly), and under other names as well [17]. To avoid confusion in this discussion, the name Laika will be used from here on in the narrative when referring to Kudryavka/Laika.

Flight preparations

Following the October decision to launch Sputnik 2, the preparation of Laika and Albina had to begin immediately, as the necessary medical procedures required a 10-day healing period. Since neither dog nor capsule would be recovered, physiological reactions had to be read by instruments and radioed to the ground. One of the trickiest vitals to record was blood pressure. To constantly monitor this function, the carotid artery had to be drawn out of the dog's neck and sewn into a flap of skin. A rubber bulb inside a blood pressure cuff attached around the neck could then be inflated at intervals to compress the artery and measure the pressure, similar to the procedure used on humans. Training then had to include having the dogs spend many hours wearing the cuff, to accustom them to this new piece of apparel.

Yazdovskiy and Gazenko, both physicians, performed the carotid artery surgery on Laika and Albina and implanted sensors in each animal to record heart activity. This second operation, which involved inserting silver electrodes in the form of small rings two-tenths of an inch in diameter beneath the skin on the chest was a simpler procedure, but it needed to be done with care or the electrodes would be rejected. Nor could it be done too far in advance of a flight, as live tissue would form around the electrodes and interfere with their ability to detect minute electrical impulses from the heart. Wires attached to the electrodes ran beneath the skin and emerged on both sides of the backbone at a distance of about $\frac{1}{2}$ to $\frac{3}{4}$ inch from each other and at the level of the first pectoral vertebra [13].

Other physiological readings that would be taken during the flight did not require surgery. Laika's movements would be recorded by a potentiometric sensor, using a wire connected to her harness and wound onto a drum by means of a spring. The drum, mounted at the rear of the capsule, maintained a tension on the wire, allowing it to play out and rewind as Laika moved about. Electrical resistance detected any difference in the length of this wire, which would indicate movement. Her rate of breathing would be measured by strain gauges on a belt fastened around her chest [18].

The intensive training of the dogs continued at the Institute of Aviation Medicine in Moscow, before advancing to the launch site. It was during the final days of training in Moscow that Laika made her debut for a Russian audience. Radio Moscow, which had been providing saturation coverage of the Sputnik satellite, mentioned in its 27 October broadcast that preparations were nearly complete for the launch of an "animal-carrying" satellite. The announcement was followed by a broadcast directly from the Institute of Aviation Medicine introducing the radio audience to Laika, the dog that would go into space. To the delight of all, she obligingly barked into the microphone.

One final heart-warming event played itself out in late October when Yazdovskiy temporarily removed Laika from the training compound at the Institute so she could meet his family. "Laika was a wonderful dog ... quiet and very placid. Before the flight to the cosmodrome [Baikonur], I once brought her home and showed her to the children. They played with her. I wanted to do something nice for the dog. She had only a very short time to live, you see" [10].

Shortly thereafter, a telegram arrived at the Institute of Aviation Medicine, with the simple but meaningful message "We are waiting" [12]. This confirmed that preparations were complete at the launch site. Three members of the biological medicine team – Vladimir Yazdovskiy, Abram Genin and Oleg Gazenko – accompanied the three dogs – Laika, Muhka and Albina – on the flight to the Baikonur launch facility. Despite the fact that Baikonur had only a short runway, they made the journey on a large, twin-engined Tupolev TU-104 jet aircraft. It was a risky move, causing Korolev to fret that his key personnel and key dogs were flying on the same aircraft, and all might be lost if there was an accident [3].

A sense of excitement

At Baikonur, preparations continued at a feverish pace. Where only weeks before they had rushed to complete Sputnik, the world's first satellite, the harried personnel now raced to launch the first biological satellite before the 5 November anniversary of the Russian Revolution. The frenzy of the workers reflected the greater excitement of the entire country – indeed, the whole world – over the October launch of the first Sputnik. In the book *Roads to Space*, Anatoliy Kirillov, deputy commander for testing, recalled one moment during those weeks that illustrated the context of their work. The Soviet press had been providing continuous updates on Sputnik, reporting the precise times when it would pass over various countries and cities. It seemed as though the whole world had thrilled at the sight of the Earth's first artificial satellite, and yet those at Baikonur who had created the spacecraft had not seen it. At that location it appeared very low on the horizon and was only visible for brief periods.

"We were in the midst of preparing the second satellite for launch," Kirillov recalled. "Several dogs, little mongrels dressed in cloth, were running around on the floor, and we lamented the fact that one of them would soon die a gruesome death in

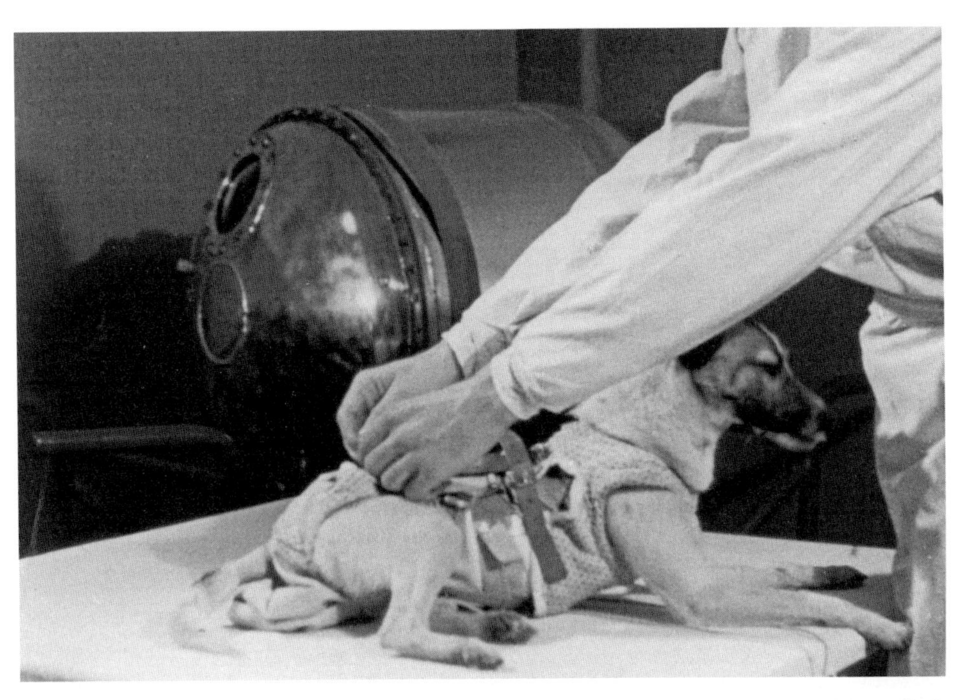

Laika being strapped into her harness and sanitation bag prior to the launch of Sputnik 2. (Photo: authors' collections)

orbit. During an interval between tests, someone came rushing into the control room shouting: 'Why are you all sitting here? The satellite is about to pass overhead. Come on outside!' ... Korolev and the members of the state commission, surrounded by test personnel, were standing in the open, waiting for the satellite to climb into the sky ... When the satellite did appear, it rose high in the sky as it moved from the south-west toward the north-east. It kept us spellbound for several minutes until it finally vanished" [3].

Pre-launch

On 31 October, after her customary 10:00 a.m. walk, Laika underwent her flight preparations. She received a sponge bath with a weak alcohol solution and an especially careful grooming. The areas around the electrode leads were painted with iodine and streptocide. At noon, she was fitted with sensors and the sanitation device, and dressed in a vest and the harness with metal restraining chains. Once the equipment had been tested in her capsule, chemical substances placed in the regeneration unit and jellied food deposited inside the feeding device, Laika was fastened in and her sensors attached to the recording equipment at 2 p.m. [12].

Nestled between two large cushions, Laika had just enough room to sit, stand, recline, or move a little forward or back. The arch-shaped heat shield went into place

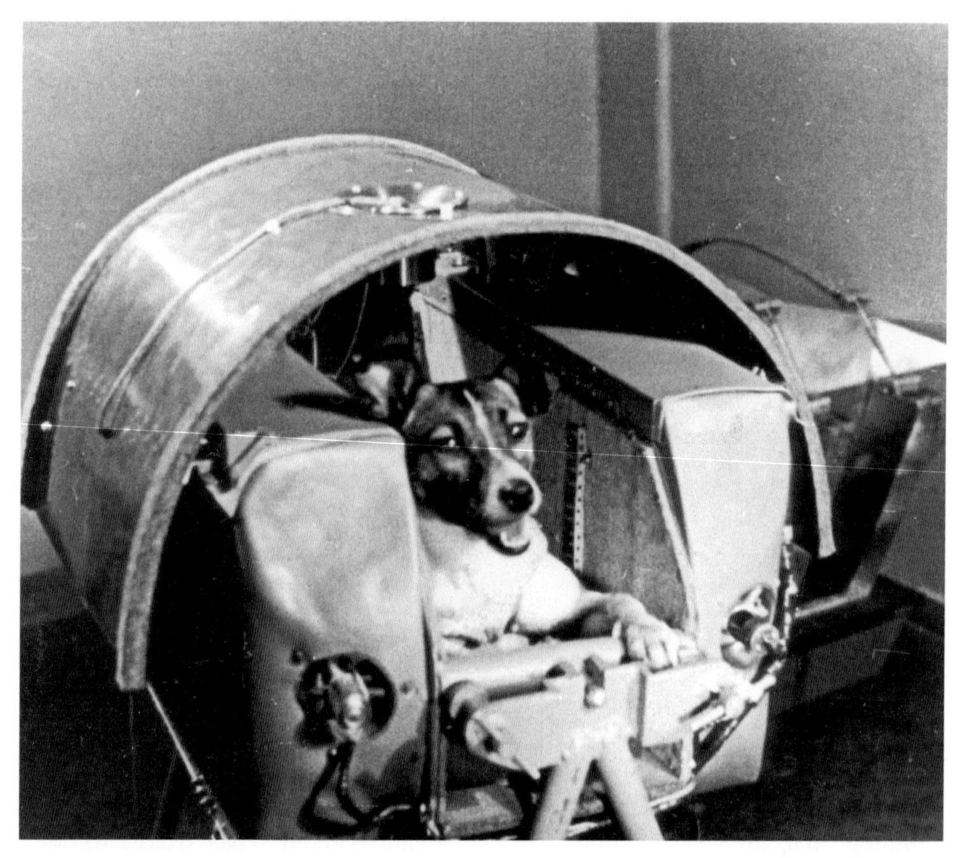

Tucked into place between her cushions and beneath the insulated heat shield, Laika was ready to be loaded into her capsule. (Photo: authors' collections)

over the top of her, then this was slid into the capsule and sealed with a forward hatch. A small, porthole-shaped window allowed in light and provided workers with a view of the interior. The capsule was then fastened in place atop the R-7 rocket.

Laika now had a 3-day wait until launch. The instruments reading her physiological signs had to be checked individually and tested in conjunction with the telemetry system to see that the readings could be transmitted over long distances. Laika exhibited her usual calm behaviour during this period. She could be observed through the porthole sleeping, feeding or reacting to a human face peering in at her. Her breathing rate held at 16 to 37 breaths per minute and her heart rate at 68–120 beats per minute. These were all within the normal ranges noted during prolonged laboratory experiments [12].

Testing also proceeded on the hermetic seal of the capsule and on such support systems as oxygen generation, carbon dioxide removal and the automatic feeding device. This phase of preparation took until 1:00 a.m. the following morning, 1 November. Night brings frigid temperatures at that time of year in Kazakhstan.

The R-7 rocket that carried Laika into space, with its distinctive gantry and payload shroud. (Photo: NASA)

An air-conditioning unit attached to the capsule kept Laika comfortable, while two assistants maintained a constant watch on her.

Shortly before launch, after Laika had already been in the capsule for 3 days, the medical staff created an unexpected stir about the pressure in the capsule. They wanted to start the flight with the capsule at normal pressure, they pleaded to Korolev, so they could observe the real changes in pressure that would occur during launch. The only way to accomplish that, they claimed, would be to de-pressurise the capsule before launch. It was an unusual request so close to launch, but they persuaded Korolev of the necessity to allow the change. However, it was all a subterfuge, because they had something very different in mind.

The capsule had air holes that were sealed with screws when not in use. Oleg Ivanovskiy, an engineer on the Sputnik 2 project, later reported that – as soon as the engineers had removed the screws – Yazdovskiy and the medical staff accosted the engineers, begging them for permission to give Laika a drink. For 3 days she had eaten only her one serving of jellied food, with no real water. Successful in their request, the physicians filled syringes with water and convinced the engineers to take them up to the capsule and squirt the water through the air holes into Laika's food trough [16].

The cabin was re-sealed, and shortly thereafter a protective cone was lowered in place to shield the capsule and instruments during launch. It would be jettisoned once in orbit.

LAIKA MAKES HISTORY

At 5:30 a.m. Moscow time, on Sunday, 3 November, the R-7 rocket carrying Laika lifted off from the launch pad at Baikonur. The powered phase of the flight would be the most gruelling for the dog. As g-forces climbed through three, four, then five times the force of gravity, while the rocket blasted a deafening noise and imparted a severe vibration, Laika reacted in fear. Her heartbeat spiked to 260 beats per minutes, approximately three times her normal rate, while her breathing rate jumped by a factor of more than 4. For the flight controllers this was no cause for alarm, however, as these reactions were consistent with the reaction of other dogs during training on the various devices that simulated launch, and with the reaction of dogs on ballistic flights.

Once in orbit, the nose cone shroud successfully jettisoned. When the first telemetry signals began coming in from orbit, a jubilant Yazdovskiy reported to the state commission, "Alive! Victory!" [10]. For the very first time, a living creature had travelled into space and was orbiting the planet.

Surprisingly, word of a dog being successfully placed in orbit had been reported to the Russian press – *before* the launch. The Communist Party maintained a bureau within the Institute of Aviation Medicine, which had already sent an article to the newspaper *Pravda* claiming a success, thus winning the race to be the first to report this historic event. The floodgates of news coverage were about to open, as the eyes of the world would once again turn to the accomplishments of the Soviet space programme [10].

Sputnik 2 achieves orbit

Sputnik 2 settled into a 140 × 1,039-mile elliptical orbit, which was close enough to expectations. The roar and vibration of launch had transformed to complete silence and a frictionless glide through the vacuum of space. At the colossal speed of 5 miles per second, it would take the satellite just over 103 minutes to circle the Earth. Laika had now entered an entirely alien and confusing world of weightlessness where the slightest movement would have caused her to lose contact with the cork floor of her capsule.

The physiological effects of prolonged weightlessness on a living being had long been a mystery. Theories existed about how they would react during extended periods in zero gravity, but these could not be verified under experimental conditions in the laboratory. Nor could ballistic rocket flights provide the answer, as they delivered only 3–7 minutes of weightlessness. During so brief an episode, it was difficult to distinguish the dog's reaction to weightlessness from its response to the preceding period of acceleration. It had become clear that at least 20 minutes of weightlessness would be necessary to isolate the effect of the absence of gravity on an animal.

To conserve vital battery power, the Tral-D telemetry system transmitted physiological data to Earth for only 15 minutes during each orbit, when the satellite passed over Soviet territory. In the first hours, telemetry indicated that Laika's vital signs were returning to normal, although at a rate three times slower than when she recovered from centrifuge tests on Earth. These findings were highly significant. They indicated that weightlessness alone did not cause any essential change in the physiological functions of a living creature. This was good news for humans heading to space.

In addition, telemetry indicated that the supply of oxygen was adequate and that the capsule had remained airtight, with no loss of pressure.

Critical problems arise

However, what quickly became apparent on subsequent orbits was that a serious problem was brewing aboard Sputnik 2. Cabin temperature had climbed far too high, reaching an unacceptable 104°F. Sensor readings, reported by telemetry, indicated that Laika was moving and barking, apparently quite agitated. Furthermore, there were indications that some of the thermal insulation protecting the capsule from solar heating had been damaged when the nose cone shroud separated. Was this causing the undue heating? Had they miscalculated the amount of heat that would be generated by allowing the spent rocket to remain attached to the satellite? And how much more heat was being generated by the elliptical orbit? The only thing clear was that the heat-dissipating screen and ventilating fan that had been installed to address the problem of overheating were not up to the task of controlling the temperature.

Approximately 5–7 hours into the flight, telemetry indicated no further signs of life emanating from the spacecraft. Laika had evidently died from stress and overheating. No word of her demise leaked out at the time. News of the successful launch of a second Sputnik was about to break like a tidal wave in the world press. Now was not the time to announce the passing of the world's first space traveller.

THE WORLD TAKES NOTE OF THE ACCOMPLISHMENT

News of the historic launch grabbed the front page of the *New York Times* on 3 November, with the dramatic headline, "Soviet Fires New Satellite Carrying Dog" [19]. No details about the canine passenger had yet been released. Other articles explained to the public such rarified space age concepts as solar radiation, cosmic rays, weightlessness and permanent space stations.

Concerned dog lovers found reassurance in the next day's *New York Times* page one headline that claimed the dog was alive and might be recovered [20]. But other articles raised doubts about whether she could be returned to Earth. The dog, name still unknown, was reported to be a breed of husky dog known as a laika [21].

No single Soviet source gave out the official word at this point. Therefore, news reports quoted whatever Soviet source they could get to comment on the event – diplomats, planetarium directors, unnamed experts. The question that overshadowed all others was not the technology involved, but whether the dog would be returned to Earth. Initial reports gave no indication, although some news articles did mention that when the Soviets used dogs in their suborbital rocket flights, the dogs had been returned by parachute.

Past criticism of the use of dogs in their ballistic rocket programme had made the Soviets sensitive to this issue, making it understandable why they were not eager to immediately set the record straight. On occasion, they had been careful to hold press conferences after successful launches to show that the dogs had survived and were in good health. When space dogs gave birth, mother and pups would often be put on display as well. Twice in October, the *New York Times* had reported about articles that had appeared in Soviet publications detailing the recovery of dogs from rocket flights using parachutes. So the idea that the Sputnik 2 dog would be recovered from space seemed a real possibility to concerned readers.

Prayers and protests for an unnamed dog

The flood of news coverage began on 4 November. Early speculation was that the dog would be catapulted from the satellite and returned to Earth. In the absence of a name for the historic canine space pioneer, some American newspapers began referring to the dog as Muttnik, a combination of "mutt" and "Sputnik." In London, when members of the National Canine Defence League met with the first secretary of the Soviet Embassy to protest the use of the dog in the satellite, he mentioned that the dog was named Limonchik (Little Lemon). The League called for a daily, worldwide minute of silence until Limonchik was brought back to Earth [22].

In the United States, the Smithsonian Astrophysical Observatory kept the public posted on when the satellite would pass over different parts of the continent. It was only viewable just before dawn, when the rays of the sun could illuminate the satellite in the still-dark sky.

A chorus of protest continued in the West. England's Royal Society for the Prevention of Cruelty to Animals received a flood of protest letters and phone calls from angry citizens. In New York, a group of outraged dog owners fastened protest signs to their pets and picketed outside the United Nations, displaying messages such as "Be Fair to Our Fellow Dogs".

On 5 November, the *New York Times* overflowed with articles relating to Sputnik 2, including a report from an unnamed Russian scientist suggesting the dog could not live much longer. Other articles stressed the importance of the information that could be learned by sending an animal into space, most notably how it would handle weightlessness and the effects of cosmic-ray particles. The dog, still being referred to as Limonchik, was reported to be the pet of a Russian scientist, and was erroneously identified as a male.

An Associated Press story out of Australia on 6 November reported a Russian source as saying that the name of the dog was really Laika. It mentioned that she had been treated to a three-course dinner at an airport restaurant before her flight to the launch site, and that she was still alive. In press articles on 7 November, Soviet scientists were still claiming to receive telemetry signals of Laika's physiological reactions and that she was in good condition. This, of course, was untrue. Although telemetry still functioned at this point, sending back valuable information on solar radiation and cosmic rays, Laika had been dead for the past 4 days. Onboard instruments would continue to transmit data until 10 November, when battery power was finally exhausted.

Meanwhile, American fascination with Sputnik went on full boil. Block-long lines formed outside New York's Hayden Planetarium to attend a programme about the Sputnik satellites and hear a recording of radio transmissions from the new satellite. Sputnik-viewing became a popular sport. Thousands rose in the pre-dawn darkness for a chance to glimpse this space age phenomenon.

One *New York Times* reporter gave an account of his vigil atop the RCA Building on 8 November. He was but one of many "moonwatchers" around the city, he explained, up before dawn for a chance to see the satellite. He caught sight of it above the southern horizon at 5:12 a.m., describing Sputnik 2 as being nearly as bright as the Dog Star, Sirius. The satellite seemed to appear and disappear, which he attributed to its tumbling motion. After a 2-minute passage across the sky, the Earth's newest satellite faded into the northeastern horizon [23].

In what must surely be the first recorded commercialization of the world's fascination with Sputnik 2, a full-page Hart, Schaffner & Marx advertisement appeared in the *New York Times* on 7 November promoting a new line of fall coats. It featured a bevy of well-coated men assisting a dog as the animal finished its descent to Earth by parachute.

Hiding the facts

The truth about Laika's death emerged incrementally. An 8 November news story claimed that telemetry continued to register the dog's physiological reactions [24]. However, another article noted that in their nightly communiqué, the Soviets had mysteriously dropped all mention of the dog, leading to speculation that Laika had become history's first victim of space travel. A statement from Moscow on 11 November reported that the radio transmitters on Sputnik 2 had ceased to function, and that all medical and biological observations had been completed [25]. The following day Radio Moscow finally confirmed the speculation: the dog aboard the satellite had died.

A *Pravda* article on the 14th gave the fullest description of the Sputnik 2 satellite and its various components, and explained how Laika had been trained for her mission. Revelations from the article about the dimensions, functions and purpose of the massive satellite were truly impressive, describing a 1,100-pound satellite, loaded with scientific instruments, not to mention a live passenger [26].

These details about Sputnik 2 came at a time when the United States was struggling mightily to launch its own first satellite, weighing a mere 3.25 pounds. In Western newspapers, tucked between the coverage of Soviet success, were demands for an urgent analysis and recrimination over the delays in the American space effort. Congressional investigations were called for, revisions to defence policy, increased spending for scientific research and improved science education.

The now-inert Sputnik 2 and the spent core of the R-7 rocket remained in orbit for 162 days, making 2,570 orbits before falling from space in very dramatic fashion on 14 April 1958. Paul Dickson, in *Sputnik: The Shock of the Century*, described the fiery demise of Sputnik 2:

Shortly after midnight on April 14, 1958, UFO sightings were reported by reliable witnesses along the east coast of the United States ... They reported a brilliant, bluish-white object moving high across the sky at incredible speed. According to reports, it suddenly turned red, and several smaller objects detached from the main object and fell into formation behind it.

Minutes later, observers on more than fifteen ships in the Caribbean reported seeing one or more brilliant objects of different colors and configurations, all moving in the same direction in the sky [27].

What these hundreds of people were actually seeing was the flaming death of Sputnik 2, the vehicle carrying the corpse of Laika.

The exact cause and time of Laika's death remained a matter of conjecture over the years, thanks to conflicting accounts provided in various Soviet publications. Some reports claimed that she had died after about a week when the satellite's batteries lost power and could no longer circulate oxygen. Others suggested that she had been euthanised with poisoned food, poison gas or a poison injection. Later, Soviet sources hinted that Laika had died after several hours when her cabin overheated – a claim validated by Oleg Gazenko during a 1993 interview with one of the authors in Vienna [28]. Gazenko revealed at the time that Laika had perished "soon after launch" due to a problem with stripped insulation and overheating, which had sent her capsule's inner temperature soaring to unsustainable levels in excess of 100°F.

Finally, in October 2002, the definitive answer emerged at a meeting of the World Space conference, in Houston, Texas. In a paper presented by Dr. Dimitri Malashenkov, one of the scientists involved in the Sputnik 2 mission, he explained that Laika had died between 5 and 7 hours after launch from heat and stress [29].

Laika's legacy

It's fair to say that in the arena of public opinion, the death of Laika overshadowed the incredible accomplishment of the launch of Sputnik 2. On the other hand, the travels of that one dog in space immediately served to galvanise scientific opinion about the near-term plausibility of a human following in Laika's tiny footprints. A living being had ridden a rocket into space and been able to function in the prolonged weightlessness of orbit, and that was monumental news.

Once the flight of Laika had been completed and the data combined with that from the animal ballistic flights, Korolev's design bureau immediately began making preparations for a manned space flight. But dogs had not yet finished their role in the

Soviet programme. More than 3 years would pass between the launch of Sputnik 2 and the first manned orbital flight of Yuri Gagarin on 12 April 1961. On 16 occasions during that period, dogs would fly on suborbital and orbital flights to aid in the development of the Vostok capsule that would carry the world's first space traveller.

So, many dogs played a critical role during the development of space travel and, like Laika, many paid with their lives. Yet none has come to symbolise that collective role of sacrifice more than a small dog from the streets of Moscow that became the first living being to travel in space. Whether the singular sacrifice of Laika was justified in the development of space flight is debatable. Certainly, her flight could have been delayed until a method of recovery existed. But that's where the politics comes in. The passage of time has given Oleg Gazenko a new perspective. "The more time passes, the more I am sorry about it," he stated after Malashenkov's revelations. "We did not learn enough from the mission to justify the death of the dog."

In November 1997 a statue commemorating fallen cosmonauts was unveiled at the Institute of Biomedical Problems in Star City, Moscow, and in one corner appears an image of Laika. On 9 March 2005 NASA gave a nod to Laika's place in history when, during operation of the Mars Exploration Rover *Opportunity*, space agency mission controllers unofficially named a patch of Martian surface in the Vostok crater "Laika".

Various measurements were taken on the Laika "soil target" and on another nearby target in the same crater named after Yuri Gagarin. It is truly fitting that the names of the first dog, and the first human being, to venture beyond our planet now have a place beside each other on another world, as well as in the history books.

REFERENCES

[1] Richard M. Bissell, Jr., with Jonathan E. Lewis and Frances T. Pudlo, *Reflections of a Cold Warrior: From Yalta to the Bay of Pigs*, Yale University Press, New Haven, CT, 1996, p. 119.

[2] Asif A. Siddiqi, "Korolev, Sputnik, and The International Geophysical Year," NASA History Office. Website: *http://hq.nasa.gov/office/pao/History/sputnik/siddiqi.html*, accessed 8 March 2005.

[3] John Rhea (ed.), *Roads to Space, An Oral History of the Soviet Space Programme*, compiled by the Russian Scientific Research Centre for Space Documentation, translated by Peter Berlin, Aviation Week Group, McGraw-Hill, London, 1995.

[4] *Space Handbook: Astronautics and Its Applications*, Staff Report of the Select Committee on Astronautics and Space Exploration, 86th Congress, 1st session, House Document No. 86, United States Printing Office, Washington, 1959. Website: *http://www.hq.nasa.gov/office/pao/History/conghand/astrussr.htm*, accessed 27 April 2005.

[5] "Soviet 'Rocket Dogs' Get Geophysical Year Role", *New York Times*, 9 June 1957: p. 7.

[6] James Harford, *Korolev: How One Man Masterminded the Soviet Drive to Beat America to the Moon*. John Wiley & Sons, New York, 1997.

[7] I.S. Balakhovskiy, L.I. Karpova and S.F. Simpura, "Providing Dogs with Food and Water under Space-Flight Conditions," pp. 379–392 in N. M. Sisakyan (ed.), *Problems of Space*

Biology, Vol. 1, NASA, Center for Aerospace Information. Translation of *Problemy Kosmicheskoy Biologii*, USSR Academy Publishing House, Moscow, 1962.

[8] "Soviet Space Medicine, Session Two," transcript of Smithsonian Videohistory Programme, interview with Oleg Gazenko, Abraham Genin and Evgenii Shepelev. Cathleen S. Lewis, Interviewer, 28 November 1989.

[9] Sven Grahn, "New revelations of frequencies for early Soviet spacecraft." Website: *http://svengrahn.pp.se/radioind/mirrradio/earlyfxs.html*, accessed April 12, 2005.

[10] V. I. Yazdovskiy, *On the Trail of the Universe: Collection of Cosmic Biology and Medicine in the Exploration of Space*, Firma Slova, Moscow 1996 [in Russian]. Russian title: *Na Tropakh Vselenniy: Vklad Kosmicheskiy Biologii I Meditsini v Osvoenye Kozmicheskovo Prostranstva.*

[11] Sergei P. Korolev, "Synopsis of Report on Development of Conceptual Design of an Artificial Earth Satellite," Report presented to the USSR Council of Ministers, 30 January 1956. Website: *http://www.hq.nasa.gov/office/pao/History/sputnik/russ3.html*, accessed 8 November 2005.

[12] M.A. Gerd and N.N. Gurovskiy, *The First Astronauts and the First Scouts of Outer Space*, Translation Services Branch, Foreign Technology Division Wright-Patterson-AFB, OH, 1963. Translation of *Pervyye Kosmonavty I Pervyye Razvedchiki Kosmosa*, Akademiya Nauk, SSSR, Izdatel'stvo Akademii Nauk, Moskva, 1962.

[13] O.G. Gazenko and S. Georgiyevskiy, "Preparation of the Animal Prior to the Experiment," pp. 353-359 in N.M. Sisakyan (ed.), *Problems of Space Biology*, Vol. 1, NASA, Center for Aerospace Information. Translation of *Problemy Kosmicheskoy Biologii*, USSR Academy Publishing House, Moscow, 1962.

[14] O.G. Gazenko, A.A. Gyurdzhian and G.A. Zakhar'yev, "Sanitary Devices in Pressurized Capsules," pp. 361-368 in N.M. Sisakyan (ed.), *Problems of Space Biology*, Vol. 1, NASA, Center for Aerospace Information. Translation of *Problemy Kosmicheskoy Biologii*, USSR Academy Publishing House, Moscow, 1962.

[15] V.N. Chernov and V.I. Yakovlev, "Research on the Flight of a Living Creature in an Artificial Earth Satellite," *American Rocket Society Journal Supplement*, vol. 29, 1959, pp. 736–742.

[16] Aleksei Ivanov (Oleg Ivanovskiy), *The First Stages*, 2nd edition, Molodaya Gvardia, Moscow, 1975 [in Russian]. Russian title: *Na Tropakh Vselenniy: Vklad Kosmicheskiy Biologii I Meditsini v Osvoenye Kozmicheskovo Prosstranstva.* Website: *http://epizodsspace.testpilt.ru/bibl/ivanovskiy/obl-i.html*, accessed 15 October 2005.

[17] Vladimir Gubarev, "Academic O. Gazenko: Wind of Cosmic Travels," *Nauka I Zhizn*, no. 7, 2001, pp. 30–37 [in Russian]. Russian title: *Akademik O. Gazenko: Vyeter Kozmicheskikh Stranstviy.*

[18] O.G. Gazenko and A.A. Gyurdzhian, "Fastening of an Animal in a Pressurized Capsule, Fabric Apparel and Distribution of Sensors for the Registration of Physiological Functions," pp. 369–377 in N.M. Sisakyan (ed.), *Problems of Space Biology*, Vol. 1, NASA, Center for Aerospace Information. Translation of *Problemy Kosmicheskoy Biologii*, USSR Academy Publishing House, Moscow, 1962.

[19] "Soviet Fires New Satellite Carrying Dog: Half-ton Sphere is Reported 900 Miles Up," *New York Times*, 3 November, 1957, p. 1.

[20] "1,056 Miles High: Russia Reports New Satellite Is Final Stage of Rocket," *New York Times*, 4 November 1957, p. 1.

[21] "Recovery of Dog Possible," *New York Times*, 4 November 1957, p. 8.

[22] "Britons Protest Dog in Satellite," *New York Times*, 5 November 1957, p. 12.

[23] "Satellite Rivals Sirius Over City," *New York Times*, 8 November 1957, p. 3.

[24] "Condition of Dog is Now in Doubt," *New York Times*, 8 November 1957, p. 3.
[25] "Radio of Second Satellite Has Stopped, Russians Say," *New York Times*, 11 November 1957, p. 1.
[26] "Soviet Issues Satellite Pictures," *New York Times*, 14 November 1957, p. 1.
[27] *Sputnik, The Shock of the Century*, Paul Dickson, Walker & Company, New York, 2001.
[28] Interview with Oleg Gazenko conducted by Colin Burgess at the Association of Space Explorers congress, Vienna, Austria, 12 September 1993.
[29] Dmitry C. Malashenkov, "Some Unknown Pages of the Living Organism's First Orbital Flight," paper presented at the 53rd International Astronautical Congress, The World Space Congress – 2002, 10–19 October 2002, Houston, TX.

7

Prelude to manned space flight

The Wallops Island Flight Test Range, now the NASA Wallops Flight Facility, is situated on an Atlantic Ocean barrier island on Virginia's eastern shore. Established under NASA's predecessor, the National Advisory Committee for Aeronautics (NACA), Wallops Island would host two crucially important spacecraft test flights using Holloman monkeys.

SAM, OR THE SCHOOL OF AVIATION MEDICINE

History will recall 1959 as the year in which Alaska and Hawaii became America's 49th and 50th states, the X-15 winged spacecraft made its maiden flight, the St. Lawrence Seaway was opened and Fidel Castro swept to power in Cuba. It would also prove a benchmark year for space exploration in the United States, particularly in regard to the announcement of NASA's man-in-space programme, Project Mercury. That same year the USAF School of Aviation Medicine (later renamed the School of Aerospace Medicine) was relocated from Randolph AFB to nearby Brooks AFB in San Antonio, Texas. Known by the acronym SAM, the medical facility was populated by scientists and researchers eager to assist both the Air Force and the newly-formed space agency in their space efforts.

NASA would come to rely heavily on SAM's staff and facilities to provide medical research on aerospace stresses that might be encountered by humans flying into space. Vital work carried out at the school would include the development of space cabin simulators, life-support systems and spacesuit design. SAM would even play a major role in the evaluation and selection of America's first astronauts.

Devising an escape rocket

When it came time for verification tests of the manned Mercury spacecraft's escape systems, NASA needed to know that the astronaut could be saved by promptly removing him from danger in the event of a catastrophic failure on the launch pad or during early ascent. The escape system that had been devised might have seemed rudimentary, but it was extremely effective. A small but powerful rocket with three exhaust bells, canted out at an angle, was attached to a 16-foot tower above the Mercury spacecraft. It could be fired during a manned launch if anything went wrong, blasting the craft free of a malfunctioning booster. But testing the complex sequence of events associated with the escape system required the use of a suitable, uncomplicated launch test vehicle.

Prior to the establishment of NASA in 1958, and realising that manned space flight was imminent, engineers and scientists at the Langley Research Center had developed a solid-propellant launch vehicle for testing a prototype Mercury capsule and its rocket escape system. According to a contemporary NASA fact sheet, the booster system "had to be simple in concept, use existing proven equipment wherever possible, be flexible so that it could perform a variety of missions, avoid the use of complex systems, be as inexpensive as possible, and be designed in such manner as to keep ground support requirements at a minimum" [1]. The Little Joe booster would emerge from Langley, having been given the series nickname because of its stubby appearance when compared with other rockets from that period.

Little Joe rockets were capable of creating conditions during initial flight through the atmosphere that would closely approximate those of the much larger and more powerful Atlas, which would become the launch vehicle used during Project Mercury's orbital flight missions.

An animal space programme takes shape

In March 1959, Air Force Headquarters directed SAM to provide biomedical support for Project Mercury launches at the invitation of NASA's Space Task Group, who needed to conduct tests of their equipment and procedures relating to an emergency-induced separation of the launch vehicle and the Mercury spacecraft. Brigadier General Don D. Flickinger, a surgeon and assistant deputy commander for the USAF Air Research and Development Command, directed that the personnel, equipment and funding necessary to support Project Mercury be transferred to the 6571st Aeromedical Field Laboratory at Holloman.

Originally, the Air Force task group had planned to conduct two proving flights of the proposed Mercury capsule using the reliable Jupiter booster, with the second flight carrying a primate passenger to qualify the environmental control system. On 1 July 1959, according to a report on Mercury primates by then executive director of the Alamogordo Space Center, Gregory P. Kennedy: "Abe Silverstein, Director of Space Flight Development at NASA Headquarters, sent a formal memo to the Space Task Group cancelling the Mercury Jupiter program." However, he added that "the concept of testing the spacecraft with primates remained, and during the same time frame that the Mercury Jupiter was being terminated, the animal test program took shape" [2]. The 6571st laboratory at Holloman was subsequently directed to provide chimpanzees for up to four planned flights.

Meanwhile, on 26 May, SAM was principally tasked with manufacturing, fitting out and testing several airtight capsules that could safely restrain and offer life support to a primate within a full-size mock-up of the Mercury capsule. Due to several factors including capsule space and weight constraints, technicians and scientists determined that it would be far more practical to train and fly 6- to 8-pound rhesus monkeys, rather than the larger chimpanzees [3].

While the geometric configuration of the capsule would duplicate the configuration of a production Mercury craft, the structural materials were a little different. This simplified model, manufactured at Langley, became known as a "boilerplate" version.

The programme would undertake two ballistic flights. The first would test the escape system at high altitude, while the second would simulate an emergency separation when dynamic pressures or air loads were at their maximum during ascent.

A monkey gets a name

One of the brighter Holloman monkey graduates, who had proved to be an attentive and industrious worker, was chosen to make the first of these short ballistic flights. Working under the patient training and care of Dr. Wade Lynn Brown, a professor of psychology at the University of Texas since 1937, this monkey was not only good at concentrating on his tests and seemed to be unfazed by loud noises and other distractions, but was also in fine physical shape. All of the monkeys at the base were officially known by numbers, but when this 7-pound, American-born rhesus monkey was nominated for the flight he was given the name Sam, the acronym for the School of Aviation Medicine.

During the flight – designated Little Joe 2 or LJ-2 – Sam would wear a complex spacesuit that had been developed at the school, which measured his physiological reactions and conditions within a fibreglass biologic capsule, or biopack.

This suitcase-size biopack, fitted with a clear Plexiglas window that would allow a camera to film the monkey's face and arms, would be fitted inside the boilerplate spacecraft. Sam would be snugly laced into a scaled-down couch, although he would have some freedom of head movement and his arms would also be left free, allowing him to operate the psychomotor unit as part of a shock-avoidance test. The couch was made from polyester resin with fibreglass cloth reinforcement, lined with $\frac{1}{2}$-inch-thick polyurethane foam. The body mould for the original couch had been produced around an anaesthetised animal, with its hips and knees flexed at the most comfortable angles. An oxygen supply would be available to the biopack, circulated by a small fan, offering a life-sustaining atmosphere for 56 hours.

Among other experiments on board was one to test the effects of radiation on film plates of barley, measuring the impact of fast particles such as protons that shoot in from the sun. Also included in the particular experiment were nerve cells from a rat and *Neurospora*, a laboratory culture of a low form of mould. A separate experiment involving eggs and larvae of the flour beetle would also study changes in appearance and mutation rate.

Sam rides a Little Joe

History's newest space voyager, his body plastered with monitoring sensors, was strapped onto the biopack couch the afternoon before launch day, after which he was given a pre-flight snack of an apple and some orange juice.

The following morning, with a spectacular launch predicted for the Little Joe, two interested spectators at hand were Mercury astronauts Alan Shepard and Virgil (Gus) Grissom. Shepard's astronaut duties involved a study of launch escape systems, so he was keen to see how well the equipment and procedures worked. There had actually been some behind-the-scenes discussions regarding the possibility of placing a man aboard one of these Little Joe test flights, but ultimately (and probably wisely) it was decided that the dynamic pressures would be far too great.

At 11:15 a.m. on 4 December 1959, the Little Joe carrying Sam was launched from the pad at Wallops Island and tore into the skies under full power. Fifty-nine seconds into the flight, with the booster's fuel fully expended, an abort sequence was initiated by timers. The motors on the escape tower fired for just 1 second but with incredible power, propelling the spacecraft away from the booster rocket. The assembly then continued to coast upwards to an apogee of 53 miles, about 15 miles less than hoped for, before the tower and rocket-motor case were jettisoned on cue from the capsule. Gravity then began acting to pull the spacecraft back down to the ground in a mild re-entry. The drogue and main parachute systems operated perfectly at just over 20,000 feet and 10,000 feet, respectively, and the capsule splashed down hard but safely into the Atlantic Ocean off Cape Hatteras, North Carolina.

Two SARAH beacons were part of the onboard recovery equipment, and they had been activated with the deployment of the main parachute, serving as homing transmitters to locate the spacecraft while it was still descending, and they continued to operate after splashdown. Two bags of bright green dye were also released after impact to aid in spotting the capsule from the air [4].

LOCATING THE CAPSULE

There was a distinct feeling of excitement in the air as the Lockheed Neptune P2V-5F headed out to sea from Naval Air Station (NAS) Chincoteague in Virginia. The nine-man crew had been assigned the task of tracking the Little Joe flight and directing one of several recovery ships deployed off the coast to the splashdown site.

Tracking Sam

In all, the air and surface support consisted of two P2V aircraft, four helicopters, two destroyers, one dock landing ship (LSD) and a sea-going tug. Two of the helicopters

Sam is inserted into his biopack container. (Photo: NASA)

were stationed at Wallops Island and the other two were located on the deck of the LSD [5].

Ralph Papa was flying with his regular P2V crew that day. An Aviation Electronics Technician serving with Patrol Squadron VP-8, he was manning an A-Type oscilloscope mounted in the bow of the Neptune, registration LC-11.

Launch of the Little Joe 2 from Wallops Island carrying Sam. (Photo: NASA)

Neptune P2V LC-11, the aircraft in which Ralph Papa's crew tracked the Little Joe launch and splashdown. (Photo: Ralph Papa)

"Our aircraft was already in the air at the time of launch," he recalls. "We were downrange, about one hundred miles east of Wallops Island, ready to detect the transmission from the capsule when it landed in the water.

"Using the specially equipped A-Scope in the Plexiglas bow, I would be able to guide our pilot Lieutenant Robert Hogg to the capsule by comparing the transmitted signal strength at the left and right of the scope's vertical cursor beam. A stronger or larger signal showing on the right side of the cursor would indicate that the capsule was off to the starboard or right side" [6].

Stationed at a mid-section radio console that day, fellow crewman Wayne Meier clearly remembers seeing the rocket's distant contrail through a small window above his desk as the P2V turned and headed further downrange, ready to vector one of the two waiting destroyers in the direction of the spacecraft. "I think I was too young to comprehend the significance of the event," he said, reflecting on his involvement. "Back in the radio compartment you kind of get lost in listening to message traffic and transmitting position reports. Ralph had the best seat in the house out in the nose cone of the aircraft" [7].

An exemplary job

According to Ralph Papa, the entire crew of LC-11 did an exemplary job that day, and he was able to lock onto the Mercury spacecraft after it had splashed down in the Atlantic Ocean just 11 minutes after lift-off.

"I guided the pilot to the capsule by keeping the signal the same size on both sides of the vertical cursor. The signal size increased as we neared until it finally blossomed in a bright round circle as we flew over it. We then passed on its location to a US naval destroyer that would pick up and return the capsule along with the rhesus monkey Sam."

Two hours after the Little Joe booster left the launch pad, crewmembers were anxiously scanning the ocean from on board the recovery ship USS *Borie*, which had been directed to the area by the P2V. They soon spotted the telltale green fluorescent dye surrounding the bobbing spacecraft. The destroyer then swept in, came alongside and hooked up while a Navy LSD, the USS *Fort Mandan*, maintained a support position nearby. Then, exercising great care, the destroyer's crew hauled their precious cargo aboard.

It hadn't been a long flight for Sam – he'd only experienced 3 minutes of weightlessness – but he and the Mercury capsule's escape systems had come through the exercise with flying colours. In fact, the longest parts of the mission for the monkey had been sitting on the launch pad, and then the 90-minute wait before his spacecraft was recovered from the rolling sea. When he was finally removed from the biopack some time later, however, he was found to be in fine shape.

Contemporary newspaper accounts report that the *Borie*'s officers decided against opening the sealed capsule in the absence of a physician, who would be able to give the monkey immediate treatment if any injuries had been sustained. Later reports, however, suggested they were actually unable to find a way of opening the spacecraft to extract Sam and had to radio for instructions.

Crew aboard the USS *Borie* capture the bobbing Mercury spacecraft. (Photo: NASA)

Sam was described as being "hungry and very responsive" once his biopack had been opened, quickly devouring half an apple, half an orange and a cup of water [5]. He would later be examined in Norfolk, Virginia, by Major Claude Green, the Aviation Medical School's senior physician for the flight. He would then be taken to NASA's laboratory at Langley AFB, where he would be treated to a joyful reunion with his female counterpart, Miss Sam, before being flown back to Brooks AFB in Texas.

There was immediate concern that Sam had not performed his psychomotor tests with the same diligence he had displayed on the ground. He had begun well enough, pulling the little lever whenever a red light blinked on, but his reactions to this test had tapered off significantly in the latter part of the flight. Even the mild shocks through his feet when he missed his cue did not seem to antagonise him into action.

THE SECOND FLIGHT

Another primate test flight loaded with a simulated ascent "accident" was scheduled to take to the Virginia skies 6 weeks after the successful conclusion to Sam's mission.

Prior to this, however, another Little Joe flight took place with less than satisfactory results. This fully-instrumented repeat flight, designated LJ-1 and with no monkey aboard, was scheduled for launch on 21 August to test a launch abort under high aerodynamic load conditions. Just 31 minutes before launch the capsule's escape tower suddenly fired due to a faulty escape circuit, carrying the boilerplate capsule to an altitude of 2,000 feet, and sending ground crews and photographers scurrying for cover. Fortunately, no one was injured, and another Little Joe proving flight (LJ-6) designed to test the spacecraft's aerodynamics and integrity was carried out on 4 October. On this occasion the non-instrumented boilerplate capsule was fitted with an inert escape rocket system. After $2\frac{1}{2}$ minutes of flight the rocket was intentionally destroyed to prove the effectiveness of the destruct system [5].

From Sam to Miss Sam

With repairs and systems overhauls effected, the earlier unsuccessful Little Joe/Mercury launch was redesignated LJ-1A and finally took place on 4 November. The purpose of this second attempt was to test the escape system under the maximum dynamic pressure conditions of an Atlas flight, or at about 1,000 pounds per square foot. The Mercury spacecraft contained a pressure-sensing device that should have initiated a planned abort 30 seconds after launch, but when this system fired the escape motor igniter, it took a few seconds for sufficient thrust to build up and the abort sequence was not accomplished at the desired dynamic pressure. The mission was written off as only a partial success.

The fourth Mercury/Little Joe test flight (LJ-1B), which was planned to overcome the abort system problems encountered on LJ-1A, would carry American-born monkey Miss Sam. The 6-pound female was subsequently transported to Wallops Island and prepared for her own mission.

There would be five main objectives manifested for this flight:

(1) To check the Mercury escape system concept and hardware at the maximum dynamic pressure anticipated during a Mercury–Atlas exit flight.
(2) To determine the effects of simulated Atlas abort accelerations on a small primate.
(3) To obtain further reliability data on the Mercury spacecraft drogue and main parachute operations.
(4) To check out the operational effectiveness of spacecraft recovery by helicopter.
(5) To recover the escape-system assembly (escape motor and tower) for a post-flight examination to determine if there were any component malfunctions or structural failures [8].

One fast, hot and crushing ride

On 21 January 1960 Miss Sam was rocketed to a nominal altitude of just over 9 miles, the Little Joe booster achieving a maximum velocity of just over 2,000 miles per hour. The extreme force of the launch pressed her back into the couch at nearly 14 times her

Miss Sam before her Little Joe flight. (Photo: NASA)

usual body weight, but she seemed to endure this with minimal physiological disturbance.

When the escape sequence was initiated, blasting the spacecraft free of its booster, Miss Sam was thrown about by the unexpected thrust of the escape rocket and squealed her displeasure. The monkey also had to contend with a higher than anticipated noise level in the capsule. For the next 30 seconds she was noticeably unresponsive before settling down and once again falling into her practised routine of pulling a lever whenever a light winked on inside the biopack.

Eight and a half minutes after Little Joe 1-B had lifted off from Wallops Island, during which time the escape sequence system and capsule landing systems functioned perfectly, the boilerplate spacecraft splashed down smoothly 12 miles from the launch site. This time, instead of being picked up by a recovery ship, a waiting Marine Corps helicopter plucked the spacecraft from the Atlantic and carried it back to Wallops Island.

Just 45 minutes after lift-off, the excited space pioneer was being extracted from her spacecraft. A medical examination soon after indicated that Miss Sam was in excellent health and had suffered little more than a very mild case of nystagmus – an involuntary wobbling of the eyeballs – after the firing of the escape rocket and again at the time of impact with the water. This reaction did cause flight physicians to reflect on an astronaut's effectiveness in a similar situation.

While only two of the six eventual Little Joe qualification test flights would carry primate passengers, the success of the Little Joe 1-B test flight meant that the next launch in the Mercury series (LJ-5) would be the first to fly an actual production line spacecraft from the McDonnell plant in St. Louis, Missouri [8].

Like Sam before her, Miss Sam was eventually returned to Brooks AFB. While little is known of what happened to her once she had been returned to the monkey colony, her mate Sam remained at Brooks for the next 11 years, after which he was transferred to the San Antonio Zoo. It is believed he died there in 1978.

OF MICE AND MEN

Duplicity played a key factor in the U.S. military's decision to announce an impressive series of engineering research and biological rocket flights in the late 1950s. While ostensibly conducting engineering experiments and research dedicated to the safety of crewmembers occupying planned manned spacecraft, one of the military's most noteworthy biological spaceflight programmes was in fact a carefully contrived and executed cover-up. It was all about spying, and elaborate hoaxes involving biomedical research would be created to fool other nations into believing there was no ulterior motive beyond science in sending dozens of powerful rockets and satellites into orbit.

A meticulous sham

Any search of books, documents and the Internet for specific details concerning a U.S. rocket flight carrying mice into space will lead researchers to the launch of Discoverer III on 3 June 1959. These sources reveal that the flight incorporated the use of a Mark I biomedical recovery capsule occupied by four C-57 black mice. The satellite carrying this recovery capsule is also correctly recorded as failing to achieve orbital velocity due to less-than-expected thrust from the Agena second-stage rocket.

While this launch carrying the mice did take place, the biological aspect of later research missions in Project Discoverer was mostly a meticulous sham, created to disguise the true purpose of the programme, which was to launch a highly-classified series of spy satellites known as Argon, Lanyard and Corona – but principally Corona. Managed by the newly-created Advanced Research Projects Agency (ARPA) of the Department of Defense and the U.S. Air Force, Discoverer was a military-backed ruse designed to fool Cold War adversaries into believing it was nothing more than a relatively benign, biological research programme.

According to spaceflight historian Dwayne A. Day, the Corona programme was managed and operated under intense secrecy. Corona was a "covert" project, meaning that its very existence was a secret. Because the rockets would be visible during

The Thor–Agena combination used in Discoverer launches. (Photo: USAF)

launch and the capsules that returned to Earth required an extensive recovery force, there was no way to keep the actual space flights a secret. Instead, the programme managers developed a cover story. At the time, Corona was one of the seven most secret projects of the United States government [9].

Background history of Project Corona

Corona was approved by President Eisenhower in February 1958 upon the recommendations of his senior intelligence advisors. The programme's primary aim was to take spy photographs of the Soviet Union and Sino-Soviet Bloc countries from orbit and later retrieve the film for processing and analysis. The spy satellite had been developed as a top-secret project, and its existence was not officially acknowledged until 1995.

Corona had originally been proposed in the late summer of 1957 as a recoverable reconnaissance satellite carrying film. The project was approved as a U.S. Air Force development programme, but lacked funding. Later in the year, after Sputnik, it was officially cancelled. On 23 December 1958, the ARPA announced the creation of Project Discoverer, and gave the Air Force Ballistic Missiles Division responsibility for the development and management of the project [10].

Officially, Project Discoverer had multiple engineering and research goals. These included biological research and the development of a stable platform for scientific observations, as well as testing ways of safely recovering space vehicles returning from space with live occupants, instruments and other scientific payloads. It would also be involved in the development of an early warning system that might be used for the detection of enemy missiles.

ARPA director Roy William Johnson said at the time that the project would be considered "very successful" if as many as three or four of the first dozen launchings put satellites into orbit. He also revealed that live mice would be placed into some of the early Discoverer satellites, and, if successful, a live monkey would be aboard on about the fifth or sixth flight [11].

Setting things in place

Technicians and scientists were duly appointed to the Discoverer programme, and they began to create a number of instrument and research payloads to be sent into orbit. These would include life-support systems for a series of flights involving the use of mice from a hardy genetic strain known as C-57. Project scientists would later become innocently involved in training and preparing monkeys for flights on more advanced Discoverer missions. Little did they suspect that their work was simply being used as part of an elaborate hoax.

The launch vehicle and spacecraft were entirely new, so the first several Discoverer firings would be diagnostic tests of the Thor–Agena guidance and propulsion systems, and not involve cameras or equipment to be used in the Corona series of spy satellites.

At the very top of the second-stage Agena A was a conical fairing, which was eventually destined to hold the camera. For the first three flights, however, the fairing would be empty except for wiring and cooling pipes for the mouse compartment.

The satellite recovery vehicle, or SRV, consisted of a gold-plated aluminium capsule complete with parachutes, an ablative shield and a thrust cone with a retro-rocket. This 160-pound capsule, known as "the bucket", was attached to the forward

Securing the nose and biological package to the Agena. (Photo: USAF)

open end of the conical fairing with two pins that would be released on signal to initiate recovery, while the aft end of the fairing was securely bolted to the Agena.

The first launch attempt was planned for January 1959. However, during a pre-launch test, a "sneak circuit" between the Agena vehicle and the blockhouse caused the Agena separation sequence to start while the rocket was still on the launch pad – and surrounded by its launch crew. Fortunately, nobody was hurt, but the Agena and Thor were damaged. The Agena was scrapped and the Thor was eventually refurbished and flew much later. This embarrassing incident, which was not publicly revealed at the time, was later referred to by the Air Force crews as "Discoverer Zero". It led to a valuable lesson that all systems should be checked end to end before a vehicle reached the pad and was filled with dangerous fuels. The Agena had been given the serial number 1019, and afterwards those involved in the programme used to say simply "1019" as a sort of high-tech mantra, a warning of what not to do.

First flight of Discoverer

The programme quickly recovered from its first try and Discoverer I, the first U.S. satellite launched from the West Coast, roared into the skies atop a modified Thor

IRBM from Vandenberg Air Force Base at 1:49 p.m. on 28 February 1959. There was no SRV loaded on this occasion; it was simply a dry run for the flights to come, and the nose cone was filled with tracking and telemetry equipment. Discoverer I was launched in a southerly direction, intended for an orbit that traversed both the North and South Poles. Use of this polar orbit meant that – because of the Earth's rotation – every point on the Earth would then be within the view of the satellite at some stage during the flight.

Based upon some initial data from a ground station, the Air Force issued a press release that the spacecraft was in orbit. However, no further signals were received. Two Air Force officers involved in the launch programme were then ordered to "prove" that it was in orbit, but not to investigate whether this was in fact true. They pieced together some dubious data and unenthusiastically presented it to their superiors. But they and many others had privately concluded that the Agena had probably suffered a failure late in its burn, and the spacecraft had most likely impacted in Antarctica. Because the Agena had to burn nearly all of its fuel to reach orbit, and because it had to be very precisely oriented during the entire burn, there was no margin of error during these early flights.

A near-polar orbit and a predicament

Discoverer II was launched from Vandenberg on 13 April, and sped to a near-polar elliptical orbit with an apogee of 225 miles, a perigee of 156 miles and a 2° angle of inclination to the Earth's axis. Like the earlier flight, the satellite was part of the 19-foot Agena A second stage. This time it not only carried 245 pounds of cosmic-ray as well as other scientific and communications equipment, but also an SRV to test ejection and recovery procedures of an instrument package from orbit.

Discoverer II became the first artificial satellite to be stabilised in all three axes with the aid of a device known as a horizon scanner. This scanner, which made a fearful sound like a coffee grinder in action, would view the horizon and fire pneumatic nitrogen jets to align the craft. On later flights (with the scanner replaced by a quieter, solid-state version) this would keep the Corona camera pointed downward to the Earth.

Inside the SRV capsule was a biomedical compartment designed to hold four mice. According to Bill Obenauf, who worked on the Corona system from 1960 to 1967 at Vandenberg, it was not on the second flight. "The mice weren't ready," he revealed, "so their places were taken by four multivibrators to put simulated heart beats on Agena telemetry" [12].

The following day, as the satellite neared Hawaii on its 17th orbit, a radar signal transmitted from the ground separated the re-entry module from the Agena. However, a timer malfunction caused the SRV to eject prematurely, and it most likely re-entered over the North Polar region.

"The capsule probably, errantly, landed on Spitzbergen Island near Russia," Obenauf stated. "No one knows for sure. It was supposed to land in the Pacific, but due to a mix-up, the Hawaii tracking station issued approximately thirty-two orbital-timer advance commands instead of twenty-two, causing the capsule to eject half an orbit too soon."

Under normal circumstances, and in a precision-planned operation, the descending SRV was scheduled to be plucked from the air at around 8,000 feet. This would take place with the aid of a trapeze-like device fitted with snare hooks mounted below a USAF C-119 "Flying Boxcar" aircraft from 6593 Test Squadron, based in Hawaii. Because they believed that the spacecraft had come down on Spitzbergen Island, Air Force officers rushed to the Arctic to search for it. But Spitzbergen had no landing strip and an air search attempt failed to find the capsule. ARPA reported the loss, saying that "analysis of radio beacon and telemetry signals ... indicates that the capsule did eject ... as predicted" [10].

"The Russians may have the capsule today," was Bill Obenauf's wry comment on the total loss of the satellite. "The thirty-two commands may have been intentionally sent so that the capsule would land near the USSR and allow the Russians to see the 'mouse' configuration" [13]. Bob Powell, the tracking station manager on duty in Hawaii, strenuously denies this claim.

According to Dwayne Day, there was another officer involved in the programme who also does not think that the capsule came down near Spitzbergen. Frank Buzard was in charge of the Corona launch programme and was one of the officers ordered to prove that Discoverer I was in orbit. Given the track record for the early programme, Buzard thinks that it is far more likely that Discoverer II failed like all the others. He also points out that – with all of that open ocean to fall into – it was unlikely that Discoverer II just happened to hit the one little bit of land in the vicinity [9].

The mice that soared

The next Discoverer/Thor–Agena flight in June finally carried four mice into space in an unsuccessful attempt to make them the first animals sent into orbit and brought back alive. They were scheduled to return after 26 hours' exposure to space radiation. Dark-coloured mice were being used as it was known that cosmic radiation could actually turn some black hairs grey. Each weighing less than an ounce, the mice were inserted into individual cages with small radio transmitters strapped to their backs to relay data on their physical reactions.

While the vehicle was sitting on the launch pad, the mice urinated on the humidity sensor inside their capsule before launch. Then all telemetry from the mice ceased. Obenauf does not recall if this delayed the countdown, but he says that there was no built-in provision to dry out the mouse chamber. "Opening up the capsule to service the mice or their telemetry would have required unfueling the Thor and the Agena, lowering the Thor to horizontal, de-mating the capsule, and disassembling the capsule in our building ten miles away – a procedure likely to create more problems than it solved – a job taking many days. The 'powers that be' were more interested in proving the Thor Agena system for future use with Corona and not wasting time on mice" [13].

Technicians working on the Discoverer mouse unit. (Photo: USAF)

The mice probably died of heat exhaustion on the launch pad, but fate had already decreed they would not survive their flight. The third Agena never achieved orbit and plunged back, crashing down into the ocean.

The first Corona camera

Although the programme leaders knew people might suspect that a recoverable spacecraft could carry a camera, they hoped that the small size of the spacecraft and the limited lifting capability of the rocket would make outside observers and Soviet intelligence analysts conclude that no camera could be made small enough. What few people outside of the American intelligence community knew was just how good American industry was at making powerful reconnaissance cameras.

The Corona camera was designed by the Itek Corporation based outside of Boston, Massachusetts. It was a revolutionary camera based upon an earlier panoramic camera suspended from high-altitude balloons. The camera lens swept past a long strip of film, producing a long and narrow image at a higher resolution than any previous camera. The initial cameras were manufactured by the Fairchild Camera and Instrumentation Company of New York, because Itek lacked fabrication equipment.

A whimsical photo showing three "deceased" rubber mice. The card reads "Sincerest condolences to thee, our departed friends of Discoverer III. From the Army Monkey." (Photo: USAF)

The first version was initially known only as the "C" model, but years later was designated "KH-1." The KH stood for "KEYHOLE," the top secret codename for satellite reconnaissance [14].

The first Corona camera was ready earlier than planned. Because the intelligence collection mission was so vital, programme managers decided to load it into the fourth spacecraft, even though the launch vehicle had not yet achieved success.

The KH-1 camera would be loaded in the fairing between the Agena and the SRV. The film spool faced the Agena, with very little space to spare. The film travelled through a twisting route within the camera, after which the exposed film would be wound onto a second spool within the SRV, forward of the camera. The SRV would then separate at a predetermined point, and the retro-rocket would fire just long enough to slow the capsule so that it would fall into the atmosphere at very high speed, headed for a pre-established recovery zone.

Lower in the atmosphere, having survived the fierce heat of re-entry, a timer would blow the SRV's parachute cover, deploying a small drogue parachute which would yank the capsule from its charred ablative shell. The drogue would then pull out a single large parachute. The capsule came equipped with a simple transmitter whose tone would change to indicate events such as the retro-rocket firing and parachute deployment, as well as a flashing strobe light and a bright cover to assist in its location by the catch aircraft.

Should the airborne recovery attempt fail, the capsule would float for a short while in the ocean after splashdown – the source of the nickname "bucket". If it was not recovered within a reasonable amount of time – around 24 hours – salt water would begin dissolving a plug on the bottom of the bucket. The water would then begin to gush in, filling the capsule and causing it to sink so that it could not fall into foreign hands. Apparently, the plugs were initially made of compressed salt, and later compressed brown sugar, which dissolved slower.

If recovered, the precious cargo of film would be removed from the capsule at Lockheed's Advanced Development Projects Unit – known as "Advanced Projects" (AP), or more euphemistically as the "Skunk Works" – in California. The buckets themselves were destined to be cut into pieces and irretrievably ditched in deep ocean water. The film was rapidly despatched to Eastman Kodak in Rochester, New York, for development under strict security. The developed film was then sent to Washington and the National Photographic Interpretation Center, or NPIC (pronounced as "en-pick"). There, photo-interpreters viewed positive copies of the film on light tables by looking through powerful microscopes.

A TROUBLED PROGRAMME

Unfortunately, the first few Discoverer missions were plagued by a number of launch mishaps – 12 in a row, or 13 if Discoverer Zero is included. Nine of those missions carried reconnaissance cameras and film. The irony of the programme was that – although publicly Discoverer was partially successful – in reality the classified Corona effort was suffering severe problems.

To some it seemed like a case of Murphy's Law – everything that could go wrong did go wrong. But others who were working on the spacecraft viewed it as a methodical, if frustrating, example of engineering development for a complex vehicle. For instance, before re-entry the SRV would start spinning for stabilisation. The initial design used small solid propellant rockets to start the spin and then stop it after the retro-rocket engine fired. These spin rockets proved unreliable despite various attempts to test them in batches and only use rockets from batches with high reliability. Finally, someone suggested switching to a nitrogen gas jet system to spin and de-spin the SRV. Two 3,000-psi bottles, each with a valve, fed four gas exhaust ports – two spin and two de-spin. They incorporated this system into the SRV and it worked perfectly thereafter [13].

Discoverer finally makes headway

It was not until the lift-off of Discoverer XIII on 10 August 1960 that the programme finally achieved its first fully successful mission. Because of the earlier failures, the managers decided to fly the spacecraft without a camera or film, and to instead fill it with diagnostic sensors. This time the recovery capsule was ejected on the 17th orbit, and safely plucked from the sea; not by a "catch" aircraft but a waiting helicopter. As on previous attempts, the only significant payload it carried was an American flag – bearing a 50th star for the new state of Hawaii. Amid much fanfare this capsule and its parachute were transported to Washington, where President Eisenhower openly told reporters it was an "historic" occasion. Then he and various generals had their photographs taken with their hands on the capsule, emphasising with wide grins the "innocent" nature of this spacecraft and programme. The press conference was impromptu and not popular with those who were concerned about the cover story for Corona. But it served a useful purpose by portraying Discoverer as just another scientific and engineering project.

According to the 22 August 1960 edition of *Missiles and Rockets* magazine, the success of Discoverer XIII meant that the much-touted monkey flight – the first of three possible flights to carry monkeys – would likely take place two flights later, on Discoverer XV [15].

Meanwhile, one surprising element of the programme disclosed to the authors by Bill Obenauf is that other mice actually flew during Project Discoverer. "A few of the later film capsules actually had a live mouse in a very small, pressurised 'can', bolted inside the bucket," he revealed. "Only a few buckets, a few launches, contained a mouse, but none of these mice ever came back alive from a few days in orbit. The machined aluminium can was used for many other non-animal experiments, like testing the effects of space radiation on transistor materials" [13].

Making plans for primates

In fact, scientific or biological research activities would be regarded as low-rated, secondary objectives for these missions. Most people who worked on Project Discoverer were totally unaware of the contemporaneous Corona programme, and they had no idea that their work was all part of a carefully-executed and elaborate cover story. At one time, it was even planned that specially-trained teams might covertly remove the biocapsules from the nose cone before the launch took place and replace them with the intended payload. Eventually, this plan was scrapped due to differences in the hardware involved. Instead, ARPA programme managers had openly begun announcing plans for sending primates into orbit aboard future Discoverer flights [16].

As an agency within the Department of Defense, the ARPA's mandate was to manage space programmes for individual military services such as the U.S. Air Force. In this capacity the agency reiterated earlier plans by the Air Force to use monkeys on future missions, in order to provide crucial data for an eventual human spaceflight programme. However, there was insufficient room in the Agena for both the Corona camera and a monkey capsule, a situation that would be discussed and resolved in secret behind closed doors.

The hardware for these proposed primate flights included a cylindrical container with an inbuilt life-support system. Designed and fabricated by the General Electric Missile and Space Vehicle Division in Philadelphia, it had an external diameter of 36 inches, an internal diameter of 27 inches and a length of 64 inches. Known as the Mark II biopack, the life-support system provided the proper level of oxygen while minimising humidity and carbon dioxide exhalation levels. It was maintained at an average temperature between 70 and 85°F, and provided adequate and controlled, regular nutrition for the animal – water and apple wedges dipped in paraffin to prevent spoiling – over an extended period. The total weight of the re-entry vehicle, including the biopack, was approximately 300 pounds. Agena No. 1025 was set aside for the first primate mission.

Too much monkey business

The animals that had been selected for the task were hardy American-born *Macaca mullata* rhesus monkeys. They were carefully chosen from a large colony maintained under a contract with the U.S. Air Force at the University of Texas Balcones Research Laboratory. The laboratory fell under the administration of the School of Aviation Medicine (SAM) at Brooks Air Force Base, in support of the Air Research and Development Command's Ballistic Missile Division, the executive managers of Project Discoverer.

Dr. W. Lynn Brown, the director of radiobiology at the University of Texas, who had been in charge of training monkey Sam for his suborbital flight, once again took on the task as head of experimental psychology for the bio-astronautics division of the Discoverer programme. He stressed a preference for training rhesus monkeys rather than chimpanzees for the flight, as he felt that chimpanzees tended "to brood and to pout," and once they'd learned to solve a problem through repetition they soon became bored with it and would no longer perform. The rhesus monkey, on the other hand, was extremely inquisitive and "continues to react to outside stimuli and continues to perform." He added that – as the rhesus monkey is from a more temperate climate – it is "physically more adaptable to the range of temperatures which will be encountered in manned space flight. He is neither excessively sensitive to heat nor cold."

In total, there were 31 monkeys in the laboratory at the time of the candidate selection, housed in individual cages. As Dr. Brown related, experience quickly showed that special latches had to be fitted to the cages that the monkeys could not unfasten. "They are so smart," he once emphasised, "that when one gets out of his cage, that one will go about the room letting out all the others."

All of the animals in the colony were given regular physical examinations, which included blood tests and analyses of their waste products. They would be weighed on a regular schedule and their growth carefully charted, while a complete clinical history of both major and minor weaknesses would be recorded for each animal. Each day they were fed a standard, controlled diet consisting of grains, fruit and vegetable products carefully selected for their nutritional value, which would be processed into $\frac{1}{4}$-pound cakes with the binding aid of molasses.

"X" marks the monkey

Those monkeys born in captivity at the Balcones laboratory were designated by a number, followed by the letter X. Those born elsewhere in the United States and purchased by the laboratory were given the designation Y. The three primary candidates selected for the first Discoverer monkey launch, in order of preference, were known as 21X, 20X and 3Y. They would not be given names, as it was felt this might induce a form of emotional attachment to the animals, which was frowned upon in animal research. However, each monkey was provided with a notarised birth certificate, certified by the attending Air Force veterinarian.

21X was a 4-pound female, born 26 January 1959, whose parents were C4 and A4. 20X was also a female, weighing 5 pounds, born 4 days earlier than 21X, and her parents were simply known as 299+ and 893. The third candidate was the 5-pound male known as 3Y, who had been born at Shamrock Farms in Middleton, New York, on 14 April 1958.

According to Dr. Brown, the criteria for selection of the space candidates, apart from a prerequisite of perfect health, was "emotional stability, performance, weight and physical condition." He added that the ideal launch weight had been determined as being "between 4 and $5\frac{1}{4}$ pounds". Based on his experience in preparing primates for space travel, Dr. Brown pointed out that the best overall performer in tests might not necessarily be the best candidate for the first flight, as the animal might react adversely under emotional pressure caused by the environmental stresses associated with space flight. The rhesus monkey selected would be the one that combined optimum performance with optimum emotional stability. Candidate 20X would prove the better performer on psychomotor tests, but 21X, whose performance levels were slightly lower, was selected on the basis of her greater dependability [17].

Supervised training begins

Early in their training programme the monkeys were introduced to a nylon restraint garment with a soft, padded chamois face protector ring, a form-fitting couch-like seat covered in foam rubber and a special training device developed by Dr. Brown, which simulated the physical layout of the biomedical capsule that might carry them into space. The garment and couch had been designed to allow the monkey sufficient freedom to move its arms and hands in order to eat and conduct the required psychomotor tests in relative comfort. In tests of similar couches conducted at the U.S. Navy centrifuge facility in Johnsville, Pennsylvania, other monkeys, restrained in a reclining position, had been subjected to stresses up to 20 g, even though the maximum gravity expected on lift-off would be 9–10 g's over a short period.

Initially, the animals undergoing training would receive a small reward pellet if they reacted to a visual signal and depressed a lever, but in the second stage of training the reward was eliminated and replaced by a mild but unpleasant 12-volt electric shock through their feet. Over a period of weeks the animal became conditioned to pulling the handle each time a red light came on. However, the monkeys quickly learned that if they held the handle down and did not release it, they did not get a shock. The training device then had to be modified so that the animal not only received a mild shock if they did not react in time to the red light, but also if they failed to release the handle.

The three prime candidates would undergo minor surgery in order to record the physiological reactions of the chosen monkey throughout his or her space flight. Tiny stainless steel wires, serving as electrodes, were anchored to cartilage in the animals' chest, back and groin. These wires were then covered with plastic insulation and led under the animals' loose skin to a central point where a single lead was brought out through a small incision under the arm. The contoured seat to which they were fastened was wired to provide connections to these leads, allowing ground controllers to receive constant EKG information via telemetry. The seat was also designed so that a chest strap, attached to a potentiometer, would provide a record of the rapidity and depth of the animal's respiration during the flight, expected to last 17 orbits.

Systems within the biopack containing the monkey would provide oxygen, proper humidity and carbon dioxide levels during the flight, as well as an average temperature ranging between 70 and 80°F. What it would not provide was anything in the way of room or comfort, with the monkey strapped into a couch with barely enough space to extend its hand to pull the lever whenever the red light came on. A camera mounted in front of the monkey would periodically photograph the animal's face in flight, working in conjunction with a strong light that would illuminate prior to a photograph being taken [18].

There was also the problem of animal waste to consider. Several complicated methods were tried until it was realised that the best solution was the simplest: a baby diaper and doll-sized plastic pants.

Conducting tests of the biopack

Tests of the entire biomedical package, including a monkey, would begin at the Vandenberg Air Force Base launch site in California. The animal was placed in the biopack and the couch secured before the biopack was sealed and the life-support system activated. After all the connections had been checked, the capsule was loaded onto a specially-modified C-47 aircraft and transported to the three-storey high-altitude chamber at Lockheed's Missile and Space Division in Sunnyvale, also in California. For the purpose of the exercise, the flight time from Vandenberg to Sunnyvale simulated the countdown time on the launch pad. Once further checks had been completed the biomedical capsule was lowered cautiously into the nose cone,

The monkey numbered 21X was the prime candidate for the first Discoverer primate launch. She is shown here with Carolyn Kingery, a haematology technician at the University of Texas radio-biology laboratory. (Photo: USAF)

to ensure correct alignment of all components and connections, and the whole assembly was raised into the altitude chamber by a hydraulic boost.

At the conclusion of the 27-hour environment test, duplicating orbital flight, the biomedical package was removed from the chamber. The recovery capsule was then immersed in salt water for several hours to simulate splashdown conditions. Later

Another primate candidate – No. 31X. (Photo: USAF)

analysis confirmed that the biopack's integrity was sound, and the monkey involved in the test had suffered no ill effects from its experience, despite having been sealed in the biopack for more than 70 hours.

The Public Information Division of the Air Force's Office of Information had been working overtime in the lead-up to the Discoverer monkey flight. By January 1961, a series of photographs showing the training of the animals had been assembled and suitable captions prepared for release to the media. Air Force Headquarters in

Dr. W. Lynn Brown (centre) laces an animal into the flight couch in preparation for a training session. (Photo: USAF)

Washington, D.C. had prepared press releases covering all contingencies such as the successful launch or postponement of the flight, even the loss of the booster at various phases of the launch. Other pre-emptory press releases covered the safe recovery or non-survival of the monkey.

The vanishing programme

Then, suddenly, the monkey flight was cancelled, and the biomedical research team was abruptly told that the programme had been discontinued.

There are many possible reasons for the cancellation. The first is that the original cover story of biological research flights operated by the military would have been wearing thin with so many successful, hugely expensive launches, and this would not have gone unnoticed by Soviet intelligence agencies. It was also a crucially pivotal time in the Cold War with a relatively new president in the White House.

The Discoverer programme had originally been envisaged as an operation lasting about 18 months through to the end of 1960, and the animal flights would have provided an excellent cover story, but ongoing delays in producing the primate capsule meant that flights would not take place until the middle of a very active Corona programme.

With crises mounting before him, John F. Kennedy began to exert his authority by changing the security classification for all military space launches in an attempt to confuse those who might be trying to track which flights had military or scientific payloads. It is also believed, with things heating up in relations between the Soviet Union and the United States, that the CIA and USAF wanted to focus primarily on reconnaissance space missions and to shed any flights without distinct military applications. It seems the importance of the Corona cameras, which the CIA had incorporated into Discovery satellites, now outweighed any possibility of sending animals in their place, and national security appears to have prevailed.

Agena 1025, set aside for the first primate mission, never flew. Held over for this flight, it is likely that it soon became outdated in a rapidly-evolving Agena programme and was no longer considered suitable for a Corona mission.

End of a mission

Discoverer XIV carried a camera, and its bucket containing a valuable payload of film was recovered in mid-air on 18 August by a C-119 over the Pacific Ocean. Post-recovery, the polymer film was removed by Lockheed technicians in Menlo Park and secretly transported to Rochester for development by Eastman Kodak. The capsule eventually ended up in the U.S. Air Force Museum in Dayton, Ohio.

Photo-interpreters later stated that images received from the 3,000 feet of film, showing more than 1,650,000 square miles of denied areas of Russian territory, were "terrific, stupendous ... we are flabbergasted" [10]. In a single space mission, Corona had provided more images of Soviet territory than the entire U-2 spy plane programme, including the very first photograph of the Mys Shmidta airfield in the Soviet Arctic, which was being used as a staging base for intercontinental bomber flights.

Ostensibly, there was nothing hidden from the public about the airborne recovery of these satellites; in fact, photographs were released showing the dramatic and skilful capture of one of these descending satellites by C-119s in a procedure that would become almost routine. It was only the true purpose of the entire operation that was

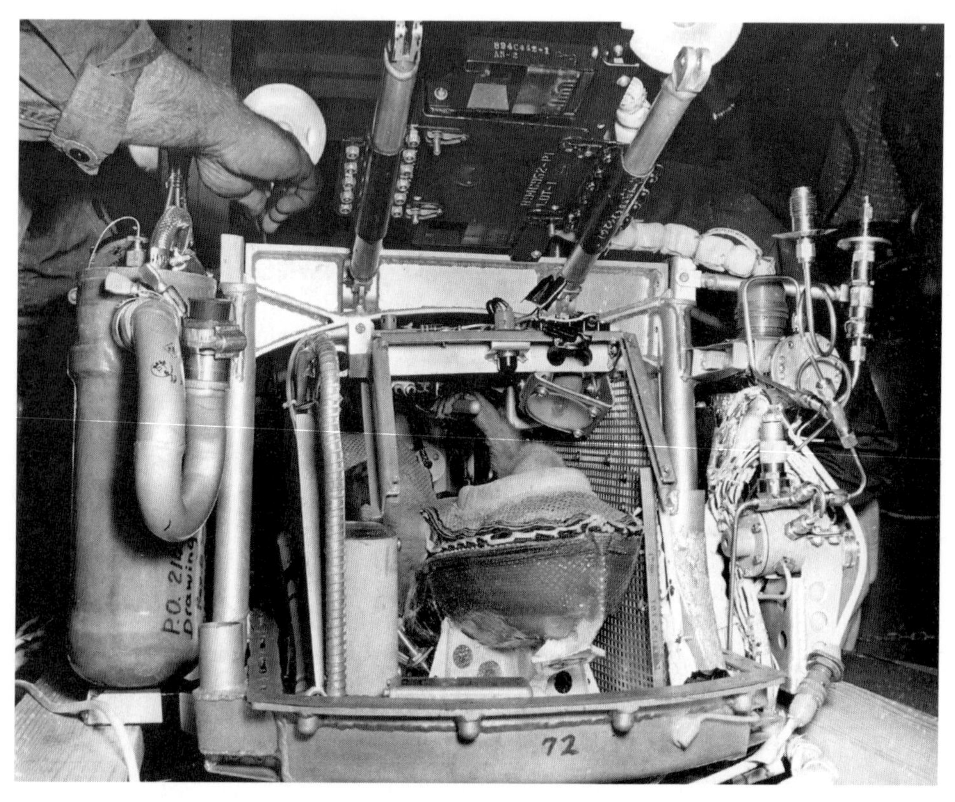

A test subject has been installed in the Mark II biopack, and the animal is already gripping the psychomotor performance lever. (Photo: USAF)

kept secret. And that was the most important part. Intelligence was now flowing in on Soviet submarine bases and missile sites, nuclear facilities and strategic military installations in China and Eastern Europe.

NASA would also benefit from the Corona programme, using similar imaging technology in the space agency's successful Lunar Orbiter programme of 1966–67.

After Discoverer XVII, Corona employed an innovative, constantly-rotating camera system known as KH-2. The spacecraft flew backwards in its orbit so that the Agena's engine nozzle could protect it from the unknown effects of atomic oxygen. The Corona camera looked down, out of the bottom of the spacecraft. The camera would be significantly upgraded a number of times during the 1960s, increasing both the resolution and the amount of film carried.

Discoverer XVIII was launched during a massive solar flare, which began and lasted for the first 13 hours of the 48-orbit, 3-day flight. Biological samples carried on this flight included *Neurospora conidia*, nerve tissue, algae, human bone marrow, eyelid tissue, gamma globulin and cancer cells. Post-flight results indicated that biological specimens could withstand solar flares with a minimum of shielding, and that thicker, lightweight aluminium shielding might prove to be better than lead.

Discoverer XIV's bucket is snagged by a C-119 aircraft under the command of Captain Harold E. Mitchell. (Photo: USAF)

Plans on hold

Meanwhile, despite the sudden cancellation of the primate programme, Lieutenant General Bernard Schriever, who ran the Air Force's powerful Systems Command, still wanted to fly small monkeys aboard Air Force rockets. He had become increasingly interested in developing a military manned space flight capability – something that he was officially prohibited from doing because manned space flight was supposed to be a civilian activity.

Part of Schriever's revised proposal involved the potential use of large, advanced Samos E-5 reconnaissance satellites to carry monkeys, possibly chimpanzees, secured inside recoverable capsules on 2-week orbital missions that would subject the animals to potentially harmful cosmic radiation, in order to evaluate the risks humans might face in protracted military orbital flights [9].

In October 1960, as a precursor to the Samos chimpanzee programme, Schriever put forward a proposal which outlined plans already in existence to fly smaller primates on Discoverer missions. By this time, however, it was a moot argument. Discoverer's programme directors had already decided to cancel life sciences research and any unnecessary work on satellite support systems for primate passengers.

The Undersecretary of the Air Force, Joseph V. Charyk, who reported directly to the Secretary of Defense, had seen very little merit in Schriever's lofty Samos E-5 proposal and recommended it be rejected. Charyk felt fully justified in stressing that biological research payloads would take up valuable space in the nose cones of missiles that could more advantageously be used for placing spy cameras into orbit.

However, Schriever was famously stubborn. He persisted, and in December 1960 his Office of the Assistant for Bioastronautics at the USAF's Air Research and Development Command issued a challenging report titled "Bioastronautics Capability and Requirements for Manned Space Operations".

The report said, in part: "Space vehicles ... will undoubtedly be substantially improved, if past experience is any guide, in usefulness, feasibility, and reliability by the presence of man [and] if the Russians got here, we've got to get there, too, in order to keep an eye on them. In the light of the foregoing considerations it is clear that failure to explore the potential of manned military space systems is a prohibitively dangerous gamble with National Security" [19].

A cover story revealed

Four months later, on 12 April 1961, a beaming cosmonaut Yuri Gagarin became the first person to fly into space, giving the Soviet Union not only considerable technological prestige, but the undeniable advantage of taking the high ground in the Space Race. Just 4 days later, the Air Force Systems Command issued an interim proposal for what was called a bioastronautics orbital space system, or BOSS for short, which revisited the Samos E-5 plan involving primate flights.

The BOSS proposal is said to have received tacit approval in June of that year, but the issue of funding would prove to be too difficult an obstacle to overcome, especially in light of the recent and successful suborbital flights of Alan Shepard and Gus Grissom. The primate project foundered, and the entire Samos E-5 programme was finally killed off by Charyk.

On 28 August 1961, a disappointed Lt. General Schriever announced that provisional plans to orbit a monkey aboard Discoverer XXII on 30 March the following year had been scrubbed, and there were no plans to reschedule the event.

The Discoverer cover story could not have lasted for ever. Nobody would seriously believe that the Air Force was launching dozens of biological, science and engineering satellites at such a crucial time. By the end of 1961, with the Corona spy camera system now in full operation, any technicians and scientists still innocently involved in biological research programmes were dispersed, much to their consternation, and moved on to other projects.

The heavily-rumoured Corona spy camera programme was officially revealed in February 1995, when President Clinton signed an executive order releasing details of its operation. Despite using a biological test programme as a cover for its true intentions, and some early launch and systems failures, Discoverer proved at the time to be an incredibly successful programme. It was beneficial in reaping results, both covert and scientific, and creating ongoing applications for both military and civilian spaceflight programmes.

SALLY, AMY AND MOE

Like their rodent cousins in the highly-classified Project Discoverer, a trio of C-57 mice also achieved a successful space flight late in 1960. On 13 October, three black mice named Sally, Amy and Moe were installed in a heavily-shielded General Electric experimental re-entry vehicle known as the RVX-2A, mounted atop an Atlas-D rocket at Cape Canaveral's Atlantic Missile Range. Specially trained for this mission by the USAF School of Aviation Medicine (hence the trio of names beginning with the letters SAM), they were surrounded by numerous scientific instruments to record their reactions, and were plentifully supplied with oxygen, oatmeal, peanuts and gelatine. A separate experiment would try to gauge the rate at which radiation in the Van Allen belts would enter human skin.

Safely recovered

After an early morning lift-off from Launch Complex LC11, the spacecraft reached an altitude of 650 miles and travelled 5,000 miles downrange over the Atlantic toward the Ascension islands, accelerating to around 17,000 mph. As well as passing through the Van Allen radiation belts, the mice were also subjected to 10 minutes of weightlessness. The nose cone containing the mice and 10 other experiments was safely recovered from the sea just 3 hours after lift-off. First reports said they seemed to have suffered no ill effects from their brief ride into space and no premature greying from their exposure to radiation.

Only one of the female mice wore a bioinstrumentation package, and physiological data telemetered for the first 17 minutes of the $\frac{1}{2}$-hour ballistic flight showed that the mouse's heartbeat was quite normal, as was her temperature and muscular movement, except for a few brief moments after lift-off.

Sally and Amy would later be mated with Moe to study any possible genetic effects of cosmic radiation on the reproductive process [20].

Tests and more tests

Once their first group of "graduates" had flown as test subjects on ballistic V-2 and Aerobee rocket flights, the Aeromedical Research Laboratory at Holloman AFB continued to operate as the principal training camp for space-bound primates. The trainers, veterinarians and scientists now had several years experience in preparing animals for flights into the largely unknown realms and hazards that lie beyond the thin shroud of Earth's atmosphere.

At any one time, around 40 Holloman monkeys were undergoing training to ride the unpredictable rockets. All were treated with care and patience as they participated in examinations and tests of their intelligence, conduct and ability to learn. Those who consistently failed to make the grade were soon identified and transferred to receptive zoos around the country.

For the remaining monkeys the training intensified. They were given increasingly harder, repetitive tasks to memorise, and there was a reward in the form of a banana-flavoured pellet if they did well in practical tests. Should they forget something or fail, however, they would receive a mild electric shock through a small metal plate attached to their feet. While not physically harmful, this tingle irritated the monkeys, and they quickly learned to avoid being zapped by doing as they were instructed. Eventually, they had memorised their exercises so well they could perform a long sequence of tasks with very little prompting.

The testing was mostly carried out while seated inside a mock-up of a Mercury spacecraft, where they were strapped into chairs opposite an instrument panel and tasked with pulling small levers and pushing buttons whenever certain lights flashed. Before long they had gone through almost every phase of astronaut training except for the control and re-entry procedures, which required precise judgement beyond their capabilities. These systems commands would be initiated from the ground.

The monkeys were surprisingly smart. One of them pulled the levers 7,000 times in 70 minutes and made far fewer errors than a visiting politician who tried his hand at the same tests. Another monkey subject was given a banana by his handler prior to an acceleration test in a rocket sled. As the monkey peeled the banana his sled suddenly lurched forward on its run, and the ripe banana was smeared all over his face. The next time he was strapped into the rocket sled and offered a banana he carefully peeled the skin off, and then shoved the fruit into his handler's face! [20].

As the Space Race intensified, coinciding with a national effort to put a man into space before the Russians, Holloman's monkeys would once again answer the call, this time at one of the world's oldest rocket launch sites.

REFERENCES

[1] Undated NASA Fact Sheet, *Project Mercury Little Joe Flight Test Program*, issued by NASA Headquarters, Washington, D.C.

[2] George P. Kennedy, *Mercury Primates* (IAA-89-741), The Space Center, Alamogordo, NM, 1989.

[3] Paul Laster, "Oral History of Brooks Air Force Base," HHM Inc., Austin, TX. History segment *Man-In-Space Program.* Website: *http://www.brooks.af.mil/history/space.html*

[4] Loyd S. Swenson, James M. Grimwood and Charles C. Alexander, *This New Ocean* (Chapter 7), NASA History Series SP-1401, 1989.

[5] Hal T. Baber, Jr., Howard S. Carter and Roland D. English, NASA Technical Memorandum: *Flight Test of a Little Joe Boosted Full-Scale Spacecraft Model and Escape System for Project Mercury*, NASA TM X-629. Declassified 28 June 1967.

[6] Ralph Papa, e-mail correspondence with Colin Burgess, 11–17 August 2005.

[7] Wayne Meier, e-mail correspondence with Ralph Papa and Colin Burgess, 11 August 2005.

[8] Joseph Adams Shortal, NASA Reference Publication, *A New Dimension: Wallops Island Flight Test Range, the First Fifteen Years*, December 1978.

[9] Dwayne A. Day, "From Cameras to Monkeys to Men: The Samos E-5 Recoverable Satellite (Part III)," *Spaceflight*, vol. 45, no. 9, September 2003, pp. 380–389.

[10] John Pike, online article, KH-1 Corona: Return. Last modified 30 March 2005. Website: *http://www.globalsecurity.org/space/systems/kh-1-return.htm*

[11] Lester A. Sobel (ed.), *Space: From Sputnik to Gemini*, Facts on File, New York, 1965.

[12] Jim Plummer and Bill Obenauf, "The Story of the Flag on Discoverer XIII." Website *http://www.livefromsiliconvalley.com/space/DISCXIII.html*
[13] Bill Obenauf, e-mail correspondence with Colin Burgess, 27 August 2005–3 March 2006.
[14] Charles J. Vukotich, "Discoverer/Corona Program," for undated Astrophile publication. (used by permission of Charles Vukotic).
[15] "Monkeys set to fly,"*Missiles and Rockets* (American Aviation Publishing, Washington, D.C.), 22 August 1960.
[16] Much of the ensuing technical information is derived from an appendix to a massive Discoverer document history collection maintained at the US Air Force Historical Research Agency at Maxwell Air Force Base, AL. The appendix has only limited citation information: "Appendix D: Notes and Editorial Background for Primate Launch."
[17] *Project Discoverer Biomedical Launch TV Caption Sheet*, Office of Information, Headquarters, Air Force Ballistic Missile Division, Air Research and Development Command, Washington, D.C., undated, circa 1960.
[18] Dwayne Day and Colin Burgess, "Monkey in a Blue Suit," *Spaceflight*, vol. 48, no. 7, July 2006, pp. 265–272.
[19] *Bioastronautics Capability and Requirements for Manned Space Operations*, U.S. Air Force Office of the Assistant for Bioastronautics, Air Research and Development Command, Washington, D.C., December, 1960.
[20] Clyde R. Bergwin and William T. Coleman, *Animal Astronauts: They Opened the Way to the Stars*, Prentice Hall, Englewood Cliffs, NJ, 1963.

8

Pioneers in a weightless world

For the staff at the Soviet Union's Institute of Aviation and Space Medicine, the second orbital biological flight, coming nearly 3 years after Sputnik 2, began very much like the first. A telegram arrived at the office of the director alerting him to the imminent launch of the satellite. Scheduled for 15 August, this flight would be the first since Sputnik 2 to take dogs back to space. With one major difference – this time they would return. The successful recovery of a satellite from orbit would mark a major milestone in the quest towards a manned orbital flight.

A CROP OF SATELLITE DOGS

The selection of the dogs for a particular flight was typically done at the last minute to utilize those animals that were the best prepared and ready at that time. A small booklet or "passport" maintained on each of the dogs included their picture and a wealth of data on how they had performed and reacted during the various phases of training. The dog trainers had also developed close relationships with the dogs and knew their behaviour patterns well. The crop of satellite dogs then in training that were considered possible candidates for this flight included: Belka, Il'va, Marsianka, Mushka, Pchelka, Strelka and Zhemchuzhnaya [1]. They were all veterans of the programme, most with years of training. Belka and Zhemchuzhnaya both had suborbital flight experience.

After careful consideration, the dogs Belka and Strelka were selected. True to the space dog norm, they were both small animals. Belka, with yellow spots on her ears and sides, was 18 inches in length, 12 inches tall and weighed 12 pounds. Strelka, though a little longer and taller, weighed the same as Belka. Dark-brown spots dotted the white fur on her head, back and sides.

Perfecting the hardware for manned flight

Shipped off to the Baikonur launch facility, along with the rabbit, rats, mice, fruit flies, seeds, and cellular organisms that would accompany them on the flight, the dogs had to wait out a 4-day launch delay caused by a faulty oxygen valve in the R-7 rocket booster. Given recent problems with the launch vehicle and the capsule, the delay may well have raised concern for the safety of the canine cosmonauts.

Belka and Strelka would fly in a spacecraft that went by the name Korabl-Sputnik 2 (spaceship-satellite 2). Korabl-Sputniks 1–5 were known in the West as Sputnik 4, 5, 6, 9 and 10. (Sputnik 7 and 8 were Venus probes.) These craft served as test vehicles for the Vostok capsule that would later be used for manned flights.

The story of the five orbital dog flights conducted over a 7-month period in late 1960 and early 1961 is really the story of the development of the orbital manned capsule, the Vostok, and the race to employ it to launch the first human into space. As early as 1957, a research plan had been developed for a piloted spacecraft that would be lifted into a low Earth orbit by the R-7 rocket. Even after the successful orbital flights of the first three Sputniks, however, an enormous amount of work remained to be done before the first manned flight could become a reality. Those early satellites had not been designed to return to Earth. Creating a spacecraft that would return from orbit, be shielded from the heat of re-entry, have a life-support system capable of sustaining a human and a means of recovery, all within the limits of the R-7's lift capabilities, was a daunting task.

Developing Vostok

Complicating matters and delaying development of the Vostok were disputes with the military over whether manned spacecraft or photo-reconnaissance satellites would get priority. However, by 1960 intensive testing of the Vostok was under way, with several flights planned for the capsule. For the staff at the Institute of Aviation and Space Medicine, the challenges included: developing the equipment to sustain life during an orbital flight and studying the physiological effects of orbital flight.

A Vostok prototype (1KP-1) was launched on 15 May 1960, with a dummy cosmonaut aboard, but was never intended to be recovered. Launched 3 days later, the first Korabl-Sputnik (KS-1) experienced orientation problems during re-entry and was instead mistakenly boosted to a higher orbit. Despite that setback, the flight of the dogs Chaika and Lisichka was scheduled for 28 July. This was planned as a 24-hour flight, having as its outcome the first satellite and the first living beings to be recovered from orbit. However, misfortune struck again when a strap-on booster broke away from the R-7 rocket during launch, and the dogs perished in the resulting explosion.

BELKA AND STRELKA ORBIT THE EARTH AND RETURN

The second team of dogs selected included Belka, a veteran of numerous suborbital flights, and a newcomer named Strelka. When they finally did launch on 19 August

1960, they flew in an ejectable sled, along with the familiar accoutrements from other dog flights: an automatic feeder, sanitation device and physiological sensors. Capsule equipment included a catapult, life-support system and parachutes. Sensors monitored the composition of the air and reported on levels of carbon dioxide, water vapour and oxygen. Mirrors and lamps installed in the capsule hatch provided illumination for the TV cameras that recorded the dogs from front and side angles.

Dog watch

For the first time, Vladimir Yazdovskiy and his biomedical staff were able to visually observe dogs during a flight. Two television cameras beamed back to Earth front and side views of the dogs. What the first images revealed, however, was not encouraging. The dogs appeared to be frozen in place. Only the medical information streaming in from telemetry assured the staff that the dogs were still alive. As with Laika, in Sputnik 2, the pulse rates of both Belka and Strelka had nearly tripled during the acceleration of launch. These indicators were now very slowly returning to normal.

After a few orbits, the dogs finally began to move. Belka became extremely agitated, barking and trying to tear loose from her restraining harness. Finally, on the fourth orbit, she vomited, then settled down. Yazdovskiy, watching the behaviour of the dogs on the TV monitor, would later report to the State Commission that a manned flight be limited to just one orbit, because there were still too many unknowns about the effects of weightlessness [2].

An onboard TV camera beamed back this image of Belka during the flight of Korabl-Sputnik 2. It took several orbits for her to adjust to the feeling of weightlessness. (Photo: authors' collections)

Despite a problem with the orientation system, after 25 hours the satellite successfully re-entered the atmosphere on the 18th orbit, the catapult system ejected the dog cabin and the dogs parachuted back to Earth some 124 miles from the planned recovery site near the city of Orsk. A recovery team headed by Arvid Pallo arrived on the scene quickly and released the dogs from their cabin. As Pallo described it, "They began to run about and jump high in the air and were visibly pleased to be back on Earth" [2].

Belka (left) and Strelka, the dogs that flew aboard Korabl-Sputnik 2, the first orbital flight of living beings successfully returned to Earth. (Photo: authors' collections)

Publicising space flights

The flight of KS-2 came during one very busy month for space achievements, and the meeting of the 11th International Astronautical Congress in Stockholm, Sweden became the venue for airing them. The Congress, which ran 15–20 August 1960, featured such luminaries as Wernher von Braun and German space pioneer Hermann Oberth. One week prior to the Congress, and within a day of each other, the USAF had launched Discoverer XIII (the first successful flight in that military series) on a 17-orbit mission, while NASA had placed a 100-foot diameter Mylar polyester film balloon – the communications satellite known as Echo 1 – into orbit. During the Congress, on August 18, the USAF would enjoy another reassuring achievement with the follow-up success of Discoverer XIV.

After a series of failed attempts, Discoverer XIII achieved the programme's first success when its re-entry capsule returned to Earth on 12 August, making it the first orbital object ever to be recovered after re-entry, beating KS-2 by 7 days.

In the days leading up the Stockholm meeting, wire service articles buzzed with hints that the Soviet representative at the Congress, Leonid Sedov, would be arriving with news of a "space spectacular" [3]. Even with the 4-day launch delay of KS-2, Sedov was able to hold a press conference on 19 August, the last full day of the Congress, to announce sketchy details about the flight. Once again the Soviets had upstaged the Americans, by not only recovering a satellite from orbit but one with living beings onboard.

Meanwhile, back in the USSR, Belka and Strelka returned to a hero's welcome in Moscow. After an "interview" on Radio Moscow, they were whisked off to a press conference at the Academy of Sciences, during which dog trainer Oleg Gazenko held the dogs aloft in a gesture of triumph. In November of that year, Strelka gave birth to a litter of puppies, one of which would play its own role in international diplomacy. Following a meeting at a Vienna summit conference, Soviet premier Nikita Khrushchev gave one of Strelka's pups to the U.S. First Lady, Jacqueline Kennedy. The dog, named Pushinka (Fluffy) would grow up in the Kennedy White House. Before the dog could be accepted, however, and reflecting the deep suspicions of those Cold War times, army experts thoroughly searched and X-rayed the little dog to check for any implanted "bugs" or a possible doomsday device.

Data obtained during the flight of Korabl-Sputnik 2 confirmed the findings from Sputnik 2, namely that animals could adapt themselves to long periods of time in conditions of weightlessness. In addition, this second, animal orbital flight also proved the capabilities of the Vostok capsule's life-support system to sustain habitation on long flights.

THE RACE TO PUT A MAN IN ORBIT

That fall, the Kremlin learned from Soviet intelligence that NASA planned to launch a suborbital manned flight in early 1961, which prompted Khrushchev to sign his "Document 10/11", an 11 October order that gave the highest priority to manned

Dr. Oleg Gazenko, trainer of the dogs, holds aloft Strelka and Belka at a press conference after their historic flight. (Photo: authors' collections)

space projects [4]. With the success of Korabl-Sputnik 2, the thinking in the Soviet space programme was that they might be ready for a manned flight as early as December, following just one more dog flight. That schedule may well have worked if not for a catastrophic event that took place at the Baikonur launch site.

The Nedelin disaster

On 24 October, during a test-firing of the new R-16 ICBM, the countdown was halted to fix a minor electrical problem. As commander of the R-16 development programme, Marshall Mitrofan Nedelin was furious with the delay and positioned himself beside the fully-fuelled rocket in order to better direct his team of technicians. However, instead of following the standard but time-consuming procedure of draining the R-16 of its highly-corrosive fuel, which would delay the launch by several days, the technicians were ordered to complete the work while the engorged rocket sat on the launch pad. During repairs, the second stage of the rocket was accidentally ignited, causing a massive explosion that killed 126 people. Some were vaporized instantly and others died of burns over the following weeks. As well as Marshall Nedelin, the victims included many of the Soviet Union's top engineers, rocket technicians and managers.

The accident brought a brief halt to the manned space programme, enough to make the hoped-for December launch of an orbital mission untenable. In a 10 November letter to the Central Committee, several of the leaders of the manned spaceflight programme requested permission to resume testing the Vostok spacecraft [5]. A positive response led to the launch of Korabl-Sputnik 3 (Sputnik 6) on 1 December. The flight would once again carry two canine passengers to continue the study of the performance of the life-support system, including regeneration, temperature-control, feeding and water-supply systems.

Pchelka and Mushka

Accompanying the dogs Pchelka and Mushka on this flight would be a menagerie of living specimens, including guinea pigs, rats, mice and fruit flies, plus plants and other biological experiments. Different strains of mice and fruit flies were used to study the effects of cosmic radiation. Fruit fly varieties included those prone to both high and low mutation, with some to be shielded by lead and some not.

As with KS-2, television cameras on KS-3 recorded the behaviour of the dogs during the flight. The Soviets called this a "radio-television" system and broadcast its signal at 83 megacycles. During the flight, the U.S. Central Intelligence Agency succeeded in demodulating this signal and getting their own glimpse of the dogs in their orbital capsule [6].

The flight of KS-3 proceeded smoothly until re-entry when the retro-rocket – meant to slow the craft – malfunctioned. KS-3 flew an extra orbit and a half before beginning its descent from orbit on an unplanned trajectory. The official Soviet media later gave the impression that the craft had burned up on re-entry as it entered the atmosphere at the wrong angle, when in fact it had been purposely destroyed. Soviet spacecraft of that period contained a self-destruct mechanism that would be activated if the satellite was projected to land outside the boundaries of the Soviet Union, where it might be retrieved and fall into the hands of a foreign government. With ground controllers fearing this might happen, the device aboard KS-3 was activated, destroying the satellite and its biological payload.

Three weeks later, the launch of Korabl-Sputnik 4, on 22 December, encountered new problems. A slightly different version of the R-7 rocket was used on this occasion, employing a third stage with higher thrust. There is considerable confusion as to just which dogs took this flight. Many sources list Kometa and Shutka, while others claim Zhemchuzhina (Zhemchuzhnaya) and Zhulka were aboard. To further confuse the issue, Oleg Gazenko claimed that Zhulka, his personal pet, flew on three occasions, but under different names, including the name Zhemchuzhina/Zhemchuzhnaya [7].

A re-entry problem with Korabl-Sputnik 3 caused a self-destruct mechanism to be activated, which took the lives of Pchelka (pictured here) and her crewmate Mushka. (Photo: courtesy of George Meyer)

The renaming of dogs for some flights was common and appears to have been fairly impromptu. It is possible that Zhulka (Mutt) and Shutka flew on this mission, but at the last minute were accorded the more glamorous names of Zhemchuzhnaya (Pearl) and Kometa (Comet). Unfortunately, these dogs were in for a short ride and a harrowing ordeal.

Siberian weather and self-destruct mechanisms

During launch, the new third stage of the rocket malfunctioned, cutting off thrust prematurely and failing to lift the capsule to orbit. Instead, the spacecraft's emergency

Mushka lost her life aboard the ill-fated Korabl-Sputnik 3. (Photo: courtesy of George Meyer)

escape system separated the capsule at an altitude of about 133 miles as it inscribed a ballistic arc across the Soviet Union. A sour mood settled over everyone back at mission control as they assumed the capsule and its canine passengers had all fallen victim to the same self-destruct mechanism that had claimed KS-3 [2].

Then, unexpectedly, a phone call came into mission control reporting that some of the homing stations were picking up a radar signal from the spacecraft, which appeared to have landed safely in Siberia near where the famous Tunguska meteor impacted in 1909. Apparently, the self-destruct mechanism had failed to do its job. This was more cause for alarm than celebration. Aside from an errant trajectory triggering this explosive device, it could also be activated by a timer. If the capsule was not retrieved within 60 hours, it would explode, killing the dogs.

What unfolded next ranks as one of the most harrowing animal recoveries on record. With precious hours ticking away, a dozen experts, led by Arvid Pallo, head of search and rescue, converged on the landing site where search planes had located the downed capsule [2]. The recovery crew commandeered a helicopter in the nearest town and flew to the site, where they had to jump from the hovering aircraft into waist-deep snow. Temperatures hovered at −45°F, and the short winter day was rapidly losing light.

A bundle of charred wires hanging from the exterior of the spacecraft hinted at the problem. The capsule containing the dogs had not ejected from the recoverable spacecraft. It would later be determined that the Vostok capsule had jettisoned itself from the rocket while the third stage was still firing. The dog capsule, which was to have been ejected during descent, subsequently failed to do so. The ejection rockets fired at the same time as the spacecraft hatch, instead of after a 2.5-second delay, causing the mechanism to jam in place [8].

Frost covering the portholes prevented the team from seeing the dogs inside, and knocking on the container got no response. The first order of business had to be disarming the self-destruct mechanism as well as the capsule's ejection rocket motor and the parachute deployment pyrotechnics. While some of the team took cover behind trees, Pallo and an explosives expert managed to disarm the charges.

But the drama was not yet over. The rapidly fading daylight necessitated a quick return to the airfield, and the recovery of the dogs had to wait until the following morning. That evening Pallo got two frantic calls from Korolev seeking information about the capsule and the dogs, information Pallo was unable to provide.

Miraculously, the following morning the dogs were recovered alive. They had spent three frigid nights in the capsule after enduring an estimated 20 g's during re-entry, followed by a hard-landing. The Russian press welcomed their successful mission and said nothing about the mishap.

THE FINAL HURDLES

The rush was on during the opening months of 1961 to perfect the Vostok capsule for its manned orbital flight, with intensive ground-testing of all systems. The imminence of an American Mercury suborbital flight pushed the Soviet development schedule. The capsule, the Vostok 3A, which weighed in excess of 5 tons, had now evolved to what it would be for the manned mission.

Ivan Ivanovich flies

For final confirmation of the capsule's systems, two more preliminary launches were scheduled, each planned for a single orbit and each carrying a single dog, a large payload of biological specimens and a life-sized mannequin cosmonaut, christened Ivan Ivanovich. The mannequin would be dressed in the same SK-1 pressure suit to be worn by the first cosmonaut.

The first of these flights, Korabl-Sputnik 4, was launched on 9 March 1961, carrying the dog Chernushka, along with 40 black and 40 white mice, guinea pigs, reptiles, human blood samples and cancer cells, plus various micro-organisms and plant seeds. Chernushka was enclosed in a pressurised sphere along with some of the biological experiments, while the mannequin rode in the ejection seat, with other biological specimens stowed in its chest, stomach, thighs and elsewhere.

The single-orbit mission took place without any of the major problems that had troubled many of the earlier flights. On re-entry, the mannequin was ejected for a parachute descent, while the dog descended with the capsule. Once again, snowy weather hampered the recovery. General Nikolai Kamanin and Vladimir Yazdovskiy led the recovery team, which travelled by plane, trucks and finally horseback. When they returned with Chernushka to a nearby village, a huge crowd of farmers and children had gathered to see the dog that had flown in space. Later, Yazdovskiy phoned Moscow with the good news of the safe recovery. The glowing success of the mission gave a much-needed boost to the Vostok programme.

One curious story came out of the Chernushka flight: after the flight, members of Chernushka's recovery team found a wristwatch fastened to her leg. This was unexpected, unauthorized and a mystery, until they noticed the inscription on the back and traced it to its owner, Dr. Abram Genin from the Institute of Aviation and Space Medicine. He told the story of the watch in a 1989 interview conducted by the Smithsonian Institution.

When Genin graduated from the military academy, he had been given a Pobeda watch, the first model produced after the war. He had grown tired of the watch and wanted to get rid of it, but the watch proved to be exceptionally rugged. "I swam in the sea with it, I dropped it on the floor, but it still worked and resisted all abuse." Just before the Chernushka flight, Genin fastened the watch to the dog's leg, hoping that he would never see it again. But, following the safe recovery of Chernushka, the watch was traced back to him and he was severely reprimanded because there was a strict inventory of what was allowed to go into space on the flight. Some journalists later made up a story that Genin wanted to check if the watch would work in zero gravity, which was not true, since he knew that it would. He was still wearing the watch at the time of the interview in 1989 [9].

Dress rehearsal for a manned flight

The final rehearsal for a manned flight occurred some 2 weeks later on 25 March, following a virtually identical script. The six key cosmonauts then in training flew to the Baikonur launch site on 17 March for additional training and to view pre-launch preparations. The dog, Zvezdochka (Little Star), and the previously-flown mannequin Ivan Ivanovich, along with a collection of biological specimens, were launched aboard Korabl-Sputnik 5, achieving an orbit similar to that chosen for the imminent manned flight. Aside from the now standard snowstorm at the recovery site that caused a 24-hour delay in the retrieval of the dog, the flight once again came off without problems.

Three days after her flight, Zvezdochka was back in Moscow for a special event. At a press conference at the Soviet Academy of Sciences, she went on display with four other dogs that had flown orbital missions, as well as a litter of Strelka's pups, born the previous November.

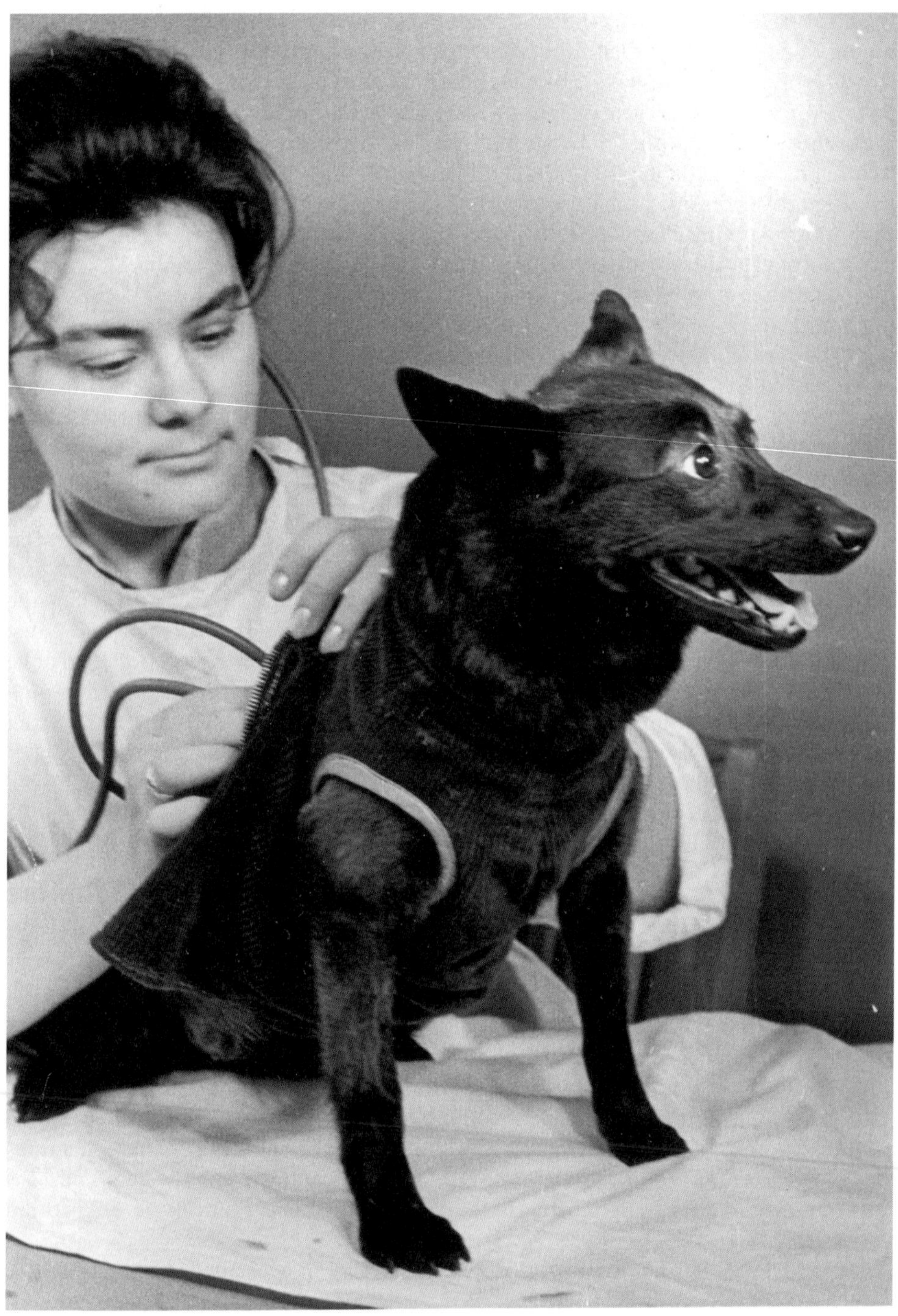

Chernushka is given a clean bill of health following her orbital flight in March 1961. (Photo: authors' collections)

Zvezdochka (Little Star) flew the last orbital mission prior to the historic manned flight of Yuri Gagarin. (Photo: authors' collections.)

Although dogs would be called upon for one final record-breaking orbital mission 6 years later, the March press conference brought a close to the space dog programme, which by then had been in operation for 10 years. Nearly four dozen dogs had flown more than three dozen suborbital and orbital flights, with 18 losing their lives. Within 2 weeks of this press conference, the Soviet space programme would achieve its historic goal of placing a human in orbit, when Yuri Gagarin was lofted into space by an R-7 rocket, strapped inside a Vostok capsule. The space age now officially belonged to human beings.

Like their suborbital colleagues, the "satellite dogs" had performed a critical role leading up to this historic moment. They had established the physiological profile of changes that occur in an organism during launch, re-entry and stays in weightlessness. Their role was also crucial in helping to develop the medical and life-support systems to be used in the first manned spacecraft.

THE FINAL CANINE MISSION

In the early 1960s, following successful manned orbital flights, both the Soviet Union and the United States set their sights on sending a man to the moon, the Soviets with

Veterans of the orbital dog missions went on display at a March 1961 press conference. Chernushka stands far right and behind her is Belka. Centre rear is Zvezdochka, fresh from her orbital mission. The proud mother, Strelka, sits front left, surrounded by her litter. (Photo: authors' collections.)

Soyuz-derivative spacecraft and the U.S. with its Apollo programme. But, both countries would take an intermediate step with a series of multi-crewed space flights that would help to develop the hardware and procedures required for a moon shot.

The passing of Korolev

From the start, the Soviet Voskhod programme was a risky venture. Aside from pursuing an accelerated schedule to outdo the American Gemini programme for new space records – first multi-crewed flight (Voskhod), first spacewalk (Voskhod 2) – problems arose with the launch vehicle and the capsule's life-support system. In addition, the programme also had to contend with the loss of two individuals critically important to its success.

During the flight of the first Voskhod, launched on 12 October 1964, Soviet premier Nikita Khrushchev was removed from power. His desire to beat the Americans had been the political engine behind Voskhod. Then a second stroke of misfortune befell the programme. Before Voskhod 2 launched in March 1965, the

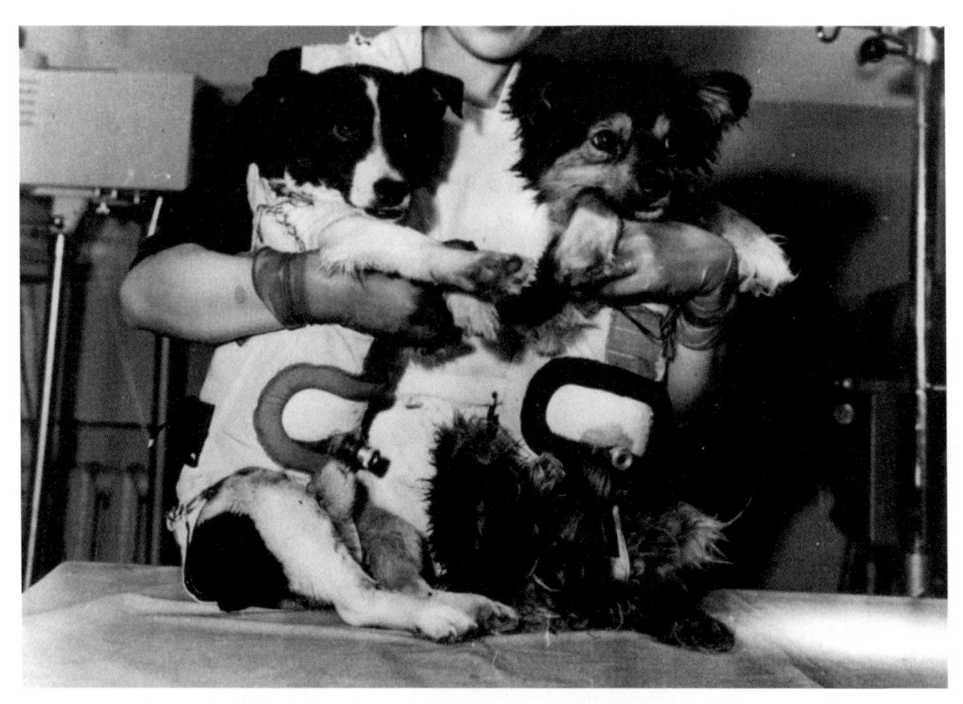

Prepared for their 22-day flight aboard Cosmos 110, Veterok and Ugolyok have feeding tubes connected directly to their stomachs. The dogs were also presented with food by mouth. (Photo: authors' collections)

guiding light of the Soviet space effort, Chief Designer Sergei Korolev, died unexpectedly at age 59. Korolev's desire to keep a step ahead of the Gemini programme had been an equally strong motivating factor in sustaining Voskhod.

Despite the loss of these advocates, several Voskhod missions still remained on the schedule: two would be long-duration flights to study the effects of long-term weightlessness and another which would utilize an all-female crew. The only factor preventing the quick launch of Voskhod 3 was concern for the performance of the capsule's life-support system. Voskhod used the same life-support system as the earlier Vostok capsule, which had been designed to support one pilot for up to 10 days. The plan for Voskhod 3 was to fly two cosmonauts for 18 days.

Twenty-two days in space

Before Cosmos 110 could launch on 22 February 1966 with the dogs Veterok and Ugolyok, the Soviets scored another stunning first with a soft-landing on the moon. Their Luna 9 probe touched down on 2 February and sent back the first surface images of the moon's barren landscape. Cosmonaut Gherman Titov would later opine to the Associated Press that dogs might visit the moon before humans [10]. But, first on the docket was the 25-day flight of Cosmos 110.

Ugolyok poses inside the container that sustained him while in orbit. (Photo: authors' collections)

The mission's biological payload included bacteria, yeast, blood serum samples, protein growths and *Chlorella*. Along with the dogs, they would be exposed to extreme levels of radiation as the craft's elliptical orbit took it within the Van Allen radiation belts. As protection, the dogs received anti-radiation medication. Food was delivered both through an automated feeding tray and through tubes directly into their stomachs.

Although the life-support system performed reasonably well, by the 20th day air quality had begun to decline, and a decision was made to terminate the flight.

Returning after 22 days in space, a new duration record, the dogs were in poor health. Dehydration was the most obvious ill effect, the dogs having lost 30% of their body weight. Additionally, the long exposure to weightlessness, as well as the period of restraint, had caused a host of other health problems that alerted scientists to the difficulties astronauts would face during long missions.

By this point, however, the capabilities of the Voskhod's life-support system had become irrelevant. The Voskhod programme had outlived its usefulness, pushed off the agenda by lunar probes and the higher priority of the Soyuz programme. All future Voskhod missions were cancelled.

The realization of the deleterious effects of prolonged stays in weightlessness became the most important finding of this final dog mission. Now that the effects of long-term space flights were being catalogued, on this flight and the American Gemini flights as well, the stage was set for research into how and why these physiological changes occurred.

It was findings such as these that would soon lead to the age of the biosatellite, flying long-duration, orbital missions devoted solely to the study of the effects of space flight on living organisms.

THE FRENCH SPACE CONNECTION

At a time when worldwide attention was focused on the higher-profile rocket flights of animals from Russia and the United States, the French had quietly set about conducting their own series of launches using live test subjects. Although very little has ever been written on the subject, and they did not have the benefit of captured German rocket engineers and scientists, the French nevertheless became the third nation to achieve space power status. They were also the third nation to send animals into space.

Rats and cats and pig-tail monkeys

Earlier balloon flights into the upper atmosphere had tested the effects of cosmic rays on rats and cats; while significant post-war progress in aeronautics and data recording systems had allowed researchers to obtain a better understanding of human physiology and dynamic stresses on pilots in the air. Drawing their inspiration from related experiments in the United States and the Soviet Union, French scientists became increasingly interested in the effects of weightlessness and cosmic radiation, seeking a better understanding of the impact these and other issues might have on warm-blooded creatures before human beings could venture into space.

In 1949, construction began on a new missile launch centre located in the Sahara Desert in southwestern Algeria, then a French colony. It was built at a remote site known as Hammaguir, 75 miles to the southwest of Colomb-Bechar and near the border with Morocco. This area provided the vital prerequisites of isolation and solitude, not just from a security aspect but in the possible event of booster failures after lift-off, which would litter the skies and surrounding area with hazardous debris.

The CIEES (Inter-arms Special Weapons Test Centre) launch facility site would become operational the following year. Eventually, four launch bases were established at Hammaguir, named Blandine, Bacchus, Beatrice and Brigitte [11].

Also in 1949, the French directorate for the study and manufacture of armaments (DEFA) proposed the construction of a new liquid-fuelled sounding rocket that would permit relatively inexpensive studies of the upper atmosphere. Originally known as Project 4213, the 20-foot rocket was developed at Vernon in northern France and given the series name Veronique (VERnon electrONIQUE).

The Veronique rockets

Initially, Veronique sounding rockets developed sufficient thrust to send a 130-pound payload to an altitude of around 40 miles. They were fuelled by nitric acid and kerosene, although later in the rocket's 9-year development turpentine would replace kerosene in the fuel mix [12].

Following test-firings at other ranges beginning in 1950, the first full-scale Veronique-N launch took place on 20 May 1952, with the rocket successfully reaching its intended altitude.

In all, eleven Veronique-N launches were carried out at Hammaguir over the following year in three series of test-firings, but several failed due to instability problems. Project scientists were also unhappy with the relatively low height the rocket could attain. The solution was easily remedied by lengthening the fuel tank to achieve additional burn time (which also increased the length of the rocket), allowing the rockets to reach altitudes of around 85 miles. Further tests of the modified NA version were successfully carried out between February and October 1954 [13].

For the International Geophysical Year of 1957–58 (known in France by the acronym AGI), the French National Defence Scientific Action Committee provided funding for the manufacture of 15 Veronique AGI-series upper-atmosphere sounding rockets, specifically dedicated to biological studies. Eventually seven of these would carry live animals.

Establishing CERMA

On 6 January 1945 the Centre d'Études de Biologie (CEBA) was established in Paris, with Robert Grandpierre serving as the centre's first director. He would set up research departments to undertake fundamental research on hypoxia (oxygen deficiency) at high altitude, and his teams would make significant progress. Ten years later, CEBA would merge with the Section Medico-Physiologique de l'Armée de l'Air (SMPAA) to create the Centre d'Enseignement et de Recherches de Médecine Aéronautique, or CERMA. Robert Grandpierre would also serve as the director of CERMA until October 1963.

When CERMA officially commenced operations in 1955, French scientists were forced to conduct experiments on weightlessness in laboratories. They did not have the luxury of suitable aircraft or rockets at their disposal, but – with the development

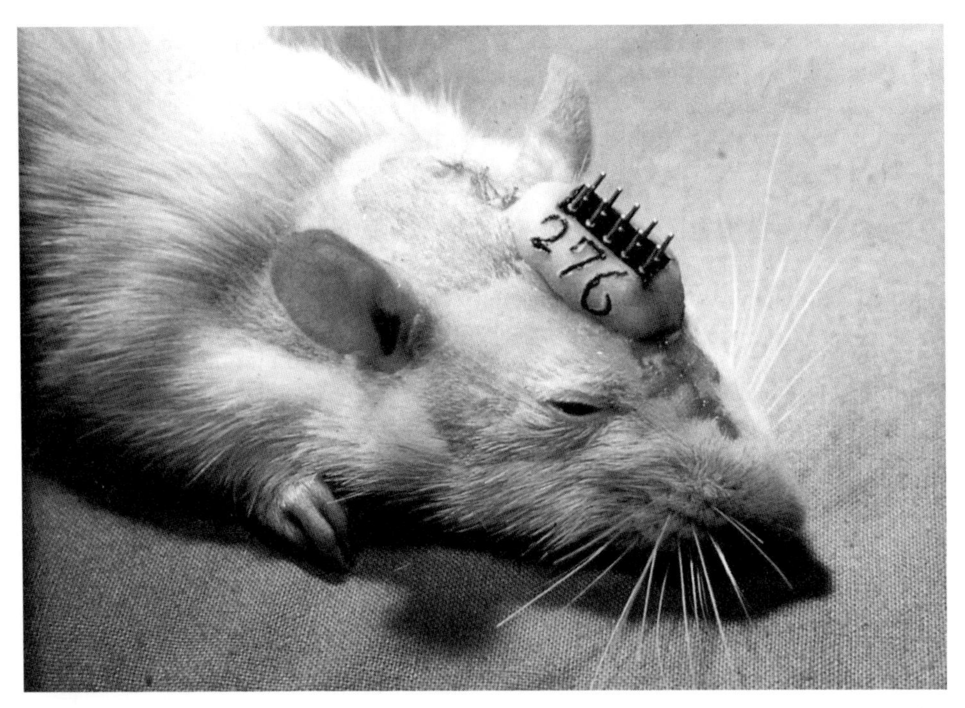

Wistar rat with an electrode surgically affixed to its skull. (Photo: courtesy Dr. G. Chatelier, CERMA)

of the Veronique sounding rocket and the creation of the Comité des Recherches Spatiales in January 1959 – new research opportunities opened up to them.

The following year, Grandpierre learned of experiments on the semi-circular canals of the inner ear being carried out by Belgian scientist Paul Ledoux. Ledoux had been successful in recording the electrical activity of the vestibular nerve of a frog undergoing rotational stimulation, and Grandpierre was fascinated by this research. He proposed to CERMA's department of sensory physiology that they undertake studies of the vestibular nerve in zero gravity, which would require the use of a research rocket able to offer up to 6 minutes of weightlessness.

Under the leadership of Dr. Gerard Chatelier, an eminently skilled surgeon, these experiments would provide the foundation of the French space biology programme [14]. Initially, following on Ledoux's work, frogs were used in laboratory experiments; however, because it was difficult to attach electrodes to the skin of the amphibians, rats were substituted. In November 1959, after several lengthy but unsuccessful attempts, Dr. Chatelier was able to surgically affix electrodes to the skull of a rat, enabling the activity of the brain cortex and reticular formation to be recorded. At the same time, other probes monitored the activity of the diaphragm muscle and recorded the electromyographic activity of the muscles at the nape of the neck, as well as the animal's heart rate, respiratory frequency and body temperature. Consideration could now be given to sending one of these rats into a truly weightless environment.

The French Veronique rocket was far less powerful than the V-2 and its American and Soviet variants that could hoist dogs and monkeys into the upper atmosphere, so the use of larger animals had already been ruled out. Under the auspices of CERMA, 47 white Wistar rats were selected for initial tests in Paris.

THE FIRST FLIGHTS

Neurophysiology tests required that each of the animals be prepared and trained well in advance of any rocket flight, which meant becoming accustomed to carrying out daily activities with electrodes surgically implanted into their craniums. The period in which each rodent could subsequently be used in the experiments was limited to between 3 and 6 months, due to the progressive polarisation of the intracranial electrodes, the aging of the rodent and necroses of the skull caused by the use of model glue in fixing the connectors [11].

Training for the rocket flight included painlessly hanging each rat by its tail inside immobilisation containers for incrementally lengthy periods. They would also be whirled around in centrifuge devices, exposing them to different accelerations and vibrations. A miniature amplifier was also developed to provide a simulated launch sound environment. Meanwhile a suitable sealed container was being developed that would be compatible with the Veronique's nose cone.

A PROGRAMME BEGINS WITH HECTOR

The first launch with a subject rat on board was scheduled for 20 February. There was a degree of trepidation about this launch, as a Veronique had exploded during the launch phase just 2 days earlier. Then, while the biological nose cone was being mated to the rocket, it was found that the rat had somehow managed to gnaw through the braided covering of recording wires in his container. The launch was pushed back to 22 February so repairs could be made and the wires repositioned. The disgraced rat was subsequently replaced, and a backup selected from the 10 who had been transported from Paris to Hammaguir.

Prior to launch, the selected rodent was fitted out in an adjustable linen vest resembling a tiny spacesuit that it had become accustomed to wearing during the training period. It was then inserted into a wide tube inside the custom-built container, suspended in the very centre by four metal hooks attached to the vest at one end and the walls of the tube at the other [15].

The installation of the biological container into the rocket's nose cone was carried out 40 minutes before lift-off. The flight plan had been carefully manifested, and like earlier Veronique firings was tentatively scheduled for 8:00 a.m. The countdown had begun 4 hours before the planned launch time.

Veronique AGI24 lifted off right on schedule from Launch Complex Blandine, but the rocket's engine fired for only 25 seconds, and the thrust was much lower than expected. As a result it only reached an altitude of 69 miles – less than half the desired

Staged photograph showing the vests used by the cats and rats in the French biological rocket flights. (Photo: courtesy Dr. G. Chatelier, CERMA)

Hector the French rat is inserted into the nose cone prior to his Veronique rocket flight. (Photo: courtesy Dr. G. Chatelier, CERMA)

Commander Brice, an engineer, and Dr. Chatelier (right) recover Hector from the nose cone of the Veronique rocket. (Photo: courtesy Dr. G. Chatelier, CERMA)

apogee. The rocket was unstable during the ascent, which resulted in a degree of pitching and rolling. Despite these problems all the recording instruments worked well, and data were collected throughout the brief journey.

Eight minutes and 10 seconds after lift-off and 28 miles from the launch pad, a parachute brought the payload back to the ground. It was safely recovered by one of two Lark helicopters at 8:40 a.m. When the capsule was opened by members of the CERMA team, they found the occupant alive and in good condition. Transported back to the Paris laboratory, the animal was kept under observation until 30 June. News of the flight was widely reported in France, and the rat was even dubbed "Hector" by the media. Some time later he achieved a further measure of fame as the namesake for a European cartoon series, "Hector the Space Rat".

Six months after his space flight, Hector was euthanised in order to study the possible effects of weightlessness on the implanted electrodes.

It would be 20 months before the next launch involving a second Wistar rat named Castor (called Beaver in some histories), who enjoyed the luxury of a back-up rodent named Pollux – the name of the other twin in the Gemini star constellation. Planned for 15 October 1962, the launch was delayed past the scheduled time due to high winds at the pad and minor technical problems.

Lift-off eventually took place at 9:39 a.m., with Veronique AGI37 achieving a peak altitude of 75 miles, although it strayed off course and travelled more than 37 miles beyond the expected recovery zone. A breakdown of VHF communications with the pilot of the prime recovery helicopter meant that the rocket's nose cone was not recovered until 75 minutes after launch. Meanwhile the rodent's container had been exposed to the intense ferocity of the desert sun, and he later died of heat prostration. Telemetry from the flight had been quite solid for the first 3 minutes of the ballistic journey, after which it was lost; although it is believed Castor travelled through a 6-minute period of weightlessness.

Pollux takes to the skies

Just 3 days later it was the turn of the second rat Pollux to be dangled from four wires attached to his vest in the biological canister and loaded aboard Veronique AGI36. Once again minor technical problems delayed the launch, which finally took place at 9:31 a.m. Unfortunately, the premature release of one of four guidance system cables connected to the end of its fins sent the Veronique on a wayward trajectory. The separation process itself was flawless, but the nose cone was well off course and finally touched down 88 miles from the launch centre. Two helicopters were immediately despatched to the recovery area, but to everyone's frustration they were unable to locate the nose cone containing Pollux or retrieve the instrumentation.

Once again, telemetry had been received for the first 135 seconds of the flight, before the signals were lost. Even though the altitude of the rocket's flight is recorded as 69 miles, the shallow trajectory could mean it was in fact significantly less, and the rodent may not have even experienced true weightlessness, especially if the spacecraft was spinning and otherwise unstable [11].

A CAT NAMED FELICETTE

In 1961, President Charles De Gaulle decided it would be advantageous to amalgamate a number of space-related authorities, thus creating the Centre National

Veronique AGI36 prior to launch. (Photo: courtesy Dr. G. Chatelier, CERMA)

d'Études Spatiales, or CNES. The following year, CNES issued a works order for an upgraded and far more powerful version of the Veronique, to be known as Vesta. This proposed rocket, 33 feet in length and 3 feet in diameter, would be capable of catapulting a far greater payload to an altitude in the vicinity of 250 miles.

Despite the loss of two of three rats, it was decided that the animal flight programme using Veronique rockets would continue while the Vesta was undergoing development. With space-saving improvements made in the biological capsule, Robert Grandpierre decided that cats could now be selected to ride the rockets.

Safe recovery

On 18 October 1963, exactly 1 year after the launch and loss of the unfortunate Pollux, a cat named Felicette was recovered following a successful flight into the very fringes of space. The female black-and-white stray was said to have been rescued from the streets of Paris by a pet dealer and later purchased by the French government as one of 14 candidates for a flight in the nose of a Veronique.

As part of their training each of the 14 cats spent time enclosed in noise boxes, allowing project scientists to assess each animal's reaction to a variety of sounds they might hear during a launch. The cats spent some time in a compression chamber and were also strapped into the cabin of a centrifuge which whirled them around at high speed to determine their tolerance to high g-forces.

Felicette came through these ordeals without suffering undue stress, despite earlier having a series of electrodes surgically implanted in her skull and brain during a 10-hour operation. The electrophysiological recordings would be far more comprehensive than those achievable with the smaller rats, taking in data on the left and left associative cortex, the right hippocampus (the part of the brain located in the temporal lobe) and the mesencephalic tissue [14].

Housed in a special biological container in the rocket's nose cone, Felicette was sent aloft from Hammaguir on her suborbital flight aboard Veronique AGI47 at 8:09 a.m. The rocket and its nose cone were explosively separated right on schedule, and the nose cone continued to penetrate the ionosphere with the impetus provided by the booster. The cat had been subjected to a force of 9.5 g's during the ascent. Having achieved a record height for the programme of 97 miles, Felicette became the first – and perhaps only – cat to experience the weightlessness of space.

Soon after, the capsule was plummeting back to Earth. Five minutes and 2 seconds after the descent had begun, the nose cone's fall was slowed when the braking parachutes deployed. The g forces acting on the cat fluctuated noticeably during the next stage of the controlled descent, peaking at around 7 g's before the main parachute blossomed. The capsule landed safely, and just 13 minutes after the launch a helicopter with the CERMA recovery team aboard touched down alongside. Felicette was recovered from the capsule without incident.

In days when most spaceflight milestones made for good headlines, the British press quickly dubbed the flight an outstanding achievement. However, the accompanying photographs of Felicette with electrodes surgically fitted to her head like a

A candidate strapped into the restraining apparatus that would be used on the rocket flight. (Photo: courtesy Dr. G. Chatelier, CERMA)

bellboy's skullcap were far from appealing to many readers, especially to those involved in the burgeoning animal rights movement.

A second cat, whose name has never been released, flew an encore mission aboard Veronique AGI50 six days later on 24 October. This time things definitely did not go to plan. On ascent the rocket became badly unstable and veered off course, staggering to a height of just 55 miles before exploding and plunging back to ground, crashing 75 miles from the launch site. An extensive helicopter search ensued and the nose cone was finally located 2 days later, together with the body of the anonymous cat.

There would be no more feline flights by any nation – unless the Chinese, in their notoriously clandestine way, placed a cat aboard one of their research rockets. In all probability, however, Felicette is the only cat ever to have successfully flown into space.

MONKEYS IN THE FLIGHT LINE

Following these biological flights using cat subjects, Grandpierre was finally given access to two of the more powerful Vesta rockets, which offered a much larger volume within the nose cone. This encouraged the CERMA team to consider the use of primates in their next series of flights.

French space cat Felicette with implanted skull electrode. (Photo: courtesy Dr. G. Chatelier, CERMA)

Selecting the candidates

Dr. Chatelier was given the task of selecting suitable candidates, and after exploring several possibilities he settled on female pig-tailed monkeys of the *Macaca nemestrina* species, which were both compact in size and docile by nature.

Members of this species are readily distinguishable by a short, 7-inch curled tail which they carry half-erect over their hindquarters – a unique characteristic and the genesis of their common identifying name. Mostly inhabiting the remaining jungle areas of Malaya, Sumatra and Borneo, these rare monkeys are calmer and more soundly oriented than other *Macaca* species such as the rhesus monkey. They also form strong bonds with their owners, and in some remote jungle habitations have been trained to climb coconut trees, twist off the fruit with their strong hands and throw it to the ground.

It would take several months to properly train the monkeys in Paris and prepare special suits for them to wear on their flights. Eventually, one was manufactured that would fasten the animal to its seat but also allow relative freedom of arm movement, which was needed for one of the planned experiments that would assess its consciousness and motor precision while in weightlessness. For this part of their training, the monkey candidates were secured in front of a panel which featured a small joystick.

As with the earlier animals, the monkey candidates had electrodes surgically implanted into their skulls. (Photo: courtesy Dr. G. Chatelier, CERMA)

When a light came on, the animal had to pull the joystick to be rewarded with a small food pellet. As the training intensified two more joysticks were added to make the task more complex, and finally five were mounted on the panel in a cross-like formation which the monkey had to pull in sequence.

Ten of the best-performing animals were transported from Paris to Hammaguir, where a final series of tests identified the two most suitable candidates, and they were prepared for the flight.

Martine lifts off

On 7 March 1967, all was in readiness to send the first of these monkeys, named Martine, on her journey into space aboard Vesta 04. Lift-off was scheduled for

Robert Grandpierre (left) and Professor Arlette Rougeul-Buser inspect the Vesta 04 nose cone. (Photo: courtesy Dr. G. Chatelier, CERMA)

6:30 a.m., but once again technical issues delayed the launch until 10:04. In accordance with the flight plan, the rocket's engine continued to burn until it reached an altitude of 22 miles just under a minute after leaving the pad, briefly subjecting Martine to acceleration forces reaching 10 g's. Residual impetus kept the Vesta shooting even farther into the skies. At an altitude of 69 miles and 100 seconds into the flight, the nose cone separated from the carrier rocket and continued to climb in the near-vacuum. Some $4\frac{1}{2}$ minutes into the flight, the capsule reached a peak apogee of 142 miles and began a quickly-accelerating descent back to Earth. At 2.7 miles from the ground, with the nose cone plummeting downwards at 31 feet per second, the drogue chute deployed, followed soon after by the main parachute. The capsule touched down at 10:57 a.m., $15\frac{1}{2}$ minutes after lift-off. Martine was recovered without problem, and a later flight report stated she "recuperated quickly and was in perfect health" [11].

Despite a few technical problems, it had been a good test of the rocket and its systems. Although the temperature in the monkey's capsule had risen from 25 to 33°C during the flight, thermostatically-controlled ventilation fans had begun operating, and Martine displayed no ill effects from the heat. Television cameras captured the monkey sitting strapped in her couch carrying out the trained task of pulling the small

Vesta 04 carries Martine into the sky, 7 March 1967. (Photo: courtesy Dr. G. Chatelier, CERMA)

Martine is successfully recovered safe and well after her space flight. (Photo: courtesy Dr. G. Chatelier, CERMA)

Dr. Chatelier photographed with Martine while the monkey was still recuperating from her flight. (Photo: courtesy Dr. G. Chatelier, CERMA)

joysticks each time a light blinked in front of her. Excellent telemetry had been recorded right up to 3 minutes before landing, the recorders and camera had worked well, and all other onboard systems had functioned as planned.

Another Vesta rocket was launched just 6 days later on 13 March with a second female monkey named Pierrette (Pebble). Planned for 7:30 a.m., the launch was also delayed, this time for 3 hours. It proved to be a totally successful mission, reaching 146 miles above the Earth, and the monkey was recovered in good physical shape. As before, all data recordings gathered by telemetry or onboard equipment were of superb quality.

Although three animals had been killed on flights through rocket malfunctions, an impressive amount of data had been collected and recorded, particularly neurological information relating to the electrical activity of the brain.

Following the Vesta flights, Robert Grandpierre left CERMA and took another position in Bordeaux. On 1 July 1967, the Hammaguir launch site would be abandoned and turned over to Algeria under an agreement ratified in 1962. A newer, far larger launch facility would then be established amid the tropical forests at Kourou, in French Guiana, South America.

Although further biological flights had been discussed by Grandpierre's colleagues and successors, these plans never came to fruition. The French programme of biological rocket flights was subsequently abandoned, apart from those later carried out in cooperation with the Russians and Americans, under the auspices of the CNES.

POLISH ROCKETRY

Even Poland flirted with the use of living specimens on rocket research flights. On 10 April 1961, just 2 days before Yuri Gagarin's history-making flight, two launches of somewhat less magnitude and global significance took place in the remote Bledowska Desert, midway between Krakow and Czestochowa in the country's south.

Known as the "Polish Sahara", and still the largest in the European complex of post-glacial dunes, the Bledowska Desert is rapidly diminishing under the vigorous encroachment of vegetation and a growing population. The area had once provided Field Marshall Erwin Rommel with a suitable training ground for his Panzer units, simulating the conditions they would encounter in North Africa. Today, the area where Polish rockets first took to the skies looks increasingly less like a traditional desert and more like a steppe, and is barely twenty square miles in area – a fifth of its size early last century.

Biological studies on mice

Standing just 57 inches tall and with a girth of just over $3\frac{1}{2}$ inches, the RM series of rockets may not have been overly imposing, but they would prove to be useful research vehicles. Developed both as a test of missile technology and a demonstration vehicle for later meteorological RASKO and Meteor rockets, they would eventually reach altitudes of around 62 miles.

The first test firings of the RM-2 were carried out on 7 September 1959 under the supervision of meteorologist Professor Jacek Walczewski, then head of the Polish rocket program. The following month a basic test postal payload was carried aboard the RM-2P rocket. In December 1960 the launch of RM-1A failed when the motor burst on ignition, but a second that month, using RM-2B, was apparently successful.

It was then decided to carry out some stomatological research (the science dealing with diseases of the mouth) using mice for biological studies. The principal researchers were Drs. Eugeniusz Gwizdek and Boguslaw Horodyski from the Krakow Medical Academy, who would later publish the results in a Polish medical journal.

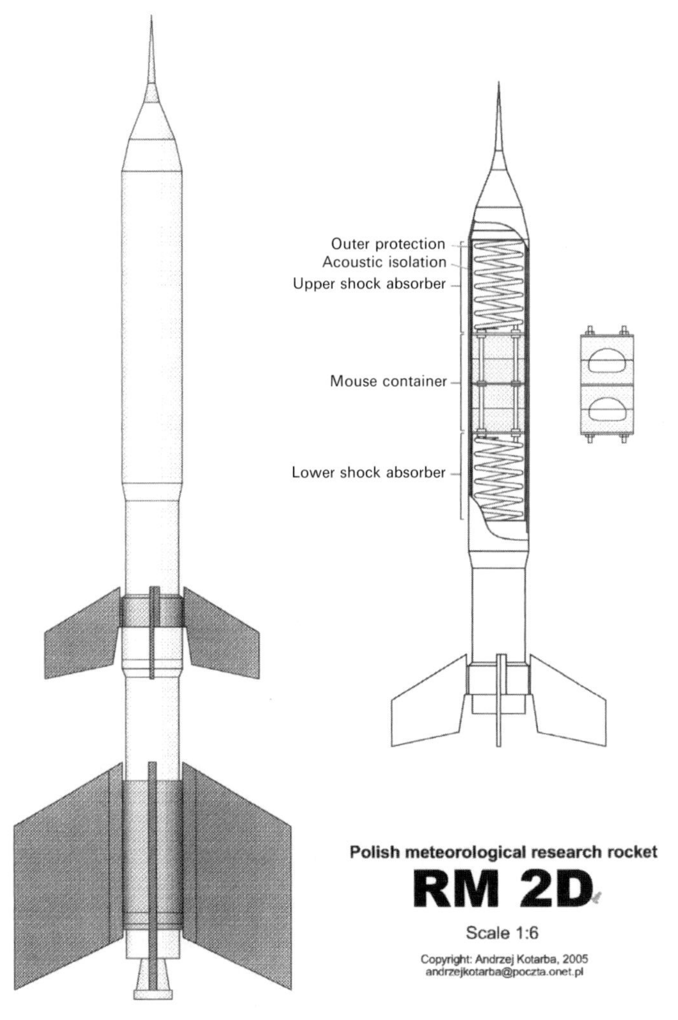

RM-2D Polish rocket. (Illustration: Andrzej Kotarba)

Two launches were scheduled for 10 April 1961 from portable, makeshift gantries in the Bledowska Desert. The first, using RM-2C, would have a flashing electronic light attached, while the second would carry two mice in a specially-designed container as well as a parachute recovery system. The container and parachute pack would add a further 9 pounds of payload to the vehicle, giving an overall weight of 35 pounds at launch.

Two unnamed mice were selected from a group of four candidates that had undergone centrifugal tests up to 20 g's, with the launch expected to produce acceleration resulting in around 17 g's. Their container, just six and a half inches long and three inches in diameter, was protected at either end by acoustic shields and shock absorbers, while carbon dioxide absorbing equipment had been installed to prevent the rodents from suffocating. The two mice were placed in separate holders: the upper holder made from moulded polystyrene foam, and the second manufactured from plastic.

RM-2C successfully rocketed out of its custom-built gantry and soared into the skies at 4:00 p.m. on the scheduled launch date. The precursory test flight had gone well. Then, just 45 minutes later, RM-2D also streaked into the air, the second-stage motor cutting out 30 seconds after launch. Now unpowered, the rocket's second stage reached an altitude of just under one mile before arcing over and tumbling nose-first back towards the desert. The parachute system functioned perfectly; just 84 seconds after launch the spent rocket with its live payload touched down in the sand. The mice were later examined and found to be unaffected and in good health [16].

While the RM launch program would continue, this proved to be the only known occasion on which live test subjects were carried on a Polish rocket. The successful flight of RM-2D and its mice passengers was said to have created a lot of interest in the Polish media at the time, but the manned launch of the first Vostok spacecraft just 2 days later largely overwhelmed any further news of local rocket research.

REFERENCES

[1] M.A. Gerd and N.N. Gurovskiy, *The First Astronauts and the First Scouts of Outer Space*, Translation Services Branch, Foreign Technology Division Wright Patterson AFB, OH, 1963. Translation of *Pervyye Kosmonavty I Pervyye Razvedchiki Kosmosa*, Akademiya Nauk, SSSR, Izdatel'stvo Akademii Nauk, Moskva, 1962.

[2] *Roads to Space, An Oral History of the Soviet Space Program*, compiled by the Russian Scientific Research Center for Space Documentation, translated by Peter Berlin, edited by John Rhea. Aviation Week Group, McGraw-Hill, London, 1995.

[3] Sven Grahn, "The flight of Sputnik-5, a.k.a. Korabl-Sputnik 2." Website: *http://www.svengrahn.pp.se/histind/sputnik5/sputnik5.html*, accessed 9 May 2006.

[4] Andrei Kislyakov, "Man's First Space Orbit," Russia Profile.org, April 2006. Website: *http://www.russiaprofile.org/culture/2006/4/12/3426.wbp*, accessed 5 May 2006.

[5] Asif A. Siddiqi, *Challenge to Apollo: The Soviet Union and the Space Race, 1945–1974*, NASA, Washington, D.C., 2000.

[6] Sven Grahn, "Sputnik-6 and the failure of 22 December 1950." Website: *http://www.svengrahn.pp.se/histind/sputnik6/sputnik6.html*, accessed 4 April 2005.

[7] Vladimir Gubarev, "Academic O. Gazenko: Wind of Cosmic Travels," *Nauka I Zhizn*, no. 7, 2001, pp. 30–37 [in Russian]. Russian title: "Akademik O. Gazenko: Vyeter Kozmicheskikh Stranstviy."

[8] Kamanin Diaries-1960, Encyclopedia Astronautica. Website: *http://wwwastronautix.com/articles/kams1960.htm*, accessed 17 May 2006.

[9] "Soviet Space Medicine, Smithsonian Videohistory Program, with Abraham Genin," Smithsonian Videohistory Program. Cathleen S. Lewis, Interviewer, November 29, 1989.

[10] "Soviets Orbit Two Dogs for Biological Study," *New York Times*, 23 February 1966, p. 1.

[11] Dr. Claude Timsit, Dr. Gerard Chatelier and Herve Moulin, presentation paper: "French Space Biological Experiments with Animals Before 1968," given at the 32nd History Symposium of the International Academy of Astronautics, Melbourne, Australia, 1998.

[12] Jean-Jacques Serra, "Rockets in Europe: Veronique and Vesta" (English version). Website: *http://www.univ-perp.fr/fuseurop/lrba_e.htm*

[13] "Veronique," Encyclopedia Astronautica (Mark Wade). Website: *http://www.astronautix.com/lvs/veronique.htm*

[14] Jean Timbal, "Space Biology at Hammaguir, 1961–1967," Aeronautical and Space Medicine Review (undated). Website: *http://www.soframas.asso.fr/Pagesweb/hammaguir/hamduk.htm*

[15] "Especialista en la Realidad" (Specialist in the Reality), author not named. Translated from website: *http://blogia.com/real/index.php*

[16] Jacek Walczewski, *Polskie rakiety badawcze* (Polish Research Rockets), Bibliotezka Skrzydtatej Polskie, Krakow, Poland, 2000.

9

Biting the hand

While most people would be unable to name any primate that has flown into space, those that do would likely give the names Ham or Enos. Both chimpanzees flew precursory flights in the lead-up to American manned space flights, with Ham preceding the suborbital flight of Alan Shepard, and Enos demonstrating that the orbital flight of John Glenn could go ahead.

ED DITTMER AND THE CHIMPANZEES

Master Sergeant Edward C. Dittmer was one of many outstanding aeromedical technicians who assisted in the bioastronautics research activities of the Air Force Systems Command at Holloman AFB. The young medical student from Laverne, Minnesota, had joined the U.S. Army in 1943 and served as a medical technician with the 770th Army Engineers in the Pacific Theatre before transferring to the Air Force medical field in 1947.

An "innovative experience"

Arriving at Holloman AFB in 1955 after a 4-year service basing in England, Dittmer reported directly to Captain David Simons at the Space Biology Branch of the 6571st Aeromedical Research Laboratory, working with Simons and the Man High balloon project during its 3-year development phase. In this capacity he assisted all the Man High pilots including Simons, Captain Joseph Kittinger and Lieutenant Clifton McClure as they made ready for their unprecedented 20-mile ascents above the Earth.

During one Man High low-level test flight near Lake Superior, it was only Dittmer's quick reactions and medical training that saved the life of pilot-elect Captain Grover Schock when a balloon car carrying Schock and the balloon's designer, Otto Winzen, suddenly plunged down and crashed to the ground. Both

Master Sergeant Ed Dittmer with a chimpanzee hooked up to a spirometer to check the animal's vital capacity. (Photo: courtesy Ed Dittmer)

men were critically injured. Schock was bleeding profusely from a deep cut across his throat when a ground-tracking pickup truck, fortuitously carrying Dittmer, arrived at the scene. He performed emergency first aid on Schock until an ambulance arrived to take the badly injured men to hospital, and his prompt, trained actions undoubtedly saved the man's life. But it was for his valuable work with two chimpanzees named Ham and Enos that Ed Dittmer is more widely known, and he says sharing this innovative experience with them was undoubtedly "one of the great highlights" of his life [1].

A demonstration flight required

By early 1961 things were moving rapidly in NASA's structured push for the first manned space flight. According to the space agency's official history, "This New Ocean: A History of Project Mercury," the technical outlook for the ambitious programme had greatly improved, even though the first attempts to launch the Mercury–Redstone combination were proving less than successful.

The end of the qualification flight tests was in sight, if only the Little Joe, Redstone and Atlas boosters would cooperate. First priority was to make sure the Mercury-Redstone combination was prepared for the first manned suborbital flights. Now, according to the progressive build-up plan, the reliability of the system required demonstration by the second Mercury-Redstone (MR-2) flight, with a chimpanzee aboard, as a final check to man-rate the capsule and launch vehicle [2].

Setting up the chain of responsibility

Meanwhile, Dr. James Henry had been appointed as an Air Force representative to a NASA committee that would establish and set in motion plans and procedures for precursory animal flights within Project Mercury. He was assigned the position of coordinator for these flights under Lt. Colonel Stanley White, a physician and the head of the Mercury medical team, and he became part of NASA's Space Task Group at Langley Field in Virginia. Henry's specific responsibilities included the establishment of an animal flight test protocol, developing operational flight plans, and overseeing the design and manufacture of the flight hardware. He would also monitor the progress of chimpanzee training at Holloman, headed by Air Force Lt. Colonel Rufus Hessberg, who had been appointed project officer for the STG's animal research programme at the 6571st Aeromedical Research Laboratory. Personnel there had begun training chimpanzees for space flight after the programme commenced in July 1959.

Initially, the 6571st set out to train and test 10 suitable chimpanzees from their colony. As with earlier programmes they began by incrementally conditioning the animals to restraint conditions that might be found in an actual spacecraft. However, their test facilities were limited. The NASA JSC publication, "Space Medicine in Project Mercury" by Mae Mills Link states that – while Holloman possessed sufficient animals, veterinarians and space physiologists – "it lacked facilities to obtain behavioural measurements of the animals."

Accordingly, arrangements were made to train several chimpanzees under contract with the Wenner-Gren Aeronautical Research Laboratory, University of Kentucky. Subsequently, Air Force personnel were transferred from the Unusual Environments Section of the Aerospace Medical Laboratory, Aerospace Systems Division, Wright-Patterson AFB. Also, arrangements were made with the Walter Reed Army Institute of Research to aid in the establishment of a comparative psychology branch at AMRL. Training of eight chimpanzees began with the use of standard operator conditioning equipment and special restraint chairs [3].

MERCURY–REDSTONE 2

Mercury–Redstone 2, or MR-2, would provide the first major test of several new design components in a spacecraft, including an environmental control system, or ECS. The Mercury craft would also be the first to be fitted with a pneumatic landing bag, designed to absorb the shock of landing on water. Manufactured from a plasticised fabric, this hole-filled, accordion-like device was attached to both the heat shield and the lower pressure bulkhead. Following re-entry, the bag and heat shield would automatically deploy and drop down about 4 feet like a large air bag to cushion the impact. That job done, the bag would then slowly fill with sea water, helping to stabilise the spacecraft in the ocean swell and keep it upright.

Training the candidates

Subsequent to working with Grover Schock on Project Mouse in Able, Dittmer became involved in working with the chimpanzee colony under the Space Biology Branch at Holloman, where he was the officer in charge. "Back then we got these small chimps from Africa – they were about a year old – and we started up a training project," he recalled. "Of course a lot of things were classified back then, so we had no real idea what we were training these chimps for, but we were teaching them to sit up and work in centrifuges, so it became quite evident that we were training them for use in missiles.

"We started out by teaching them to sit in these little metal chairs set about four or five feet apart so they couldn't play with each other. We dressed them in these little nylon web jackets which went over their chests and we could then fasten them to their chair. We'd keep them in the chairs for about five minutes or so and feed them apples and other fruit, and we'd progressively put them in their seats for longer periods each day. Eventually they'd just sit there all day and play quite happily."

Each of the chimpanzees was kitted out with one of these nylon "spacesuits", and would quickly become accustomed to wearing them. A diaper would also be worn beneath the nylon suit during lengthy training exercises [4].

All chimp candidates were progressively conditioned for flight durations up to 24 hours. (Photo: USAF, New Mexico Museum of Space History)

After the chimpanzees had become familiar with sitting in the steel chairs, Dittmer's team began securing them in individually moulded aluminium couches, much smaller versions of those that astronauts would one day occupy in Project Mercury. Once they had become comfortable with being strapped into the contoured couches the animals were introduced to a device mounted across their lap called a psychomotor, which was specifically designed to test their reflexes.

Trick or treat on the training machine

Apart from participating in tests of the spacecraft's life-support systems, one of the main chores assigned to the MR-2 chimpanzees would be to push levers on the device in sequence throughout the brief suborbital flight, in order to prove that astronauts could adequately perform similar tasks.

Three lights with three levers below were located just above the chimpanzee's couch. The first light was a red "continuous avoidance" signal, and it glowed all the time. The second was a white light, which would come on when the test subject pushed the lever below it. If they didn't do this every 20 seconds they would get a mild electric shock through metal plates attached to the underside of their feet. The third light was blue, and glowed for 5 seconds at irregular periods every 2 minutes. The lever beneath this had to be pushed before the light went out or the chimp would receive a shock.

Ham and Enos are shown practising on psychomotor machines. (Photo: USAF, New Mexico Museum of Space History)

In actual flight, this test would begin at lift-off and continue right through the flight, transcending periods of high g-loads and acceleration, weightlessness and re-entry.

Throughout their training, a corps of veterinarians closely monitored the health and well-being of the chimpanzees, keeping close track of their skeletal development with periodic exams and X-rays, as well as ensuring they were free of any parasites, and offering regular checkups of their heart and muscular reflexes. Diet and diet supplements were an important part of these tests, so the animals were given small doses of antibiotics stirred into their favourite cocktail – liquid raspberry gelatine. In fact, some of the primates enjoyed the diet and attention so much they began to pack on some excess weight, eventually washing themselves out of the programme by exceeding the weight limit of 50 pounds.

CHIMPANZEE SUBJECT 65

One of the standout primates, officially known as Subject 65 but affectionately called Ham, was a 3-year-old male, light-faced *Pan satyrus* chimpanzee. Ham is said to have

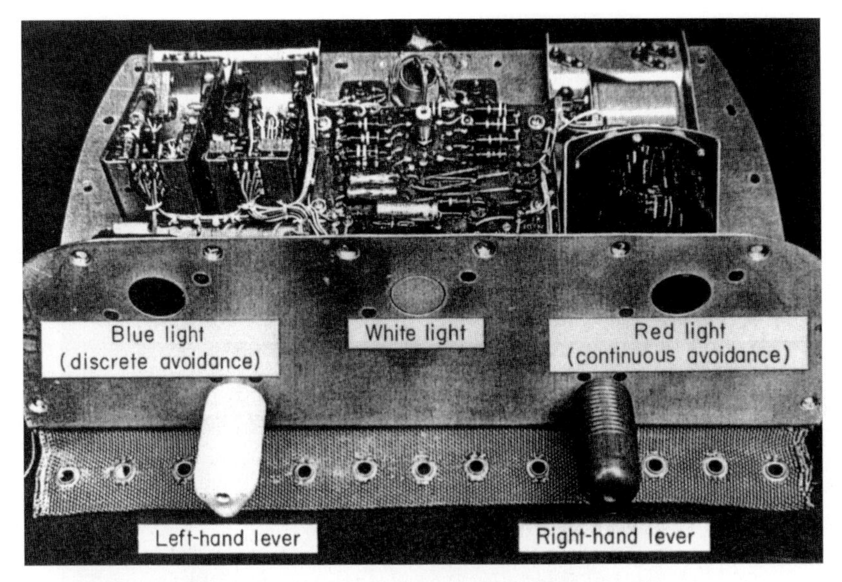

The MR-2 psychomotor panel. (Photo: NASA)

been born in July 1957 amid the dense rainforests of the French Cameroons in Equatorial Africa, and taken at a very young age by animal trappers. He had then been transported to the United States and procured by the U.S. Air Force from the now-closed Rare Bird Farm in Miami for the sum of $457. At the time of his purchase he weighed just 19 pounds [5].

Each of the chimpanzees delivered to Holloman had received an identifying number and pet name, so the colony was soon populated by chimps bearing such names as Caledonia, Duane, Elvis, George, Little Jim, Minnie, Paleface, Pattie, Roscoe and Tiger.

Ham was originally called Chang, but it was soon decided that this might prove mildly offensive to the Chinese, so his name was changed. According to popular history, he was named for the Holloman Aeromedical Research Laboratory, but as Ed Dittmer wryly points out: "Our lab commander at that time was a Lieutenant Colonel Hamilton Blackshear, whose friends called him Ham, so there may have even been a dual purpose behind that particular name."

"I had a great relationship with Ham," Dittmer recalls with obvious fondness. "He was wonderful; he performed so well and was a remarkably easy chimp to handle. I'd hold him and he was just like a little kid. He'd put his arm around me and he'd play ... he was a well-tempered chimp" [1].

Choosing the best candidates

Through their tests, the veterinarians at Holloman eventually winnowed the original field of 40 flight candidates down to 18, and a further evaluation brought this number

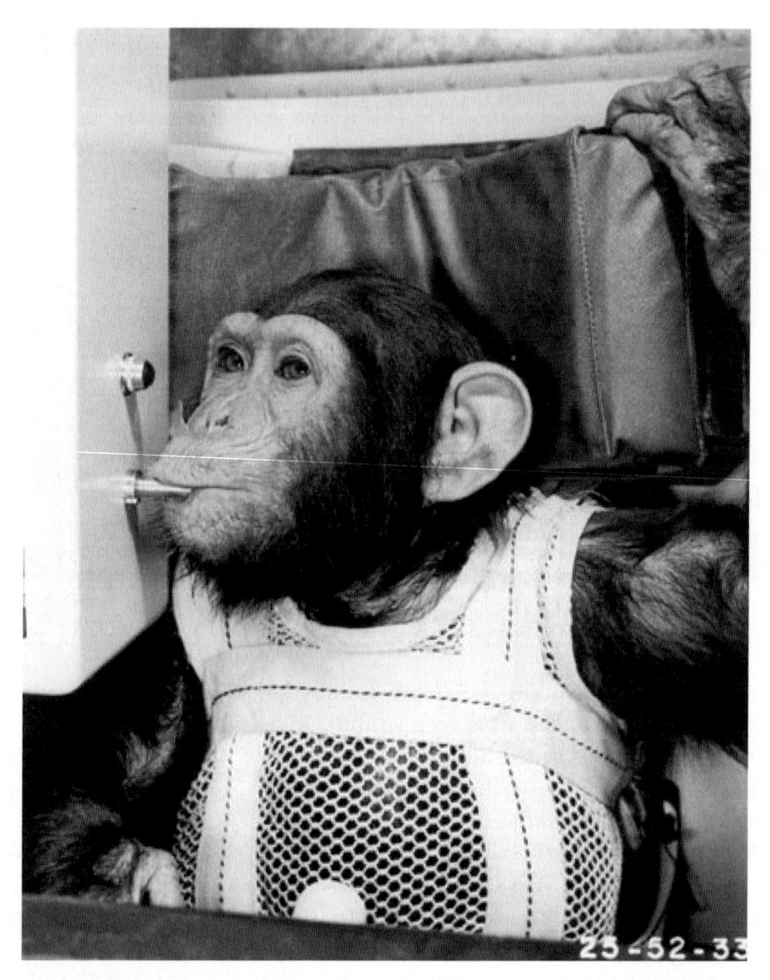

The reward for a good job – Bobbie Joe enjoys a drink of water. (Photo: USAF, New Mexico Museum of Space History)

down to 4 females and 2 males, including Ham. On 2 January 1961 the 6 chimpanzees, accompanied by 20 of Holloman's medical specialists and animal handlers, left Holloman and were transported to the Cape in Florida. Their arrival, 29 days before the launch, was needed to help stabilise the animals and to enable the preparation team to practise with the animals and the flight apparatus. They would also monitor spacecraft systems checks involving the chimpanzees.

Once at the Cape, the animals were divided into two separate colonies of three primates; each group occupying one of two large trailer vans parked inside a fenced-off compound adjacent to Hangar S. The Air Force had actually procured seven of these large vans; four were arranged to provide two animal training facilities and separate "living quarters" containing four cages. A fifth van was fitted out as a medical unit containing a surgical table and X-ray unit, while another became the

Subject 65, better known as Ham. (Photo: NASA)

transfer van for transporting the chimps out to the launch pad. The seventh trailer was used as a portable office and sleeping quarters for those on night duty. The two training vans each contained a mock-up of a Mercury spacecraft, psychomotor testing cubicles and flight preparation areas. It was felt prudent to separate the chimps into two training schools as a sensible precaution against common human diseases or illnesses, such as colds, mumps and measles, which could easily delay a flight.

"We did not want to be wiped out at the launch site and left with no candidates," explained programme official Lt. Colonel Rufus Hessberg. "What we were doing was

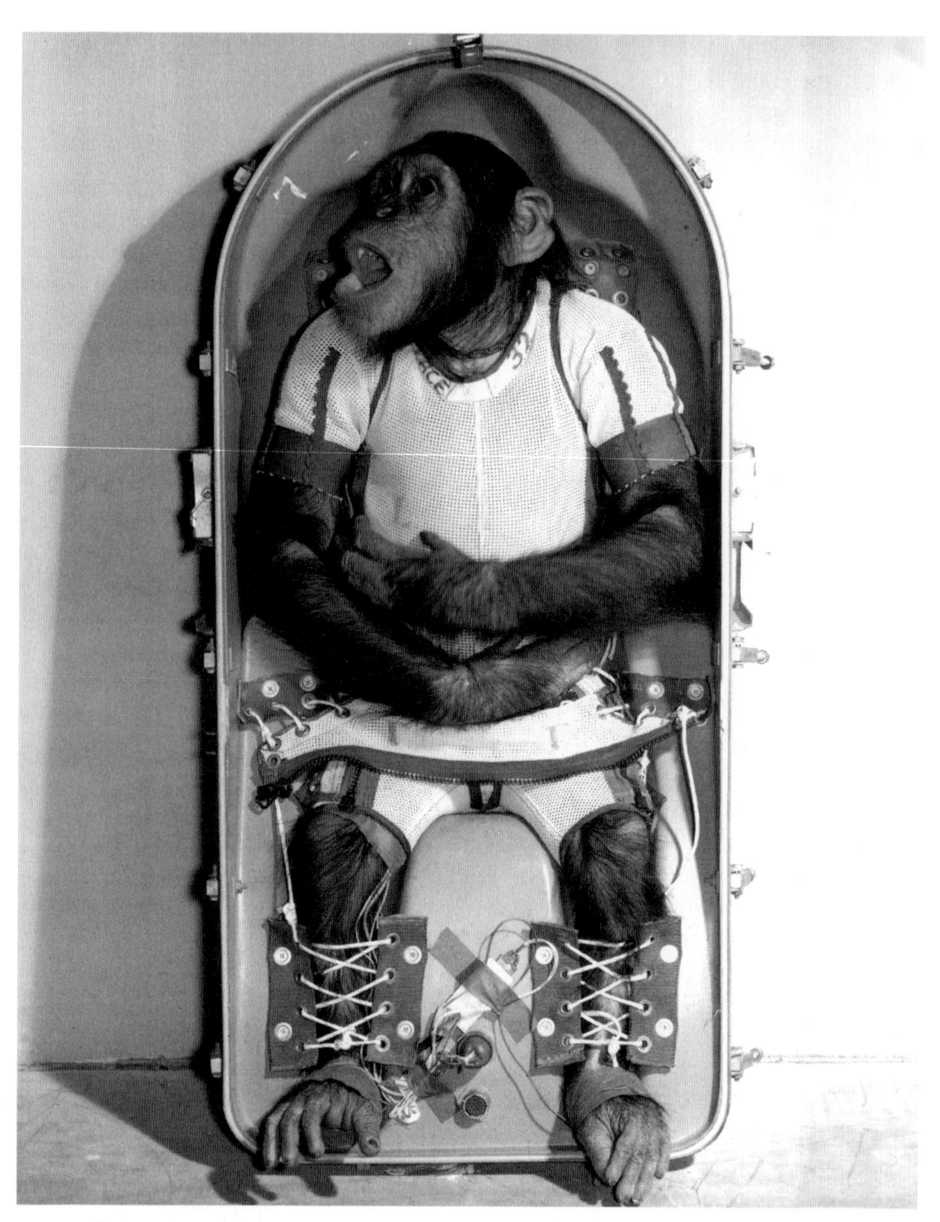

Ham is shown in the bottom half of his flight couch. (Photo: NASA)

similar to the care that was later taken with the astronauts. You have a back-up for each astronaut and they are fed from different kitchens with different foods and different people preparing it to preclude any possibility of an infection break-out" [6]. Even the handlers assigned to one group were not allowed to mingle with those tending to the other trio.

Five simulated countdowns would be conducted by members of the animals' preparation team. These consisted of exercises in preparing the animal and couch, and then proceeding to the top of the launch gantry. One of these countdowns would be used for a telemetry check, and another for a spacecraft pressurisation check. Radio frequency compatibility checks would be carried out on another of the countdown tests, and the remaining two would involve simulated flights.

Ham is given the task

The ultimate selection of the prime and backup candidates for the flight was left to James Henry and John Mosely. "We didn't know which chimp would be going until the day before launch," according to Dittmer. "There were six of them that were selected and they were all good, but Ham easily stood out as the best of the bunch." After Henry and Mosely had conducted extensive physical and training tests on all six chimpanzees, they selected Ham because of his good mood and alertness at the time, with Minnie (a.k.a. Number 46) as his backup. "The veterinarians picked Ham for his exceptional physical well-being and for his work under the actual test conditions," Dittmer added [1].

Weighing in at an age-average 37 pounds, Ham was destined to become the heaviest animal ever sent on an experimental flight, either by the Americans or the Soviets.

INTO THE UNKNOWN

Nineteen hours before the planned MR-2 lift-off, the two chimpanzees were being prepped in advance of the mission in one of the trailer vans, even though only one would fly the next morning. Launch day certainly had the makings of a favourable anniversary; exactly 3 years earlier America finally became a space nation when Explorer 1 had soared into the heavens. But Ham's spacecraft would not follow Explorer 1 into orbit; the Redstone's velocity was insufficient to send a Mercury capsule on anything other than a ballistic flight. A lot hinged on his flight, however, which would use the same combination of spacecraft and booster planned to send the first American into space later in the year. If the primate subject survived the flight and all the physiological results were nominal, the medical team would then clear the way for an astronaut to follow.

Ham prepares to make history

Early the next morning Ham sat down to breakfast with his handlers. While astronauts would enjoy a traditional pre-flight meal of steak and eggs, this spread was more to a chimp's taste. He quickly made his way through some cooking oil and flavoured gelatine, half a fresh egg, half a cup of baby cereal and a couple of spoons of condensed milk.

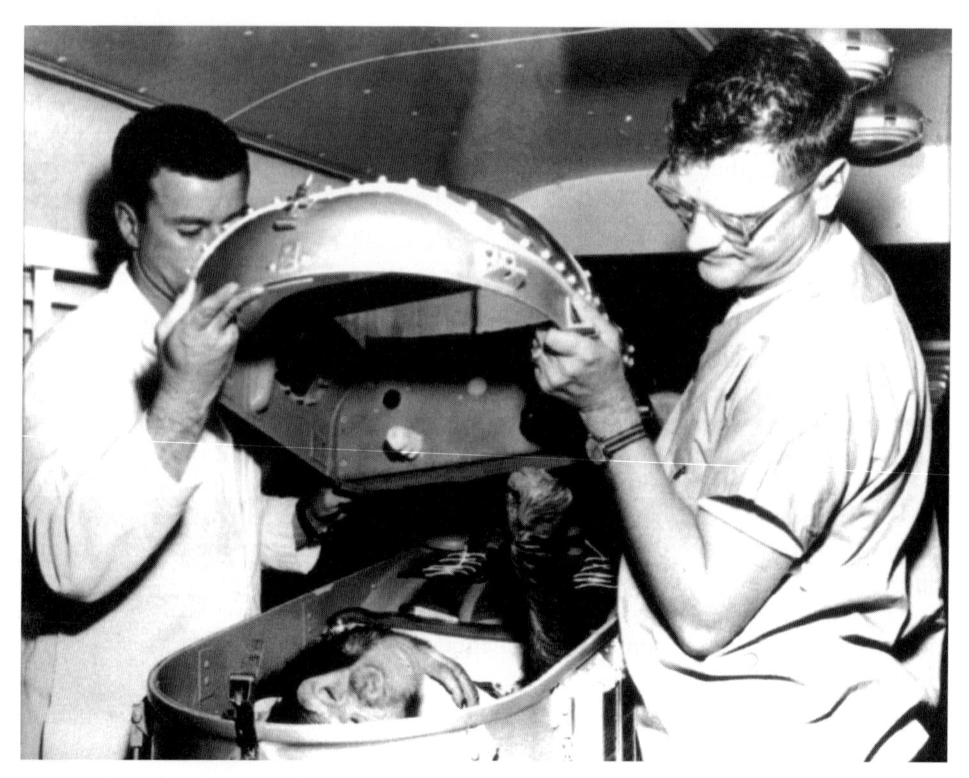

Ed Dittmer (right) helps to fit the lid onto Ham's container prior to the chimp's insertion into the Mercury spacecraft. (Photo: NASA)

Seven and a half hours before the anticipated lift-off time, both animals were once again given a thorough physical examination ahead of more sensor and psychomotor tests. Then the two chimps were dressed in their diapers, plastic waterproof pants and spacesuits and firmly strapped to their couches. Once all the tests had been completed, Ham's open container was loaded into a transfer van and driven out to the launch pad.

"My particular job at this time was to do with the medical aspects of Ham's flight," Dittmer recalled. "I'd put the electrodes on for the EKG in-flight as well as the respiration sensor and I'd hook those electric shock plates to his feet." Ham's arms had been left free so he could perform his assigned psychomotor tasks. "Then we dressed him and secured the lid before conducting pressurization tests. After making sure there were no leaks we checked Ham's physiological readouts while he pulled away at his levers" [1]. To help ensure the animal was entirely comfortable with all his equipment, the psychomotor he would be using in space was the same unit on which he had been practising his routine.

Ninety minutes before lift-off, Ham was still deemed the better behaved and more animated of the two candidates, so at this final hurdle he won his place in spaceflight history. He did not know it, but Ham was about to get what legendary flight controller Gene Kranz would later describe as "a hell of a ride" [7].

MR-2 and a primate passenger

Just before 6:00 a.m. on 31 January, Dittmer and task scientist Captain William E. Ward gingerly stepped out of the transfer van carrying Ham's container and made their way over to the gantry elevator. Ever so slowly, they were taken up to the gantry's third level where the Mercury spacecraft stood open and ready to accept its primate passenger. The two men inserted and locked down the biological container, after which Dittmer took a last look at his charge through the small window above Ham's face before the hatch was secured at 7:10 a.m. "It looked like he was smiling at me," he reflected with affection [1].

The Redstone's engine was expected to fire for approximately 140 seconds and achieve a speed of around 4,400 miles an hour. With all the rocket's fuel consumed, the Mercury spacecraft would separate explosively and reach a final altitude of about 115 miles. On ascent it was thought that Ham might experience around 9 g's, followed by 5 minutes of weightlessness after separation of the capsule from the Redstone as his spacecraft arced over at the apex of its flight. The three retro-rockets attached to the craft's heat shield would then fire in rapid succession, after which the retropack would be jettisoned and the automatic stabilisation and control system (ASCS) would manoeuvre the craft into a heat shield-down orientation prior to re-entry.

Delays and more delays

The launch had originally been scheduled for 9:30 a.m., but a series of engineering delays would frustrate flight controllers. First an electronic inverter began to overheat and the temperature started to rise in Ham's container. Flight Director Chris Kraft ordered the environmental and electrical control unit shut off to reduce the temperature to a more comfortable level for the chimp. Once the problem was thought to have been resolved the countdown resumed, but then the problem recurred. Once again technicians made their way out to the launch pad, finally giving the all-clear an hour later, as concerns were being raised about a band of inclement weather moving towards the Cape. Then, just as the crews began evacuating the gantry, the pad elevator jammed and had to be repaired. Once that had been done the pad area was cleared amid the blaring of klaxons and the countdown resumed at 11:00 a.m.

LIFT-OFF!

The Redstone booster finally roared into the air just 5 minutes before noon, arcing out over the Atlantic in what appeared to be a flawless launch. Soon after, however, flight telemetry indicators began to tell a different story. A faulty valve in the Redstone booster meant the fuel pump was injecting too much liquid oxygen into the engine, causing the rocket to over-thrust and accelerate faster than expected. As a result, the Redstone's trajectory was slightly steeper than predicted, throwing out splashdown calculations. The liquid-oxygen fuel was depleted a few seconds early, causing an abort sequence to begin. The escape system fired, tearing Ham's Mercury spacecraft

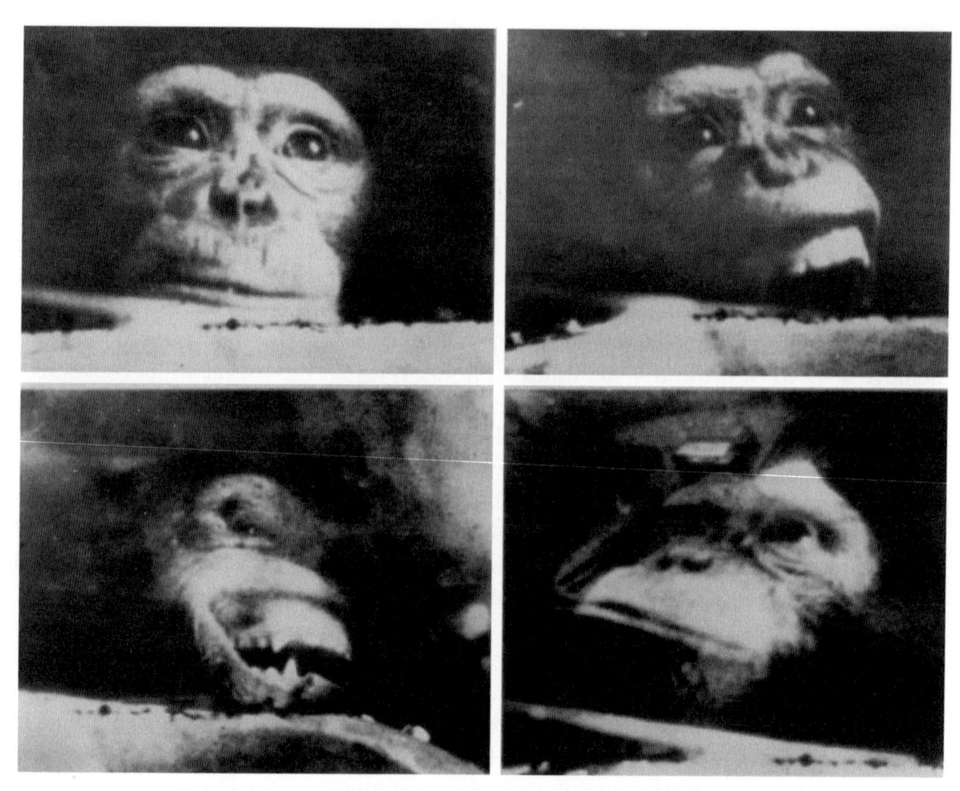

In-flight photographs of Ham taken during his suborbital mission. (Photo: NASA)

away from the spent booster with g-forces far greater than anticipated. Inside his pressurised container the world's newest spacefarer was subjected to a sudden, unexpected kick in the back of around 17 g's, causing him to temporarily forget his psychomotor duties. As the capsule flew into weightlessness he finally responded after a couple of small electrical jolts had zapped the bottom of his feet to remind him of the task at hand. Cameras recording Ham's movements then showed him starting to pull his levers.

Monitoring the flight

As Chris Kraft and his control team continued to monitor the progress of MR-2, he was informed that the fuel problem and resultant over-acceleration would carry the spacecraft an extra 42 miles higher and about 124 miles farther downrange, adding two more minutes of weightlessness to the mission. Of more immediate concern to Kraft was the fact that a faulty relief valve had caused the spacecraft's pressure to suddenly plunge from 5.5 to 1 psi. Fortunately, this would not impact on the occupant, as Ham was safely sealed in a pressurised container with an independent air supply. However, this life-threatening anomaly would certainly have created immediate concerns for an astronaut. Added to this was the unhappy fact that the retropack was prematurely jettisoned when the launch tower had been discarded after firing. As a consequence, the spacecraft would re-enter excessively fast and splash down even farther downrange.

William Augerson, a physician on duty in the Cape blockhouse, was monitoring Ham's physiological progress. He reported that despite all the onboard dramas everything was going well, and Ham was performing his tasks just as he had been trained. Weightless for more than 6 minutes, he only received those two small shocks throughout the entire journey for neglecting to push the correct levers on time. In this respect it was an almost-perfect rehearsal for a manned mission, proving that a human could easily perform manoeuvring tasks even when things did not go according to plan during any phase of the flight [7].

Heading for a splashdown

As MR-2 plunged backwards toward the sea, where a fleet of eight recovery ships was waiting, Ham began to experience a crushing 14.7-g load. Then, at 21,000 feet, a 6-foot drogue chute automatically deployed, which in turn dragged the 63-foot main parachute from its stowage at 10,000 feet. A SARAH (search and rescue and homing) beacon had been activated earlier when the escape tower pulled the capsule away from the spent booster. Tracking aircraft were circling the anticipated drop zone, continually monitoring this signal and guiding recovery ships to the expected impact area.

Seventeen minutes after lift-off the spacecraft smacked down hard in rough seas beyond the Atlantic missile range. The landing bag had deployed as planned, which helped to minimise the shock of hitting the water. On splashdown the main parachute was automatically jettisoned, some fluorescent green dye was released into the water to aid visual sighting and a high-intensity light began flashing at the top of the spacecraft. On impact with the water, a rim of the lowered heat shield snapped back so violently onto the hull that it breached the titanium pressure bulkhead in two places. Sea water then began to infiltrate the spacecraft. A cabin relief valve had also jammed open, and this allowed even more water to seep in. Just to compound the problems, the heat shield then tore loose from the bottom of the landing bag and sank. MR-2 slowly began to settle ever deeper into the tumultuous seas.

Recovering the capsule

Meanwhile the dock landing ship USS *Donner* was proceeding at flank speed to the reported landing area, together with U.S. destroyers *Ellison* and *Manley*. Twenty-seven minutes after splashdown a Navy P2V Neptune aircraft finally made visual contact with the floating spacecraft in the rough seas, but the long overshoot of nearly 132 miles meant that the closest recovery ship, the *Donner*, was some 60 miles away. It was nearly an hour before a helicopter, despatched from the ship, arrived at the scene.

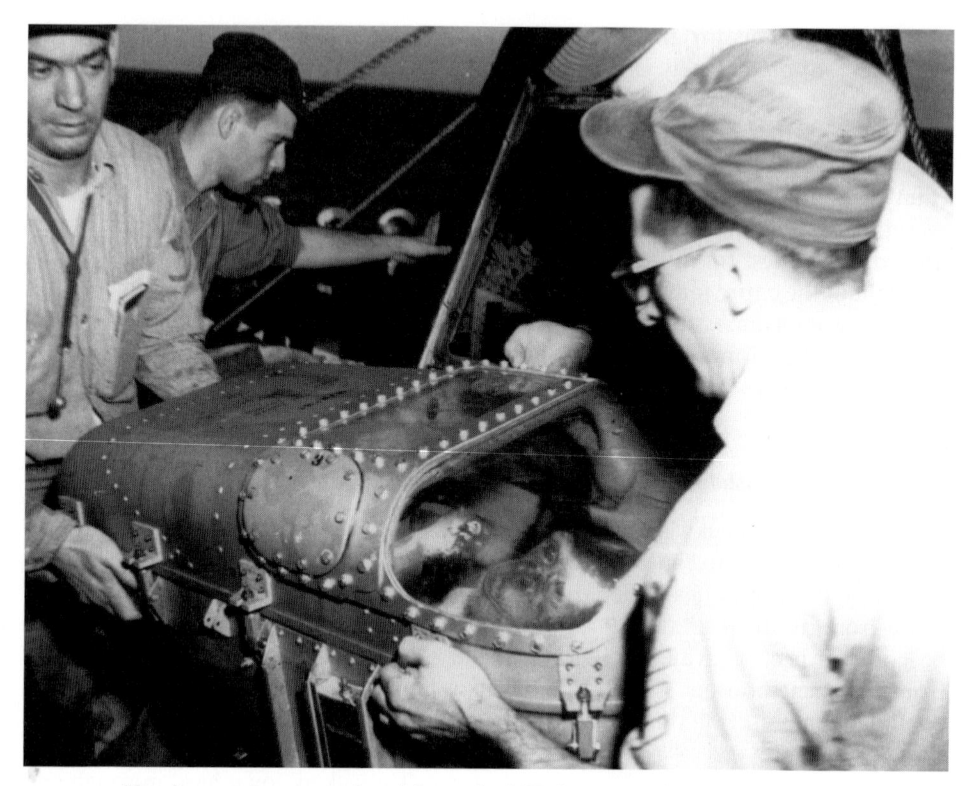

Ham's container is retrieved from the MR-2 spacecraft. (Photo: NASA)

Once they were hovering overhead, pilots John Hellriegel and George Cox alarmingly reported that the spacecraft was tilted on its side in a 7-foot swell, sitting much deeper than expected in the water. With no time to spare, trained frogmen quickly jumped out of the helicopter and attached cables to fixed points on the wallowing spacecraft. At 2:52 p.m. Hellriegel applied full power and slowly raised MR-2, streaming water, into the air. The *Donner* was now close at hand, so the precious cargo was transported to the ship and carefully deposited onto its deck, where willing hands had soon secured it in position. This good news was relayed to Cape Canaveral nearly 3 hours after lift-off [8].

It was later estimated the spacecraft had taken on some 800 pounds of water, but was otherwise in good shape. Meanwhile, Cape doctors were deeply concerned that Ham might have been injured during the crushing forces of take-off, or the hard splashdown. Half an hour later Ham's sealed container had been extracted and was resting on the ship's deck. One very confused and annoyed chimpanzee could be heard squealing his discontent from within. Oxygen was quickly fed in through a small hatch, which cleared the condensation within Ham's container, and he could be seen balefully peering out. Just 9 minutes after landing on the *Donner* the container was opened up and a relieved Ham was carefully unstrapped from his couch. Happily enough the water had not infiltrated his enclosed container; he did not know how close he had come to sinking ignominiously to the bottom of the Atlantic. He was obviously pleased to be free of the spacecraft.

A little shaken but safe

A little wobbly in the legs and slightly dehydrated, Ham was otherwise in excellent physical condition, although he had suffered a small bump to his nose during the rough splashdown. The experience had obviously not affected his appetite, as he eagerly devoured an apple in a post-flight snack.

After going through a medical examination in the ship's sick bay and a welcome-back photograph taken with the *Donner*'s commander, Richard A. Brackett, Ham was loaded onto the helicopter and transported to a forward medical facility at Grand Bahamas Island for further checks. The next day he was flown back to Cape Canaveral where hordes of reporters and photographers were eagerly waiting beside Hangar S for a glimpse of America's newest space hero.

Ham was quick to show his displeasure at this noisy, unwanted intrusion into his living space. He became agitated, bared his teeth and screamed at the melee of strangers. His handlers finally took the fretting animal back into the familiar surroundings of his van in order to quieten him down, but when Ham re-emerged he threw another tantrum as the crowd surged closer, some popping flashbulbs in his face. The handlers tried to get the reluctant chimp to pose next to a Mercury training capsule but he rebelled; he didn't want to go anywhere near the darned thing. America's latest astrochimp was definitely not impressed by his newfound fame!

BACK HOME AGAIN

The unhappy chimpanzee was finally returned to the chimp colony and familiar surroundings on 4 February. Over the next 2 years he was kept under scrutiny while performing assigned tasks in order to determine if there were any residual effects from his historic journey into space.

Enough of the glory

Ham later trained for a second mission, but it seems he'd had his taste of glory and showed very little enthusiasm for another ride on a rocket. Another chimp would make the flight. In 1963 Ham was transferred to the National Zoological Park in Washington, but he had a lot of trouble adjusting as he had spent so much time in the company of human handlers. During his time at the zoo he made a few well-received guest appearances on some television shows, and even had a cameo role in a movie featuring stunt rider Evel Knievel. However, the good life of retirement soon began to show in Ham's physique; by 1967 his pre-flight weight of 37 pounds had almost doubled.

On 25 September 1980 the famed space chimp was loaded onto a C-130 transport aircraft by men of the 156th Tactical Airlift Squadron, North Carolina Air National

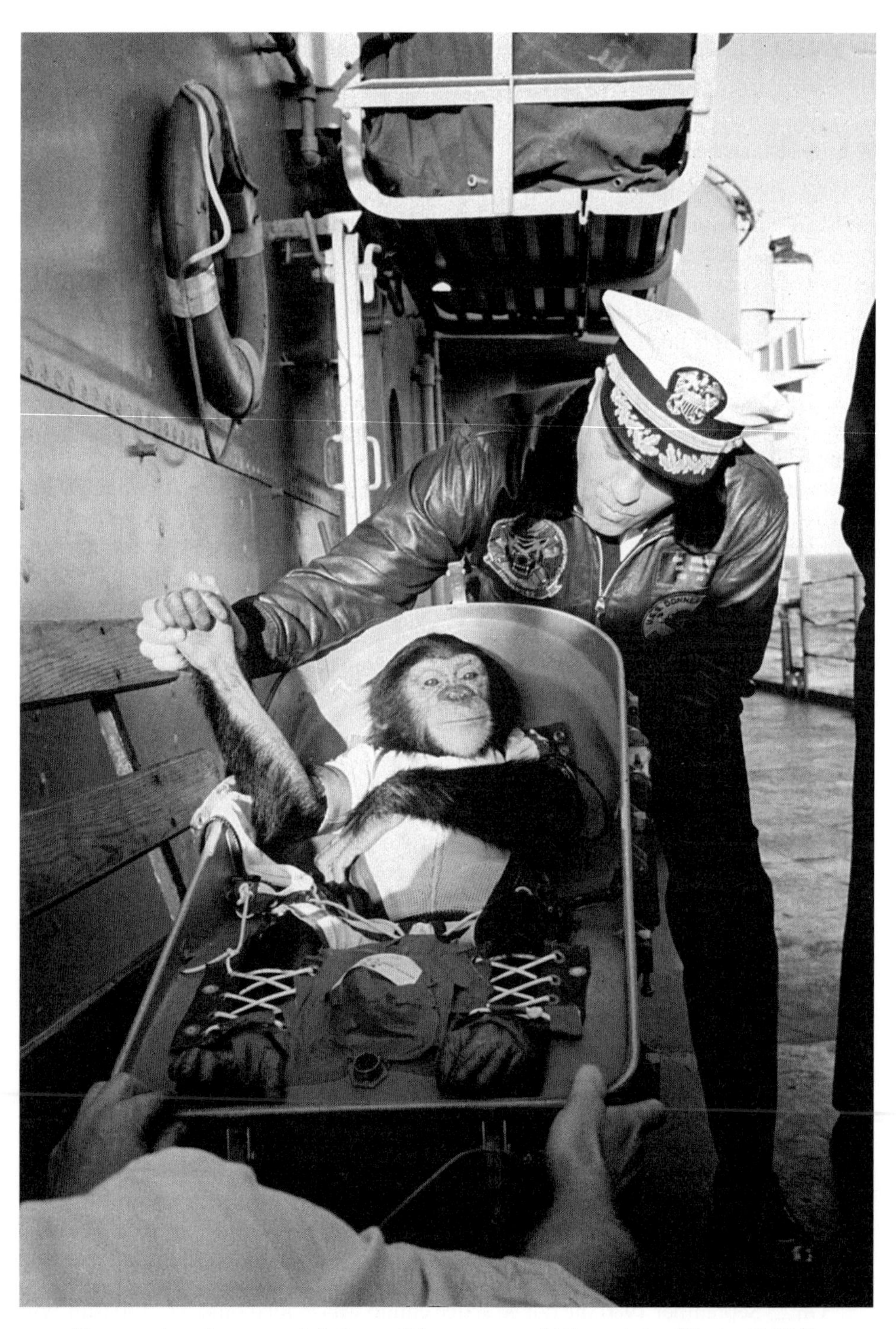

Ham receives the congratulations of the recovery ship's captain. (Photo: NASA)

An aging, overweight Ham in 1973. (Photo: NASA)

Guard, and taken to the North Carolina Zoological Park, Asheboro, on long-term loan from the U.S. Air Force and the National Zoological Park. By this time his weight had ballooned to an estimated 175 pounds. He would spend $2\frac{1}{2}$ years at the facility.

Death of a true space pioneer

On the morning of Monday, 17 January 1983, Ham's day had seemed quite normal. He was in an enclosure with two female chimps half his age named Maggie and Terry. All three ate their Purina Monkey Chow biscuits as normal. At midday Ham was separated from the females to an adjacent cage, which was normal practice when the keepers went to lunch and would not be in the immediate vicinity for some time. After lunch the chimps were placed together again, but later all three were shifted to the adjacent cage to allow for cleaning. According to a report prepared by the zoo, Ham was "a little lethargic but nothing noticeable for that time of the day. At 3:45 p.m. the keepers were checking the animals and Ham was slumped over with his back against the cage wall and his head bent toward his crotch. After taking necessary precautions to assure he was not comatose he was removed from the cage and pronounced dead" [9].

At the relatively young age of 26, Ham had succumbed to an enlarged heart and liver failure. On 19 January, under the terms of the agreement, his body was delivered to the Department of Veterinary Pathology, Armed Forces Institute of Pathology at Walter Reed Army Medical Center in Washington, D.C., where it was immediately photographed and necropsied by a team of veterinary pathologists [10]. A cosmetic post-mortem was performed in case it was decided to preserve Ham's body. There had been some well-intentioned calls to stuff and display his body in the National Air and Space Museum in Washington, in a similar fashion to his predecessor Able. However, the proposal was overturned. Ham's skeleton would be retained for ongoing examination, so it was cleaned of any soft tissue by the lengthy process of placing it into a Dermestid beetle colony at the Department of Vertebrate Zoology within the Smithsonian Institution. His other remains were respectfully laid to rest in front of the International Space Hall of Fame in Alamogordo, New Mexico, following a brief ceremony on the afternoon of 28 March. Dr. John Paul Stapp officiated at the ceremony.

Astronaut Alan Shepard, by then a retired rear admiral, was cordially invited to the ceremony for Ham by the centre's Ryita Price, who diplomatically said in the invitation: "I don't know if you're an animal lover or not – or how much you feel our space program owes the primates who first proved man could survive space. I do know that you had to cope with a lot of jokes and sometimes 'unfunny' humor about the situation. And, perhaps, now would be a good time to give a timely and dignified response to these innuendoes and, at the same time, create some goodwill for the United States space program" [11].

Shepard declined the invitation, citing the fact that he was limiting his personal appearances "to only a handful of occasions in the interest of other pursuits." It is possible that he still begrudged Ham preventing him from becoming the first person in space [12].

A much beloved chimpanzee

These days there is a memorial plaque above Ham's final resting place, and it reads:

World's First Astrochimp HAM
Born: 1955, Cameroons, Equatorial Africa
Died: 18 Jan. 1983, North Carolina Zoological Park, NC
Ham was trained at Holloman AFB, Aeromedical Research Laboratory. His name is an acronym for Holloman Aero Med. Ham's training culminated on Jan. 31, 1961, by riding in a capsule perched atop an 83-foot Redstone rocket launched from Cape Canaveral, Florida, reaching a top speed of 5,800 mph and an altitude of 155 miles. He was recovered at sea 420 miles down range from the launch site. Ham proved that mankind could live and work in space.
Dedicated March 28, 1983

Meanwhile Ham's female understudy for the MR-2 flight, Minnie, would perform backup duties for another chimpanzee named Enos, after which she became part of the chimpanzee breeding programme for the Air Force. Nine offspring would result, and Minnie would also serve as a surrogate mother to other young chimps in the Holloman colony. The last surviving "astrochimp" of America's early space programme, she died of natural causes on 14 March 1998, aged 41 years.

UNDERSTANDING ENOS

The United States stood a mere 3 weeks away from launching their first astronaut Alan Shepard into space when some dramatic news flashed around the globe on 12 April 1961. The first, brief bulletin announced that a Soviet cosmonaut named Yuri Gagarin had just completed a single orbit of the Earth in his Vostok spacecraft. He had returned safely to Earth with an in-flight promotion to the rank of major in the Soviet Air Force, as well as a far more illustrious title – that of First Person in Space.

Beaten to the punch

The news of Gagarin's flight stunned people everywhere, particularly the citizens of North America. Not only would Shepard miss out on being the first person to fly into space, but his mission was a less-complex ballistic flight. The United States was not even planning a manned orbital flight until the following year.

The sad truth is that Shepard had been ready and willing to fly before Gagarin. Originally scheduled for 24 March, politics played a dominant role in his flight being postponed to allow for a further test launch. In January of 1961 a presidential advisory group known as the Wiesner Committee had submitted a scathing report to the Kennedy administration on the state of spaceflight activities and preparedness for spaceflight activities. One of the group's principal recommendations was to immediately delay the first manned Mercury–Redstone flight. George Kistiakowski, one of the committee heads, even went so far as to say that launching a Mercury astronaut too soon would provide him with "the most expensive funeral man has ever had" [2].

At that time, the director of Project Mercury was Dr. Robert Gilruth. He was also the enigmatic and influential head of NASA's Space Task Group at Langley, with considerable and lengthy experience in aerospace technology. He was understandably outraged at many of the report's recommendations and justifiably felt that he and NASA Administrator James Webb were in a far better position to comment on the state of the space agency's preparedness to launch the first astronaut into space. Despite their protestations, many elements of the report were adopted. One resulted in Wernher von Braun and his team being advised that a further unmanned precursory flight would have to be inserted into the schedule. It would take place on 24 March – the date originally set aside for the first manned mission.

The news did not entirely upset von Braun; in fact, he had been privately lobbying for another test-firing as he harboured concerns about an unexplained "chatter" in the guidance system of the Redstone booster. He had been pressing for at least one flawless launch before placing an astronaut at the top of the rocket. The additional flight was designated MR-BD (Mercury–Redstone Booster Development), and the first manned shot was pushed back until 25 April, depending on the success or otherwise of MR-BD.

Defeat, and the road to recovery

The unmanned development flight was executed with very few problems, and approval was finally given for an astronaut to be launched on the next Mercury–Redstone flight. However, there had been considerable activity in the Soviet Union with the launch of Korabl-Sputnik 4 on 9 March. This orbiting satellite carried a frisky little dog named Chernushka, and the unmistakable pointers to an imminent manned launch were there for all to see.

Nineteen days after the successful test of the unmanned MR-BD, the Soviet Union launched Yuri Gagarin into space. NASA, and the seven Mercury astronauts, were devastated.

In May the successful mission of Alan Shepard helped turn America's gloom into elation, and his history-making flight was repeated 2 months later by that of Gus Grissom. However, more bad news was in store that year for NASA; on 6 August cosmonaut Gherman Titov flew Vostok 2 on a daylong, 17-orbit circumnavigation of the world. Like Gagarin, he too made a successful landing by parachute, having been automatically ejected in the final phase of his spacecraft's descent.

AN AMERICAN IN ORBIT

Following the suborbital missions of Shepard and Grissom, the American public was keen to have one of their astronauts in Earth orbit. This was something NASA officials desperately wanted as well, but they felt it necessary for a Mercury spacecraft to undergo a full dress rehearsal or "shakedown" flight before they could send a human into orbit with relative confidence.

The space chimps go back into training

As there is virtually no gravity at orbital height, NASA was keen to ensure that a human would be able to perform simple tasks when weightless for lengthy periods. It was also thought that orbiting astronauts, seeing the world passing so quickly beneath them, might even lose their normal concepts of up and down, speed and direction, and become confused. Once again, to resolve many of these questions, Holloman was asked to provide suitably trained candidates and prepare them for an orbital space flight.

The chimpanzees would be tested for their ability to remember increasingly complex commands in set patterns. By comparison with the simpler Redstone psychomotor panel, the orbital panel was far more elaborate, involving colours and symbols. Technicians could illuminate three symbols – circles, triangles and squares. When two symbols the same shape lit up, each subject had to press a button below the non-matching third. Every time they got it right a banana-flavoured pellet popped out

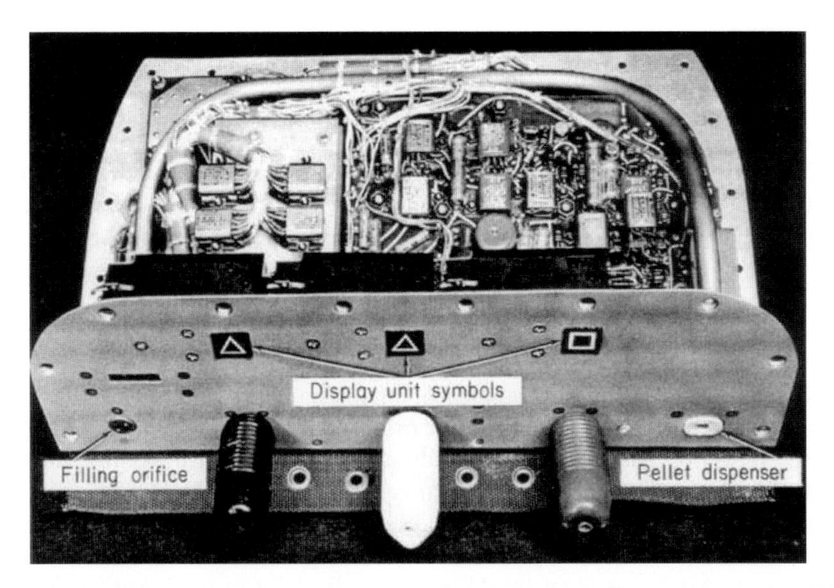

The MA-5 psychomotor panel was more complex than the earlier training units. (Photo: NASA)

of a tube near their mouth. "They had to hit the lever so many times for a drink of water and so many times for a banana pellet," Ed Dittmer pointed out. A small green light would also illuminate during other tests. If the chimp noticed this and pressed the button to turn it off within 20 seconds, the reward was a drink of water or fruit juice from another tube near their mouth. Any time one failed to perform a task a small but annoying electrical shock would tingle the chimp's feet – an unmistakable sign that something had not been done correctly.

The primate candidates also learned how to turn lights on and off by hitting right- and left-hand levers, and were taught to count by pulling another small lever exactly 50 times. As they grew conditioned to the test sequence the animals would reach their own count of 49 deliberately slowing as they approached the end of their task, and then place a hand under the tube for their banana pellet reward on the last pull. Trainers were amazed at how quickly their playful charges learned this trick. The astute chimps hardly ever got it wrong.

It was important that the subjects memorised the order in which they did their tests, as one of them would be performing them in orbit. If spaceflight had no effect on a primate's ability to do simple tasks, then there was no doubting man's ability to do the same [1].

Enos: man or chimpanzee?

On 13 September, the unmanned mission that would precede the chimp flight began when MA-4 was launched with a mechanical "simulated astronaut" wired into the spacecraft, consuming oxygen and expending carbon dioxide at the same rate as an astronaut as part of extensive systems checks. MA-4 was only planned to last for a single orbit; this was accomplished and the spacecraft was successfully recovered after splashdown.

On 29 October 1961, the 6571st Aeromedical Research Laboratory at Holloman Air Force Base delivered three of their best-trained chimpanzees to the space base at Cape Canaveral as candidates for MA-5, accompanied by 12 handlers. They would link up with another two chimps and eight handlers already at the Cape. As well as space veteran Ham, the animals involved were Duane and Jim (named after project veterinarians Duane Mitch and James Cook), Rocky (named for boxer Rocky Graziano) and Enos. But, it seemed that Ham, who had already endured one ballistic flight, was not at all keen to go on a second mission. He tackled his tests with little enthusiasm, much to the disappointment of his handlers, although 2 days before the flight he still remained one of the three prime candidates. However, another one of the three soon proved to be an outstanding and intelligent candidate, although he tended to become a bit of a handful at times.

Enos (which means "man" in both Greek and Hebrew) was a feisty, 4-year-old chimpanzee who had been captured in East Africa's French Cameroons and purchased by the Air Force from the Miami Rare Bird Farm on 3 April 1960. Before being named he was simply known as Number 81, and he had undergone 1,263 hours of training at the University of Kentucky and the Air Force Aeromedical Field Laboratory. For about a quarter of this time he was restrained in a contoured couch, much smaller but similar to that the astronauts would use on later manned missions. Technicians would also take Enos and the other chimps up in jetfighters so they could get used to the sensations of take-off acceleration and noise during high-speed flight. One veterinarian for the mission described Enos, then 5 years and 5 months old, as "quite a cool guy and not the performing type at all" [6].

One troublesome primate

Although he was definitely the star pupil this time, the 39-pound Enos sometimes acted up and was far less tolerant of humans than his chimp colleague, Ham. "No one ever held Enos," Ed Dittmer recalled. "If you had him he was on a little strap. Enos was a good chimp and he was smart, but he didn't take to people. They had the wrong impression of him; they said he was a mean chimp and so forth, but he just didn't take to cuddling. That's why in any pictures you ever see of Enos you don't see anyone holding him" [4].

While he was adjudged the ideal mission candidate, there could be a very naughty side to Enos, who had displayed back at Holloman a remarkable propensity for dropping his diapers and brazenly stroking his genitalia whenever reporters paid a visit to the sheet metal building where the chimpanzees were housed. This enclosure was right next to Hangar S, where the Mercury astronauts worked and trained, and where their offices were located. To minimise disruption for the animals visitors were discouraged, but every so often someone would want to see the chimp colony. On one occasion a visiting politician somehow pulled a few VIP strings and was given a tour of the facility by a reluctant pad leader, Guenter Wendt. In his autobiography,

Enos enjoys a drink. Note the restraining strap on his arm and the wary look on the face of his handler. (Photo: NASA)

The Unbroken Chain, Wendt recalls the politician going right up to the cage holding Enos and gesturing to the scowling chimpanzee, who was squatting down on his hands. Wendt knew Enos well enough to know what was coming and cautiously backed away, moments before Enos brought up a handful of steaming faeces and flung it straight at the politician. "I don't think we ever had him as a guest again," was Wendt's wry comment [13].

A chimp behaving badly

Like the other two chimpanzees, Enos was subjected to an intense physical and came through with flying colours. It was decided that he had met all the selection criteria and would be the one to take a seat aboard the three-orbit MA-5 mission.

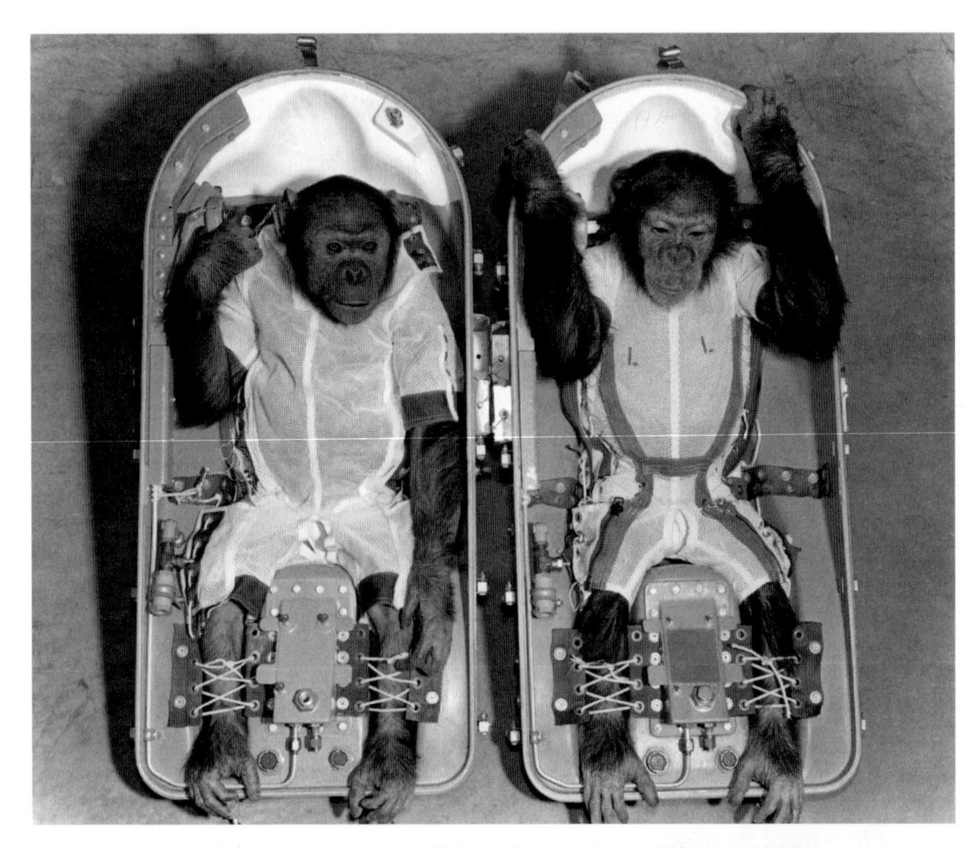

Enos (left) with his spaceflight colleague, Ham. (Photo: NASA)

With his selection, and the associated media fame it incurred, the veterinary staff looking after Enos was resolute in their determination to put an end to their charge gleefully masturbating (which had earned him the unenviable title of 'Enos the Penis'). When it became known he was doing this back at Holloman they had tried fitting him with an external catheter, but Enos had very quickly discovered that he could remove this unwanted impediment to his enjoyment. Then the doctors tried an internal catheter. That, too, was quickly disposed of by Enos. In desperation, they came up with a balloon urethral catheter, which they thought would solve the problem as it would prove extremely painful to remove. It worked – at least for a while.

ENOS IN ORBIT

Originally scheduled for lift-off on 7 November 1961, the MA-5 flight was delayed a week before being indefinitely scrubbed on 11 November, the result of a hydrogen peroxide leak in the spacecraft's manual control system. Following this, there were some passionate calls from within and outside of NASA for the MA-5 flight to be abandoned, and instead launch John Glenn on America's first orbital flight before the end of the year. It was argued that this might take some of the sting from Gherman Titov's daylong mission 4 months earlier. But good sense prevailed; it was crucial that additional flight data be gathered and systems tests be carried out before an astronaut could be launched on top of the notoriously-unstable Atlas rocket. Chief of the Mercury medical team, Air Force Lt. Colonel Stanley White, said at the time it would be "extremely hazardous" to eliminate the chimp flight at such a crucial time. "The MA-5 mission is more than a matter of just checking the spacecraft," he explained. "So far we have had experience with just one [unmanned] Mercury shot in orbit, and that for only one trip around the world. We need another shot now, for three orbits, so we can be sure that everybody in the system will have a chance to do his job" [6]. Two of the four previous Atlas launches had ended in failure, so there was still much work to be done in making the booster a safer launch vehicle as well.

The reluctant chimponaut

The MA-5 launch finally took place. Four and a half hours before the planned 7:30 a.m. lift-off from Launch Complex 14 on 29 November 1961, Enos had been woken and given his breakfast. Following this, dressed in his diapers and nylon flight suit, he was led out to a panel van which carried him across to Hangar S, where he was given a thorough veterinary examination by his personal physician, Captain John Mosely. Next his handlers escorted Enos out of the hangar and into a waiting, air-conditioned medical van which would transport him out to the launch gantry, following much the same procedure as the astronauts. Inside the van, sensors were taped onto his skin and the balloon catheter was inserted. He was then placed onto his couch and securely fastened down with his lands left free, allowing Enos to carry out his in-flight tasks. The lid was then closed, the life-support systems activated and connections made to enable physiological data to begin flowing. Dr. Henry was informed that the chimp was ready to fly, and the capsule containing Enos was carefully lifted out of the medical van, carried up in the gantry elevator and inserted into the Mercury spacecraft.

Countdown and lift-off

With the countdown held at $T - 30$ minutes, technicians had to reopen the hatch of the spacecraft to correctly position an on–off telemetry switch, which caused an 85-minute delay. It was later joked by flight controllers that Enos had probably been talking things over with Ham and had flipped the switch because he did not want to go on the flight.

In the final 2 minutes before lift-off, Enos began pulling his levers in response to the signal lights. He finally soared into the skies at 10:08 a.m. after a few minor technical hitches had further delayed the launch. On ascent he was pressed back into his contoured couch with a maximum 7.2 g's. But, Enos had sustained this force often inside centrifuge rides, so he knew it was only a temporary discomfort. Atlas 93-D propelled the Mercury spacecraft out to the northeast, and it finally settled into an

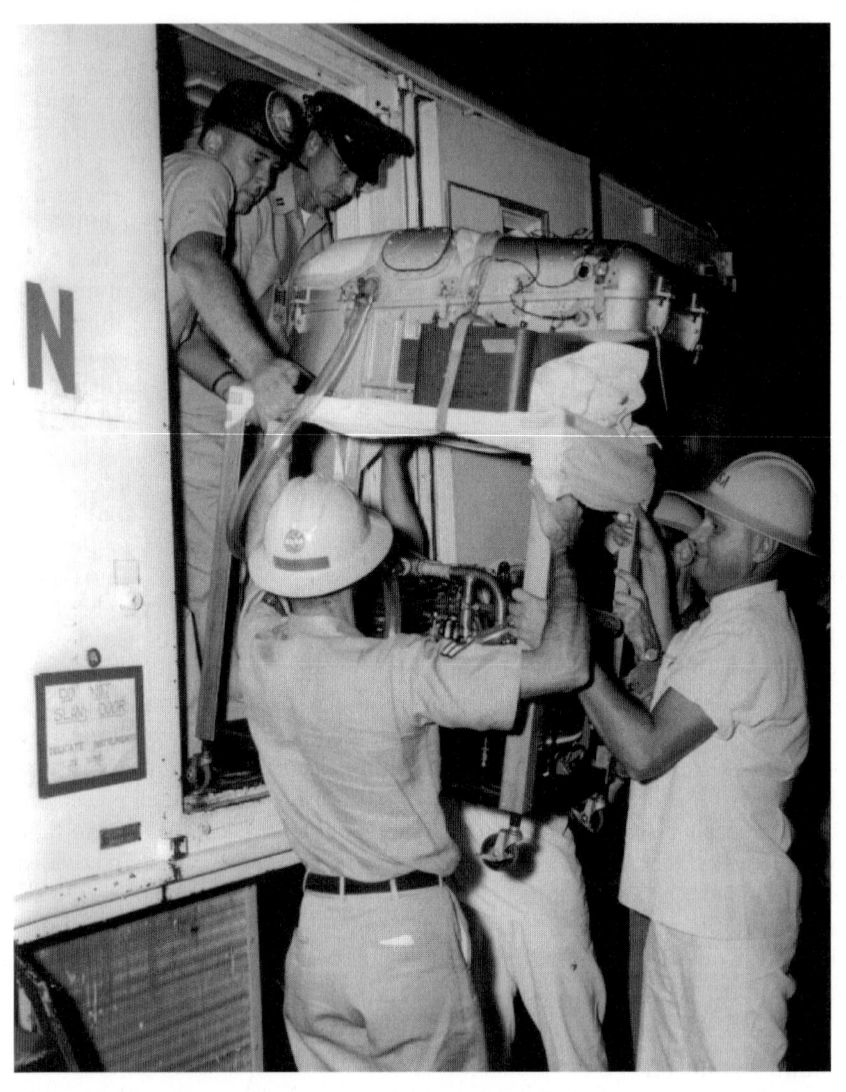

Ed Dittmer (right) assists as the container with Enos in his flight couch is gently lifted from the transfer van at the launch pad. (Photo: courtesy Ed Dittmer)

elliptical west-to-east orbit of 99×147 miles, inclined at an angle of 37° to the equator. It was almost right on the money. All of the systems worked well during the early part of the planned three-orbit flight, and Enos happily resumed his set tasks. He won 13 banana pellets on the 50-count lever, and drank just under a litre of water through a tube. As well as carrying out four basic tests, Enos also had some oddity problems to solve. In these tests, three symbols would flash on the display panel, and he had 15 seconds in which to press a lever beneath the symbol that he thought was different from the other two.

A small wiring malfunction on one of the tests meant that he received some undeserved electric shocks to his feet, but for the most part he did everything that was asked of him. Despite an overheating problem in his spacesuit which would also frustrate the Mercury astronauts, Enos remained calm and stuck to his programme. Everything seemed quite natural to him after the extensive training, and he kept happily jiggling levers and pushing buttons as his Mercury spacecraft whirled around the Earth. Not all his time was spent on mentally and physically exhausting tests; quite a bit of time had been pre-allocated to rest periods, which had featured in his training. His handlers could not help but smile when they discovered that Enos had heeded his training and did not hit any levers during these rest periods.

A voice from orbit

During the flight a pre-recorded message was played from within MA-5 to simulate voice contact with an astronaut. "CapCom, this is astro," said the taped message. "Am on the window and the view is great. I can see all the colours and can make out coastlines." When he heard about this, a delighted President Kennedy told reporters at a post-flight White House press conference that "the chimpanzee took off at 10:08. He reported that everything is perfect and working well" [7].

The wiring malfunction which gave Enos the pesky shocks was part of a more severe problem, which eventually caused the Mercury spacecraft to begin tumbling as it neared Muchea, Western Australia, during its second orbit. According to Arnold Aldrich, the remote site flight controller for the Space Task Group, the problems had begun at the end of the first revolution. "The spacecraft was acting up and its thrusters weren't all working. So there was a lot of concern that it wouldn't he able to hold the right position for retrofire.

"That set up a scenario during the second revolution where the spacecraft went east after passing over the east coast and each site looking if these thrusters were working right or they weren't. In general, the reports were that it was not working right ... You had to have the retro-rockets pointed in the right way when you fired them, so it would slow the spacecraft in the right manner and it would then dip into the atmosphere.

"There was also a thermal problem in the spacecraft ... but, as I recall, that condition had occurred and then seemed to get better. I'm not sure we would have ended the flight based on the thermal conditions, but this attitude control and thruster and propellant usage during the second revolution was quite a concern" [14].

It was later discovered that the thruster problem resulted from a stray metal chip clogging a fuel supply line, which caused the spacecraft to drift from its planned attitude. Ground controllers decided the problem could only get worse and Flight Director Chris Kraft was concerned that there might be insufficient fuel remaining at the end of a third orbit to achieve correct attitude control during re-entry. Acting on the advice of his control team he decided to end the flight early by setting procedures in motion to fire the retro-rockets. His decision came just 12 seconds before he would have had to commit to a third orbit.

One very irritated space traveller

Enos, meanwhile, was one very annoyed chimpanzee. During his work periods he had been pressing the buttons in the correct sequence, but all he was getting for his diligence was a series of increasingly annoying shocks to his feet. In frustration he began banging away at all the buttons, but he still kept getting zapped. In desperation he turned his attention to a different source of annoyance, grasping his internal balloon catheter and ripping it out. This action must have really hurt him, but it seems he was beyond caring, as he then began fondling himself in front of the camera in retaliation for all that had happened to him.

With the decision made to shorten the flight by an orbit, the necessary commands were relayed up to the spacecraft.

"We had to send retrofire from Point Arguello, from California," according to Aldrich. "And we did. We sent them, and the little lights came on in the console. In fact, we had retro 1, 2 and 3 on the console and it came on 2, 1, 3, but they all came on, which was what was needed to happen" [14].

Retrofire, and the journey home

The spacecraft slowed under the influence of retrofire, and then dipped back into Earth's atmosphere, heading for the planned recovery area some 200 miles south of Bermuda. The crew of a P5M search aircraft eventually spotted the craft and its main parachute descending through 5,000 feet and relayed its position to the recovery destroyers USS *Stormes* and *Compton*, 30 miles away. The aircraft continued to make a wide circle around MA-5 for the next 9 minutes, reporting on the descent and splashdown, and then remained in the area until a recovery ship had moved in. Altogether, the flight had lasted 3 hours and 21 minutes. An hour and a quarter later the MA-5 spacecraft was safely plucked from the sea by the crew of the *Stormes*.

Once the spacecraft had been secured on deck the hatch was explosively blown, the sealed capsule removed and opened, and an excited but overheated Enos was extracted from his couch. The temperature inside his airtight capsule had been measured at 106°F, but he soon cooled down and devoured two oranges and two apples, obviously far more appealing than the low-residue pellet diet he had been placed on prior to the flight.

Facing the press

According to the official history of Project Mercury in NASA's *This New Ocean* [15], the *Stormes* dropped Enos at the Kindley Air Force Base hospital in Bermuda, where the mission's chief veterinarian, Captain Jerry Fineg, conducted an extensive evaluation of the chimp's health and post-flight condition. As he reported, "the chimp was walked in the corridors and appeared to be in good shape. His body temperature was 97.6 degrees; his respiratory rate was 16; his pulse was 100. Apparently re-entry, reaching a peak of 7.8 g, had not hurt him. His composure at his 'press conference' surprised the correspondents. One reporter remarked that Enos, unlike Ham, did not become 'unhinged' with the popping of the flash bulbs" [2].

On the first day in December Enos was transported back to the Cape, where he underwent a further battery of physical tests. The following week he made a joyful return to the chimp colony at Holloman.

PAVING THE WAY FOR JOHN GLENN

Enos's flight, and the way he readily performed his tests in 181 minutes of weightlessness, proved that a human should also be able to conduct any tasks required of them in Earth orbit. His epic flight certainly paved the way for John Glenn to make America's first manned orbital flight 3 months later.

A one-time space traveller

Enos never sat on top of another rocket. He retired quite happily to the Holloman Air Force Base before his death from illness on 4 November 1962. Pathologists at Holloman concluded that the chimp's death was unrelated to his space flight the year before. "It wasn't due to any aspect of the flight," agreed Ed Dittmer. "He had developed shigellosis, which is a dysentery they get from Africa." Unfortunately, shigellosis resists antibiotics. "Once it gets hold of them it's just impossible to treat," Dittmer added. "And that's what he died of" [1].

Extensive research has failed to determine what became of the remains of Enos. Sadly, it seems, his remains were simply discarded, with some organs sent to interested laboratories for analysis. Nowhere is there a permanent memorial to honour Enos's achievements or to mark his final resting place, as there would be for Ham when that pioneering space chimp died of old age some 20 years later.

Results of the chimpanzee flights

According to Dr. James Henry, the respective flights of Ham and Enos had crucially demonstrated several things:

(1) Pulse and respiration rates, during both the ballistic and orbital flights, had remained within normal limits throughout the weightless state. The effectiveness of the animals' heart action, as evaluated from the electrocardiograms and pressure records, was also unaffected by the flights.
(2) Blood pressures, in both the systemic arterial tree and the low-pressure system, were not significantly changed from pre-flight values during 3 hours of the weightless state.
(3) The two primates' performance of a series of tasks involving continuous and discrete avoidance, fixed ratio responses for food reward, delayed response for a fluid reward and solution of a simple oddity problem was unaffected by the weightless state.

(4) Primates trained in the laboratory to perform during the simulated acceleration, noise and vibration of launch, and entry were able to maintain performance throughout an actual flight.

Furthermore, Dr. Henry and his project group were able to draw the following conclusions:

(1) The numerous objectives of the Mercury animal test programme were met. The MR-2 and MA-5 tests preceded the first ballistic and orbital manned flights, respectively, and provided valuable training in countdown procedures and range monitoring, as well as recovery techniques. The bioinstrumentation was effectively tested and the adequacy of the environmental control system was demonstrated.
(2) A 7-minute (MR-2) and a 3-hour (MA-5) exposure to the weightless state were experienced by the primates in the context of an experimental design which left visual and tactile references unimpaired. There was no significant change in the animals' physiological state or performance as measured during a series of tasks of graded motivation and difficulty.
(3) The results met programme objectives by answering questions concerning the physical and mental demands that the astronauts would encounter during space flight, and by showing that these demands would not be excessive.
(4) An incidental gain from the programme was the demonstration that the young chimpanzee can be trained to be a highly reliable subject for spaceflight studies [3].

"It was quite clear that the space effort at the beginning had to take the approach of a great expedition of exploration and adventure, and that research requirements should wait until the engineering problems had been solved," Henry told space researcher and author Shirley Thomas several years later.

"In view of this, the Mercury animal flights were, in my opinion, an unexpectedly elegant and complex piece of combined physiological and psychological experimentation" [15].

GOLIATH AND SCATBACK

While the civilian space agency NASA was involved in launching Ham and Enos on precursory missions for later manned flights, the Air Force was also involved in preparing two monkeys of its own for developmental test flights. And the launch vehicle involved, as with the flight of Enos, was the Atlas ICBM.

The mighty Atlas rocket

Developing the Atlas rocket for the U.S. Air Force had been a protracted, but top-priority task. It began in the late 1940s, when the Air Force appointed a company known as Consolidated Vultee Aircraft Corporation (more commonly called Convair) to produce a proposed long-range missile, the hardware part of Project MX-774. The Air Force planned to create a rocket that would fulfil a demanding role as America's first intercontinental ballistic missile (ICBM).

Project MX-774 was shelved in 1947, although several vehicles were test-fired over the next 2 years by Convair, who had continued to research the test project using residual funds. In January 1951 the Air Force turned once again to Convair, awarding the company a substantial contract to produce a rocket-powered ballistic missile, then code-named Project MX-1593.

The single-stage test-bed prototype produced by Convair (which became the Convair division of General Dynamics in 1954) was initially known as the XSM-16A, later redesignated the X-11 (Atlas). The contract called for the manufacture of 12 missiles, 3 of which would be used purely for captive or static test-firings [16].

A rocket to carry men into space

While specifically designed to carry nuclear warheads, the Atlas would later achieve a far more agreeable reputation as the formidable steed that propelled four of NASA's Mercury astronauts into orbit.

The liquid-fuelled Atlas was manufactured at Convair's Kearny Mesa plant in San Diego. Powered by rocket-grade RP-1 (highly refined kerosene) with a liquid-oxygen oxidiser, it proved to be a vehicle of amazing contradictions. Weighing around 267,000 pounds when fully fuelled, the missile's gleaming aluminium skin was actually thinner in parts than a modern compact disc; so thin in fact that – without the use of pneumatics to keep it erect – an unfuelled Atlas would have quickly crumpled under its own weight. When empty, the booster's tanks had to be filled with nitrogen gas at a positive internal pressure of 5 psi in order to maintain both rigidity and integrity.

A fame of sorts would later attach itself to a by-product specifically developed to protect the Atlas booster's thin aluminium skin in the highly-corrosive salt air it would encounter at Cape Canaveral. Industrial chemists and engineers at the Rocket Chemical Company plant, also located in San Diego, were commissioned by nearby Convair to manufacture a water displacement formula that would protect the Atlas rockets. After a series of trials, they finally met with success on their 40th attempt. The successful water displacement formula was numerically identified as WD-40, which quickly became so popular commercially that the company changed its name to WD-40, and their product remains in common use half a century later.

Test flights

Formal production of the Atlas commenced in January 1955, with captive test-firings carried out at Edwards Air Force Base the following year. A 12-month flight test of eight single-engine prototypes then took place at Cape Canaveral, beginning in June 1957. The first Atlas-A launch on 11 June ended in spectacular failure, resulting from a problem in the booster system. The errant missile was remotely destroyed by the range safety officer, who would also press the red button to detonate explosives aboard a second wayward Atlas-A launched in September.

A successful launch and ballistic trajectory finally took place on 17 December, coinciding with the 54th anniversary of the Wright Brothers' first successful flight. The following year, during a mixed launch programme of successes and failures, an entire Atlas (now produced with three engines) reached orbit, creating communications history by beaming down a pre-recorded Christmas message from President Dwight Eisenhower.

An awesome amount of power was needed to hoist a fully-fuelled Atlas off the launch pad. The now-upgraded, 82-foot ICBM, officially deployed in September 1959, was powered by two large booster engines, which produced a combined 360,000 pounds of thrust. They flanked a smaller, single-sustainer engine that developed a further 57,000 pounds of thrust. All three main engines would ignite explosively at the moment of lift-off, followed $2\frac{1}{2}$ seconds later by two small vernier engines mounted above the sustainer. The two booster engines would cut out 140 seconds into the ascent, while the sustainer engine would continue to burn for a further 130 seconds [16].

Overcoming a bad reputation

With good reason, the Atlas had earned an unwanted reputation as a notoriously volatile vehicle. Around the time that John Glenn made America's first manned orbital flight, one out of every three Atlas launches was ending in catastrophic failure, including the one that preceded his mission. However, all four manned orbital missions using the Atlas were successfully launched into space, and the booster's integrity grew as it slowly evolved through an alphabetical series of modified configurations. The four manned Atlas flights all took place using the D configuration.

THE SAD SAGA OF GOLIATH

At 14:55 GMT on 10 November 1961, just 2 weeks before the orbital mission of the chimpanzee Enos, an advanced E-model Atlas rocket lifted off from Launch Complex 13 at Cape Canaveral. This was an Air Force research and development flight planned to reach an altitude of 650 miles and hurl an instrumented nose cone pod 5,000 miles downrange to a splashdown in the South Atlantic.

As well as undertaking the milestone 100th launch of the mighty booster, Atlas 32-E was also carrying a primate passenger. Somewhat paradoxically, the petite, $1\frac{1}{2}$-pound squirrel monkey had been given the name Goliath by his Air Force handlers. In a time-honoured spaceflight tradition, an acronym had also been attached to the biological flight, and Goliath was flying as a living part of Project SPURT – or Small Primate Unrestrained Test.

Victim of a failure

In addition to the 6-inch monkey, who had been strapped into a life-supporting tubular canister, the Atlas nose cone (SP Pod 13) was also home to an experimental assortment of other life forms – fruit flies, insect eggs, bread mould and viruses. If the pod was recovered from the ocean as planned, these would later be examined to establish the detrimental effects, if any, on living organisms passing through the lower section of the lethal Van Allen radiation belts that surround the Earth.

The Van Allen belts, held in place by the Earth's magnetic field, are roughly in alignment with the equator and comprised of charged, high-energy particles from the sun. Deadly in heavy concentration, the first of these belts extends from around 1,500 to 3,000 miles high, while the second, about 4,000 miles thick, begins at about 8,000 miles altitude, rising to nearly 55,000 miles above the Earth. In the central midst of the Van Allen belts, radiation levels can be measured as high as 100 roentgens an hour, sufficient to cause radiation absorption sickness in a human being in just 1 hour.

Unfortunately, Goliath's pioneering flight into space did not last very long at all. The countdown and lift-off proceeded normally, but 15 seconds into the launch a failure occurred in the main sustainer engine and the booster's trajectory became erratic. Thirty-five seconds later the range safety officer concluded that the flight was unrecoverable, and the fuel-engorged Atlas could present a possible hazard if it came down in the wrong area. It was his job to make an educated determination based on safety factors and he reluctantly pressed a red button on his console. The Atlas booster, the nose cone and its biological cargo were torn apart in a massive, controlled explosion.

A MONKEY CALLED SCATBACK

The Air Force would try again the following month. This time, the booster would be an F-series Atlas, and the rhesus monkey chosen for the flight was a much larger, four-pound *Macaque malatta* called Scatback. The reason for using this football terminology to name the primate has been forgotten. The monkey would be monitored by recorded telemetry throughout the suborbital flight to learn more about the effects of launch acceleration and weightlessness. The flight was also expected to yield vital information during a dynamic re-entry that was planned to be far more severe than any human passenger might soon endure. Disappointed by the ill-fated flight of Goliath, project scientists were eagerly waiting to examine this second primate for any physiological changes or ill effects following the capsule's passage through the Van Allen belts.

Atlas 6F was meticulously prepared for the second primate launch. The modified booster now featured a separable side-mounted capsule known as SP Pod 6 that would enclose Scatback.

Lost at sea

At 03:32 GMT on 20 December 1961, the Atlas F lifted off from Launch Complex 11 and – despite a recorded hydraulic failure – managed to reach an altitude of around 600 miles as it rapidly headed downrange. Five thousand miles from the launch site, as it soared over the planned Ascension Island target zone, Scatback's capsule separated

Lift-off of an Atlas F, similar to the one used to launch Scatback. (Photo: USAF)

from the fuel-depleted booster and descended by parachute after an apparently rough but successful re-entry through the atmosphere.

Unfortunately, the capsule's radio beacon failed to operate as designed, badly hampering the search by helicopter recovery teams, while high seas and whitecaps in the splashdown area compounded the problem. Eventually, after a number of unsuccessful sweeps over a wide area, the search for Scatback's capsule was declared futile, reluctantly abandoned and poor Goliath is forever listed as lost at sea.

Further failures

Although they carried no animals, there were also attempts to launch capsules laden with biological specimens into space on ballistic flights during a NASA project known as BION (Biological Investigations of Space). Each of the payload capsules carried bacteria, mould, fresh human blood, grasshopper nerve fibre, barley seed, sea urchin eggs and amoebae. It was hoped to loft these capsules up to 1,165 miles into the Van Allen radiation belt and allow them to sail through a meteor storm before re-entry and recovery. The resultant data would be used to determine what sort of hazards might be involved and to aid in devising appropriate safety precautions on later manned flights.

On 15 November 1961, a 62-foot, solid-fuelled Argo D-8 Journeyman rocket was launched from Point Arguello in southwest California, carrying the BIOS-1A capsule. Unfortunately, the four-stage rocket was torn apart by dynamic forces shortly after lift-off, and the capsule plunged into the Pacific. The second attempt would also fail; this time the rocket veered off course during its ascent and the separated BIOS-2A payload parachuted into an unpatrolled stretch of ocean. Both capsules were lost.

REFERENCES

[1] Telephone interview conducted by Colin Burgess with Edward C. Dittmer, Sr., 21 June 2005.

[2] Loyd Swenson, James Grinwood and Charles Alexander, *This New Ocean: A History of Project Mercury*, NASA SP-4201, NASA, Washington, D.C., 1998.

[3] Mae Mills Link, *Space Medicine in Project Mercury*, NASA SP-4003, NASA, Washington, D.C., 1965.

[4] Alamogordo Space Center Oral History Program interview with Edward Dittmer, conducted by centre curator George M. House, 29 April 1987, Alamogordo, NM.

[5] USAF News Release (70-17-R), *Space Pioneer Remembered*, by S/Sgt. Mike Meservey, for Air Force Missile Development Center, Holloman AFB, Alamogordo, NM, 18 February 1970.

[6] Clyde Bergwin and William Coleman, *Animal Astronauts: They Opened the Way to the Stars*, Prentice-Hall, Englewood Cliffs, NJ, 1963.

[7] Chris Kraft, *Flight: My Life in Mission Control*, Penguin Putnam, New York, 2001.

[8] *USS Donner LSD 20 Recovery Ship MR2 with Space Chimpanzee Ham*, article reprinted from *The Gator News*, Amphibious Force, U.S. Atlantic Fleet, Vol. XIX, Little Creek, VA, 3 February 1961. Author not known.

[9] Report on death of chimpanzee Ham by North Carolina Zoo, January 1983. Author not known.

[10] Undated 1983 letter (HDQ/SGV) to Colonel Robert L. Flentge from Colonel William R. Cowan, USAF, MC, regarding autopsy on space chimpanzee Ham.

[11] Letter from Ryita Price, Alamogordo Space Center, to Rear Admiral Alan B. Shepard, 8 February 1983.

[12] Letter from Rear Admiral Alan B. Shepard to Ryita Price, Alamogordo Space Center, 18 February 1983.

[13] Guenter Wendt and Russell Still, *The Unbroken Chain*, Apogee Books, Burlington, Ontario, 2001.

[14] NASA JSC Oral History interview with Arnold Aldrich, conducted by Kevin M. Rusnak on 24 June 2000, Johnson Space Center, Houston, TX.

[15] Shirley Thomas, *Men of Space* series (Vol. 7, chapter on James P. Henry), Chilton, Philadelphia, PA, 1965.

[16] *The Atlas Launch Vehicle: How It Boosted America into the Space Age*, report commissioned by the American Society of Mechanical Engineers, 1965. Author not known.

10

Cosmos/Bion: The age of the biosatellites

In March 1966, two dogs just back from an orbital flight aboard the Soviet Union's Cosmos 110 satellite were put on display at a press conference to show that they had survived their venture into space. Such press conferences had long been standard procedure during the 15-year period of the programme that launched numerous dogs on suborbital and orbital flights. It made for an impressive show to have the cute canines prance and pose for the news photographers, demonstrating that space flight was safe and that the Soviet Union did not mistreat its canine cosmonauts. However, a different scene unfolded on this occasion.

The two dogs – Veterok and Ugolyok – had just been whisked back to Moscow from their landing spot in central Russia after a 22-day flight, and they appeared weak and tired. Ugolyok tried to nap during the press conference, while Veterok wobbled on her feet and could only stand with the help of a laboratory technician. Although exhaustion might have been expected after their ordeal, their fatigue masked more troubling indications of how long-duration space flight affected animal physiology.

DOGS SPEND 22 DAYS IN SPACE

The Soviet Union's space dog programme had effectively ended 5 years earlier, in 1961, when a dog named Zvezdochka took a final, orbital test flight in the Vostok capsule, just 1 month before a similar craft carried the first human into space. Now the Soviet Vostok missions and the American Mercury missions had given way to the multi-crewed flights of the Voskhod and Gemini programmes. Prior to the flight of Cosmos 110, the Soviet Union had already flown two, 1-day Voskhod missions and had plans for longer-duration flights, including a next-in-line 18-day mission. However, they had been particularly concerned with the poor performance of the Voskhod's life-support system and whether it would be capable of supporting a human crew for this long [1].

For one final curtain call, Cosmos 110 put dogs back in their critically important role of testing equipment and animal capabilities prior to humans undertaking the same risk. Although the primary goal was to test the life-support system of the Voskhod spacecraft, paramount among the mission's secondary goals was a study of the effects of radiation on living beings. Cosmos 110 was placed into a highly elliptical orbit that reached a height of 560 miles, taking it well within the lower range of the Van Allen radiation belt. Onboard instruments measured a radiation rate of 500 mrad a day, compared with rates on Soviet manned flights of 20–80 mrad a day [2].

Twenty days into the flight of Cosmos 110, concerns about the life-support system were borne out when capsule air quality deteriorated to such an extent that it forced an early return of the mission. The dogs landed safely, were quickly retrieved and hurried back to Moscow.

The effects of space flight

Fifteen years of suborbital and orbital dog flights had not prepared Soviet scientists for the physiological condition of the dogs. A report issued 2 months after the flight stated that they had suffered from dehydration, loss of calcium, weakened circulation and muscle atrophy, as well as difficulty in readjusting to walking in Earth's gravity. Full motor functioning did not return to normal until 8–10 days after their return. The report concluded, "Prolonged space flight and the development of methods to combat unfavourable effects of such flights have raised new problems for space medicine" [3].

These findings gave an indication of the gap existing in the understanding of the effects of long-duration stays in space. Vostok and Mercury missions had shown that even brief stays in weightlessness could cause such physiological problems as dehydration, a redistribution of fluids and post-flight problems with balance. These temporary effects were quickly overcome once a spacefarer returned to Earth and readapted to gravity.

But the Voskhod and Gemini programmes were sending space explorers on longer flights. The first Voskhod mission in October 1964 had already carried aloft the first physician to travel in space, Dr. Boris Yegorov, who monitored the functioning of the crew and took in-flight blood samples. The American Gemini flights had produced some unexpected physiological reactions: namely, loss of bone and muscle density, bone calcium, and red blood cell mass, plus post-flight orthostatic intolerance (a drop in blood pressure causing faintness upon standing) [4].

Clearly, evidence was building that long periods of time spent in space caused a range of physical problems that were not fully understood. With longer flights looming in the near future, it was obvious that considerable basic research needed to be done to determine more precisely the effects of spending weeks and months in space.

STUDYING THE BIOMEDICAL PROBLEMS OF SPACE FLIGHT

The Cosmos 110 mission had been the first project of a Soviet research agency called the Institute of Biomedical Problems (IBMP), created in 1963 as a spin-off from the venerable Institute of Aviation and Space Medicine (IASM) [5]. The IASM was an Air Force agency created originally as the Institute of Aviation Medicine (IAM) to study the physiological effects of aircraft flight. In 1951 some of its research shifted to the study of the physiological effects of rocket flight, using various animals, plants and biological specimens as test subjects. The Soviet Union's famous space dogs were trained here for their pioneering rocket flights in the 1950s and 1960s. In 1960, to better reflect the increasingly equal status of their aviation and space research, the name of the Institute was changed to the Institute of Aviation and Space Medicine.

In the wake of the manned Vostok missions (1961–1963), when there was wide public support for space research and considerable funding as well, it no longer seemed practical to maintain the pairing of aviation and space medicine. So the Institute of Biomedical Problems (IBMP) was born out of the IASM. As Oleg Gazenko explained in a 1989 interview, "The military was tired of having chickens, geese, and plants on the premises; it seemed rather weird to them" [5].

"Two groups of scientists and engineers were here at the beginning of this Institute," explained Yevgeniy Ilyin in a 2002 interview, when he was deputy director of the IBMP [6]. These included "military doctors and engineers from the Institute of Aviation and Space Medicine and radiobiologists and physicists from the Institute of Biophysics of the Ministry of Health." Beginning with a staff of about 250, it would eventually grow to include 2,500 personnel.

The creation of the IBMP reflected a growing realisation of the need to give greater emphasis to the role of life sciences research in space. Although the IBMP worked on particular missions, first Cosmos 110 then Salyut 1, its goal was to address the long-term problems of space biology and space medicine, not just the immediate problems associated with a particular flight. The Institute studied the effects of space flight, but also engaged in long-term research relating to safety procedures and the development of support systems for long-duration flights.

In the early 1970s, in a significant commitment to basic biomedical research, the IBMP began work on a series of biosatellites – work that would ultimately extend for over 23 years, draw in numerous international partners and greatly advance understanding of the effects of space flight on living organisms. Dr. Yevgeniy Ilyin, who had worked on Cosmos 110, was put in charge of the programme, a post he would hold until its conclusion in 1997.

Life sciences comes to Ames Research Center

This same awareness of the need for a greater emphasis on the life sciences had dawned in the U.S. as well. Although NASA would struggle for most of its first decade, through scientific review panels, Congressional hearings and internal bickering, before finally elevating the status of life sciences research within the massive NASA bureaucracy, it did implement one important change. In 1965 the lead role in life sciences research became centralised at NASA's Ames Research Center, at Moffett Field, California [7].

NASA's thinking had also turned to biosatellites. In December 1966, the same year that Veterok and Ugolyok returned weakened from their 22-day flight in Cosmos 110, NASA began its own series of satellite launches, known as the Biosatellite programme, to study the biological effects of prolonged weightlessness and radiation exposure. In the U.S. and the Soviet Union, biosatellites would become the platform for the next major role animals would play in space research.

NASA'S BIOSATELLITES

Once it had been established that animals and humans could survive in weightlessness for short periods of time, attention shifted to the effects of long-term stays in space. NASA's three Biosatellite missions, 1966–1969, were specifically designed to study the cumulative effects of prolonged weightlessness and other spaceflight factors, in particular cosmic radiation.

The problems that had already shown up on the Gemini flights might be compounded by longer missions. What new physiological problems might arise for astronauts who spent months in space? How would body systems react to prolonged weightlessness? How would it affect the inner ear, which provides our sense of orientation, or the ability to focus on tasks? How would orbiting in a satellite affect circadian rhythms? What about exposure to radiation in space? Would radiation exposure in conditions of weightlessness differ from exposure on Earth? [8].

Space biology gets more scientific

Although biological experiments had previously been flown on high-altitude balloons and on suborbital and orbital rocket flights, controls had generally been inadequate to determine what caused the effects observed in the biological specimens. "Previous animal flights had been 'canary in a mine shaft' type of flights," according to Ames Research Center's former deputy director of life sciences, Kenneth Souza. "Animals were sent up simply to see if they could survive" [9]. This series of biosatellites would be NASA's first programme devoted exclusively to controlled biological experimentation.

The Biosatellite spacecraft measured 8 feet tall $\times$ $4\frac{1}{2}$ feet in diameter and included an automated laboratory for biological experiments. Electrical heating helped sustain proper temperatures. Battery power and bottled air maintained conditions on board the first two flights, while the third flight used a fuel cell and a means of carbon dioxide removal. At the conclusion of the flight, a re-entry vehicle separated from the satellite and returned to Earth with the experimental capsule.

Original plans called for three Biosatellite flights, of 3-, 21- and 30-day durations. But, from the start problems plagued the programme. Biosatellite 1, launched 14 December 1966, failed to return to Earth, forcing a change in the schedule. Biosatellite 2 would now be a 3-day mission, carrying the same experiments that

Biosatellite 1 spacecraft mated to a Delta booster with cover off. (Photo: NASA-KSC)

had been on Biosatellite 1. The 21-day mission, which was to have carried rodents, was abandoned because of mounting project costs. Biosatellite 3 remained as a 30-day primate flight [10].

Bad weather at the mid-Pacific recovery site cut short the Biosatellite 2 flight, but it successfully exposed its 13 plant and animal experiments to 45 hours of microgravity. Amoebae and eggs of the American leopard frog were used to study development, growth and cellular structure. To study the interaction between weightlessness and radiation, researchers used parasitic wasps, flour beetle pupae, and the larvae and adults of the fruit fly.

Although changes were noted in some of the specimens, results showed that short-duration space flight did not substantially disturb normal cellular processes. Results from the radiation biology experiments were less conclusive, in that they did not establish any deleterious interaction between radiation and weightlessness. The most significant finding of Biosatellite 2 may well have been that NASA determined that it could successfully operate an unmanned biology laboratory in space. These were meagre results for a programme that had grown to be very large and very expensive.

The first primate biosatellite

Problems continued – and got worse – with the third and final flight in the series, Biosatellite 3 launched 28 June 1969 carrying a male pig-tailed monkey (*Macaca nemestrina*) named Bonnie. Some very ambitious scientific goals had been set out for this 30-day mission. Meanwhile, costs for the Biosatellite programme had risen dramatically, and because the first flight had been lost and the second cut short, pressure was on to maximise the scientific return from this one remaining flight.

The research objectives were to study "space flight effects on brain states, behaviour, fluid and electrolyte balance, metabolism, and the cardiovascular system" [10]. To accomplish all of this required that the monkey be heavily instrumented. In all, 33 different channels of physiological data would be monitored with sensors, leads and catheters implanted in the animal's body. Additional pre-flight surgeries to prepare the animal for the flight included: testicular biopsy, incisor tooth extraction, tail amputation and anal suturing. Four ground control animals were identically instrumented.

Problems with Bonnie's health began to develop after the 5th day in orbit. He stopped eating and drinking. A decline in his body temperature and heart rate had become serious enough by the 8th day to require the early termination of the flight. Immediately, after satellite recovery, attempts were made to revive the animal, but he was not responsive. Bonnie died 8 hours after recovery. He suffered marked dehydration, having lost 25% of his body weight. Although ultimately the cause of death would be attributed to "over-instrumentation", at the time it was thought that the weightless environment alone had led to the death [10].

The death of Bonnie raised the uncomfortable possibility that long-term exposure to microgravity could cause serious, perhaps fatal, medical problems for astronauts. It was a problem that certainly required further study. But, with the end of the Biosatellite programme, NASA had no immediate orbital platform for biomedical research. In fact, Bonnie's death heightened opposition to the use of primates in space research and contributed to the abandonment of another promising programme. Since 1968, NASA had been studying a project called the Automated Primate Research Lab, which planned to fly two adult pig-tailed monkeys on a 60-day orbital flight. Following the loss of Bonnie in Biosatellite 3, this programme was never developed [10].

Political pressure on NASA had been growing in the U.S. Congress and within the scientific community regarding its handling of life science research [8]. The concentration on manned flights during the 1960s was thought to be at the expense of sufficient basic biological research that could better ensure the safety of astronauts on future flights. The death of Bonnie only seemed to emphasise that point. NASA was urged to

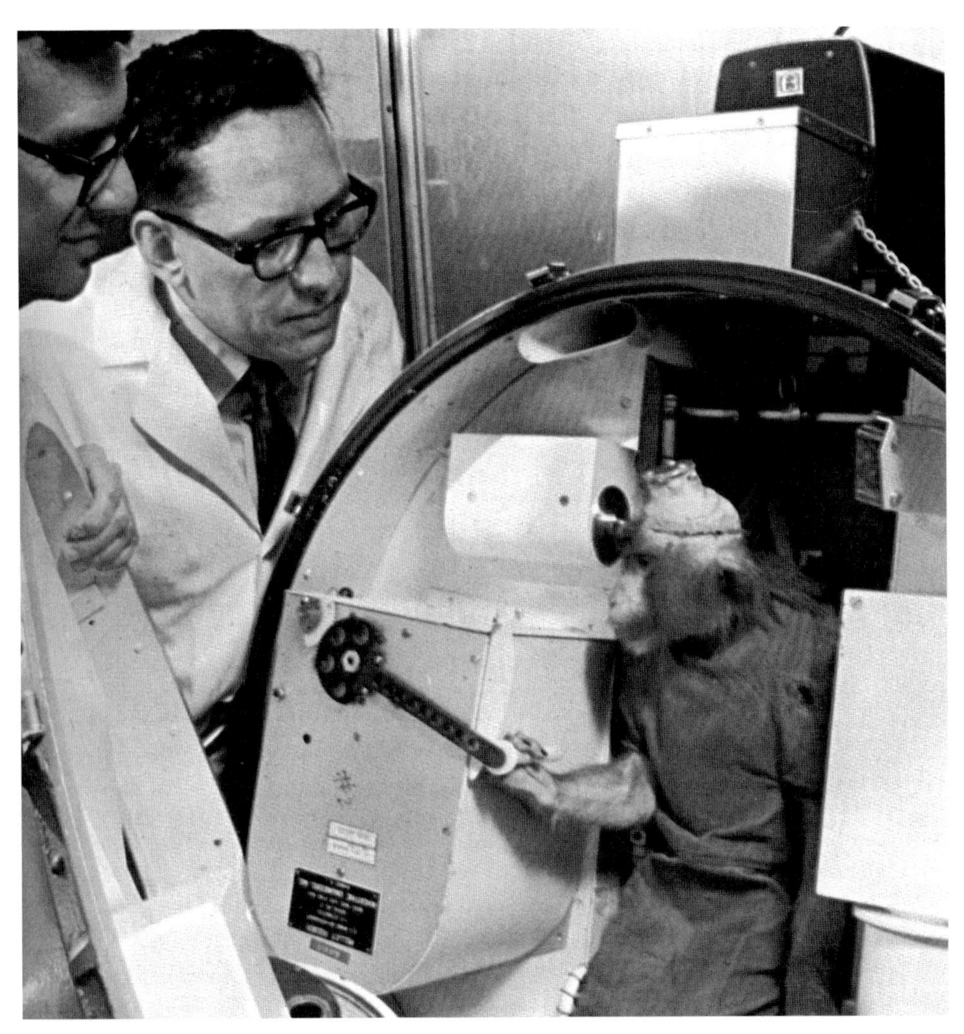

Bonnie gets loaded into his capsule for a flight in U.S. Biosatellite 3. He suffered from dehydration during the flight and died 8 hours after recovery. "Over-instrumentation" contributed to his death. (Photo: NASA)

create a central office of life science to better coordinate and advocate for life science research.

In November 1969, the President's Science Advisory Committee issued a report [11] stressing the need for NASA to develop a programme of basic biological research by beginning ground-based experiments and developing a team of experts. A 1970 report from the Academy of Sciences [12] echoed the complaint about insufficient basic research and made specific recommendations for change, including placing all life science programmes under a single director. NASA implemented many of these recommendations.

In the 1970s the United States and the Soviet Union began working together on a satellite programme – along with a host of other international partners – in a series of missions that would give pre-eminence to the basic biomedical research needed to prepare for the long-duration manned flights then on the drawing board.

COLD WAR COLLABORATION

The spirit – if not the practice – of cooperation in space had been around at least since the International Geophysical Year in 1957. In 1962, an agreement between NASA and the Soviet Academy of Sciences established several areas in which the two agencies would cooperate, including space medicine/biology and animal research. The joint working groups established to explore these areas met with some success, not least of which was that they established a spirit of cooperation and collegiality between U.S. and Soviet scientists.

To a large extent, the record of U.S.–U.S.S.R. cooperation in space can be plotted along a curve of Cold War tensions. With relations warming in the early 1970s, the stage was set for another NASA–Academy of Sciences agreement in 1971 that outlined a broad range of cooperative space activities. In May 1972, the agreement was formalised at a Moscow summit meeting between President Richard Nixon and Premier Leonid Brezhnev.

The 1972 "Agreement on Cooperation in the Exploration and Use of Outer Space for Peaceful Purposes," once again identified space biology and medicine as a main area of potential cooperation [13]. Because of the emphasis on long-duration space flights in the Soviet space programme, the United States was especially interested in an exchange of information in this area. Also identified as likely areas for cooperation were ground-based simulations of spaceflight conditions and animal research.

The U.S. and Soviet scientists who formed the Joint Working Group for Space Biology and Medicine began formal meetings in 1971, and for the first of several such meetings discussion focused on sharing information about manned spaceflight, biomedical and life-support issues. The Americans, from Ames Research Center, the Johnson Space Center and NASA headquarters found the meetings very productive and got on well with the scientists from the Institute for Biomedical Problems, including the Institute's director Oleg Gazenko and his deputy director for research Abram Genin.

It was at one of these working groups in 1974 that Russian scientists surprised the Americans by offering to fly U.S. experiments on board a Soviet satellite. The Soviets had already sent up two successful biosatellites in 1973 and 1974. Eastern European scientists had participated on those flights, but now the invitation was being extended to scientists in the U.S. and Western Europe.

In an oral history cameo in *Life into Space* [10], Ames Research Center's deputy director of life sciences, Joseph Sharp, recalled that fateful encounter in 1974 that launched 20-plus years of fruitful collaboration between the U.S. and the Soviet Union in the field of biomedical space research. Along with David Winter, NASA's director of life sciences, Sharp was attending the fifth meeting of the Joint Working

Group on Space Biology and Medicine in Tashkent, Uzbekistan, when the Russians unexpectedly made their offer to fly U.S. experiments aboard the next in their series of Cosmos biosatellites, scheduled for launch the following year.

The Soviets, who always believed in getting maximum use out of their hardware, had been using a modified Vostok capsule for their biosatellite flights, the same basic capsule that first sent a man, Yuri Gagarin, into orbit in 1961. It was simple, dependable and inexpensive. With a capacity of 850 pounds, the payload section of Vostok consisted of an 8.2-foot sphere, with attached battery pack and support modules.

The Soviet offer came at a particularly good time for the Americans, as they were faced with a gap in NASA programmes that would leave life science investigators without available satellite space for some 7 years, until the launch of the space shuttle.

The Apollo lunar-landing programme had concluded in 1972. Skylab, America's first space station, had been home to three crews of three astronauts for increasingly lengthy science missions, ending with an 84-day manned mission, before the station was abandoned in 1974. Biological experiments on the final flight had yielded no useful information because of a power failure. Evidence of the possible health risks associated with space flight that had accumulated from astronauts aboard Apollo and Skylab had raised many questions that life science investigators were eager to answer. Plant and animal experiments aboard the Soviet satellite would provide this opportunity.

What made the offer even more appealing was the low cost. The Soviets would pay the entire cost of building, launching and recovering the satellite, while the U.S. only had to cover costs associated with the experiments they flew [13].

Joseph Sharp recalled the occasion: "There had been zero preparation for that back in the States. Dave [Winter] didn't know whether he had the authority to accept the offer, whether he had the budget, and what the scope was. I can remember walking for hours in the garden near the meeting place, discussing the offer. We finally talked ourselves into it, and Dave accepted" [10]. The offer was simply too good to pass up.

American participation in Cosmos/Bion

The Soviets labelled almost all of their satellites, military and scientific, with the Cosmos designation. As soon as a satellite achieved orbit, it acquired the number next in the satellite sequence. The first Soviet biosatellite on which the U.S. flew experiments was Cosmos 782. Only after the demise of the Soviet Union did the U.S. enter a more equal partnership with Russia for the cost of these missions, signing a legal agreement for the first time. At that point Cosmos became a truly joint satellite programme and was given the new name of "Bion". Early Cosmos flights are sometimes referred to as Bion, or the Cosmos/Bion combination is used.

The flight on which the U.S. had been invited to fly experiments was scheduled for mid- or late 1975, putting preparation of experiments on a fast track. Dr. Richard Simmonds, an Air Force veterinarian assigned to NASA's Ames Research Center, served as project manager. Descriptions of the experiments had to be delivered to the Russians by December 1974 and the experimental flight hardware by August 1975.

This mock-up of a later Bion capsule (Bion 6) shares the same basic configuration as the capsule used on the early Bion flights. (Photo: State Scientific Centre of Russian Federation, Institute for Biomedical Problems of the Russian Academy of Sciences)

U.S. experiments had to be completely self-contained, meaning they could not rely on the spacecraft for power, data recording or life support. They were limited to $\frac{1}{2}$-ft^3 containers, designed specifically for the Soviet satellite [10].

The experiments were very simple, according to Harold Klein, director of life sciences at Ames Research Center. "They were not anything like the later experiments. There were no announcements, no big peer reviews. We just did them because the spacecraft was available, specimens were going to be available, and we saw a chance to get some work done" [10].

Monthly phone conferences with Soviet scientists at the Institute of Biomedical Problems (IBMP) to coordinate planning became weekly conferences as the deadline neared. On several occasions in 1975, Simmonds and other Ames personnel travelled to Moscow to establish working relationships with their Soviet counterparts and coordinate planning [14].

The U.S. would fly four experiments on this flight but would also be receiving samples of tissue from onboard Soviet rat experiments. Because the Americans would not be permitted to be present at launch or recovery, they made another trip to Moscow in September to train Soviet technicians how to handle U.S. experiments.

The NASA experiments were delivered to Moscow just weeks before the launch date. Along with experiments from France, Hungary, Poland, Romania and Czechoslovakia, they would be installed in the satellite by Soviet technicians.

Common fruit flies (*Drosophila melanogaster*) such as the female (left) and male shown here have always been involved in biological space research, from early balloon flights into the upper atmosphere to orbital space flights and beyond. (Photo: courtesy Dr. John Locke, University of Alberta)

Experiments on Cosmos/Bion 782

More than 20 species of plants and animals flew on Cosmos 782. The USSR supplied 25 male Wistar rats (*Rattus norvegicus*), which flew in individual cages. Some 500 minnow eggs (*Fundulus heteroclitus*), of different ages, flew in an onboard centrifuge, while an equal number experienced weightlessness. These would be compared with control groups on the ground. Other organisms on board included fruit flies, carrot tissue and cultured carrot cells.

Experiments were arranged in three tiers in the spacecraft, with the centrifuge on top. Fruit flies, fish eggs and the carrot experiment rotated in the centrifuge at from 1.0 to 0.6 g, depending on their radial position. The stationary middle layer contained experimental organisms identical to those in the centrifuge, except that they experienced microgravity. The rats were housed in the lower level. Five of the rats had been implanted with telemetry transmitters that sent body temperature readings to Earth.

The objectives of the U.S. experiments were to compare the effects of microgravity and artificial gravity on the genetics, aging, growth and development of biological specimens. Radiation exposure was also measured for some of the specimens.

Cosmos 782 launched from the Plesetsk Cosmodrome on 11 November 1975 and landed in Siberia $19\frac{1}{2}$ days later. Bad weather in the recovery area forced an early return. This was not the first time that the Soviets had been forced to deal with winter conditions in a Siberian touchdown, but the recovery of the Cosmos biosatellite posed special challenges.

The rat cages for Cosmos 782, consisting of individual cylinders for the rats, with waste collection below. This hardware was used on the first four collaborative Bion flights. (Photo: Dr. Delbert Philpott, courtesy NASA Ames History Office)

For investigations of how animals adapt to weightlessness, it was critical to evaluate them very shortly after their return, before they began to readjust to normal gravity. For this purpose, the Soviets airlifted a field laboratory to the recovery site within hours, including two huge inflatable, insulated tents, along with engineers and scientists, and portable generators capable of maintaining comfortable temperatures even in the midst of a Siberian snowstorm, blowing 70-knot winds [15].

Twelve of the 25 rats underwent autopsies at the landing site, while the remaining rats and other specimens were prepared for return to Moscow. Under the Soviet bio-specimen sharing programme, tissue samples from the flight animals were distributed to scientists from other countries. NASA would later adopt this same approach for its Shuttle/Spacelab experiments with rodents.

Ames's project manager Dr. Richard Simmonds and several other U.S. investigators spent several days in the laboratories at the IBMP preparing specimens for transport back to the U.S. In a 2006 interview, Simmonds recalled his return trip from Moscow. In Frankfurt, New York and Denver when he changed flights, he had to go onto the tarmac to oversee the transfer of the specially-built refrigerated box that held the specimens, which was the size of an office desk. In both American cities, he distributed some of the specimens to university investigators who would be studying the samples. The rest returned with him to Ames [14].

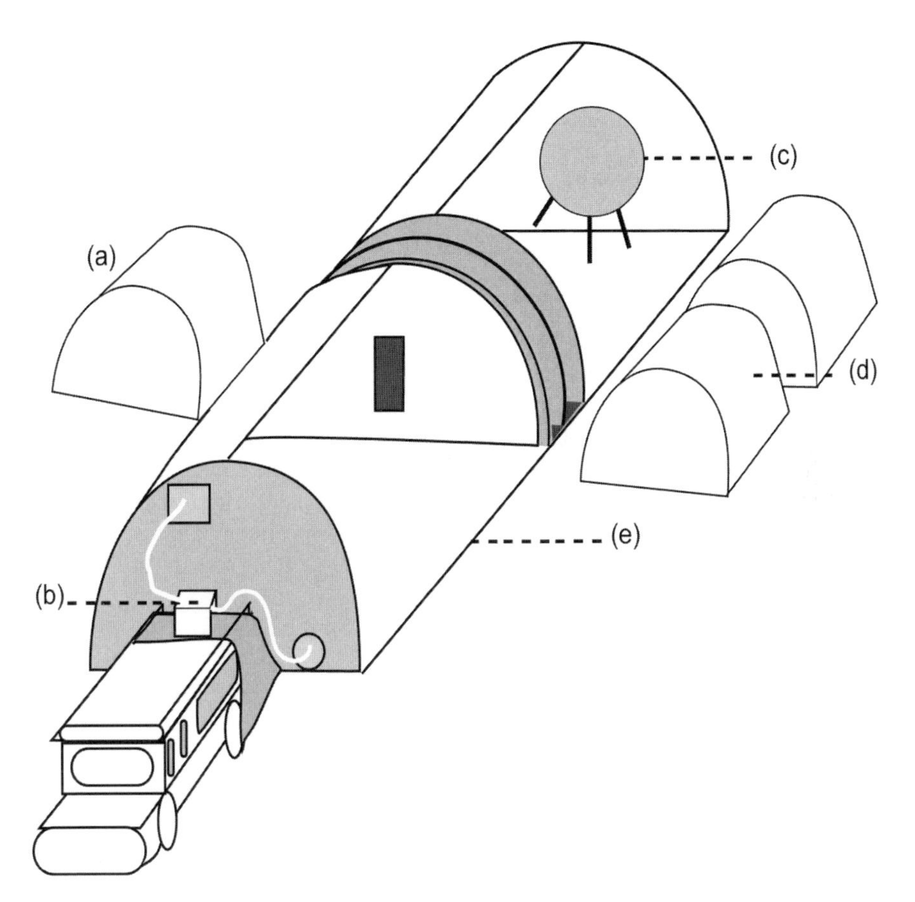

Soviet/Russian inflatable field laboratory airlifted to the recovery site for retrieval of biological specimens from the Cosmos/Bion satellite: (a) lavatory; (b) diesel pump; (c) biosatellite; (d) generators; (e) lab area. (Illustration: NASA, ARC, *Life into Space*)

Cosmos 782 findings

In September 1976, American and Soviet representatives gathered in Moscow for a symposium to discuss their findings. Kenneth Souza, who had replaced Simmonds as the Ames project manager, led the U.S. team. Researchers had been able to confirm that organisms experienced stress due to weightlessness and not to other flight factors. Physiological changes induced in the rats by weightlessness were reversed after a recovery period back on Earth, indicating that human space flights of this length would be safe [13].

Cosmos 782 was hailed as a great success, particularly as an example of international cooperation and in how effectively it had maximised the scientific returns from the mission.

The research findings of all investigators were published in *Final Reports of U.S. Experiments Flown on the Soviet Satellite Cosmos 782* [16].

COSMOS 936 AND 1129

One critically important feature of the Bion missions was their frequency. Flights were scheduled 2 years apart, giving researchers a dependable vehicle for a continuous series of experiments. Both exhilarating and demanding, it caught up the Americans in a continuous cycle of planning and preparation. Even before the September 1976 symposium in Moscow to discuss the findings from Cosmos 782, preparations were already well under way for the next launch in the series, scheduled for August 1977.

The procedures for U.S.–Soviet collaboration established for Cosmos 782 set the standard for future flights in the series, allowing things to run remarkably smoothly for these two Cold War adversaries. The procedure unfolded like this: Soviet review of U.S. experiment proposals; joint planning meetings in both countries; frequent teleconferences; the delivery of U.S. biological specimens and equipment to the Soviet Union; the training of Soviet technicians to handle U.S. experiments; the Soviet loading and recovery of experiments; NASA scientists analysing and preparing post-flight specimens in Soviet laboratories; the distribution of specimen samples to investigators; a year for tissue analysis and the writing of articles. This process culminated in a Moscow symposium for the presentation of findings.

A similar selection of biological specimens flew in the next two Cosmos satellites: rats and fruit flies on Cosmos 936 (3–22 August 1977) and rats, Japanese quail eggs, and carrot cells on Cosmos 1129 (25 September–14 October 1979). One problem with biological experiments in space is that there are generally too few samples, making it hard to draw conclusions. The Cosmos programme allowed for the replication of experiments on subsequent missions for verification of the findings.

Novel experiments on rats

Cosmos 936 again featured onboard centrifuge experiments, but this time on the rats. Of the 30 rats on this mission, 10 rode in 2 onboard centrifuges to test whether artificial gravity would counteract the effects caused by microgravity. Several sets of laboratory control experiments were conducted to provide comparative data. An experiment to test the effects of space radiation on the retina also used rats as subjects.

NASA had better accommodated itself to the schedule of flights by now. Fourteen U.S. experiments flew on Cosmos 1129, with a total of 40 scientists from 18 universities and research organizations participating in mission experiments. A more comprehensive study of re-adaptation was developed, which called for a number of flown rats to be autopsied at various times following their recovery, from 7 hours to 29 days. The Soviets even allowed American investigators to bump Eastern European scientists in competition for the most desirable specimens from the flown rats – the long bones of the legs and the calf muscles [9].

One novel experiment added for Cosmos 1129 dealt with rat embryology and featured a rodent mating chamber. Two male rats were held separately from five females. On day 2 of the flight, a partition opened between the two sections and the rats were allowed to intermingle and mate during the remainder of the flight.

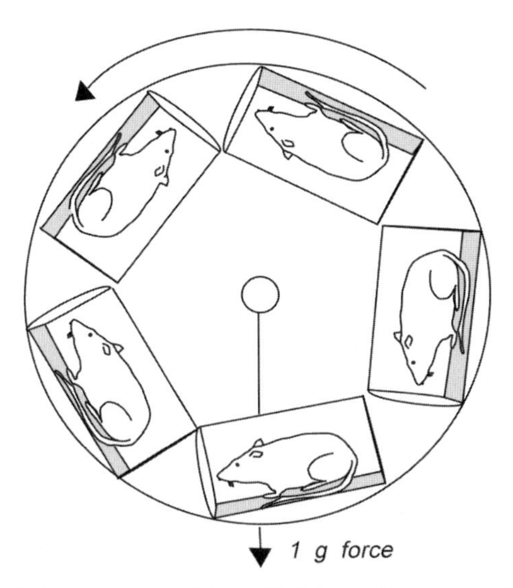

Some rats on Cosmos 936 were exposed to artificial gravity on a centrifuge during the flight. (Illustration: NASA, ARC, *Life into Space*)

The rodent mating chamber carried on Cosmos 1129: (a) female chamber; (b) feeding station; (c) remotely activated partition door; (d) male chamber. (Illustration: NASA, ARC, *Life into Space*)

Results from Cosmos 936 indicated that creating onboard artificial gravity did ameliorate some of the negative effects of weightlessness. For instance, some of the muscle and bone anomalies occurring in weightless rats did not occur in centrifuged rats. Uncentrifuged flight rats aboard Cosmos 1129 suffered decreased bone formation, volume, density and strength [10]. Although all rats carried in the mating chamber remained fertile and active, no births resulted from the flight.

Aside from the valuable scientific findings being generated from the Cosmos experiments, NASA also realised other benefits in the areas of personnel, technology and procedures. Opportunities for biological experimentation in space had never been so rich, and this helped to create a corps of life science professionals in NASA and at American universities who were developing an expertise in this area that would be valuable in future Cosmos flights and in the Shuttle programme [17].

In writing the final report on Cosmos 1129, Kenneth Souza, chief of the Ames Life Sciences Cosmos Project, also noted that "a low-cost systematic approach to the development, testing, and utilization of experimental hardware has been established which will be applied to the preparation of US biological experiments for flights aboard the Space Shuttle" [18].

In fact, the Cosmos and Shuttle programmes would serve as valuable complements to each other. U.S. life scientists and engineers worked on both projects, which resulted in hardware and technology developed for one programme later flying on the other, while experiments flown on one programme could similarly be replicated on the other [13].

COSMOS CONTINUES DESPITE COLD WAR

The fact that the U.S.–U.S.S.R. collaboration on Cosmos/Bion continued during the 1980s is an indication of the value both counties attached to the programme, especially considering the context of that decade. Cold War tensions between the U.S. and the Soviet Union ratcheted up a notch following the Soviet invasion of Afghanistan in 1979. Political repercussions from that event terminated all bi-national cooperation on space programmes, with the exception of the Cosmos/Bion programme. In fact, the 1980s would not only see four Bion missions, but the two nations would dramatically expand their collaboration, and monkeys would be added to the experimental payloads.

The motivation for NASA to continue the partnership was clear: the scientific payoff was great and the cost was low. NASA spent less than $1 million on each of the Cosmos flights in the 1970s. The per-flight cost would slowly climb to $3.7 million by the end of the 1980s; but this was still bargain pricing [13]. NASA estimated that a similar, uncrewed biosatellite programme conducted by NASA alone would have cost 10 to 20 times as much [19].

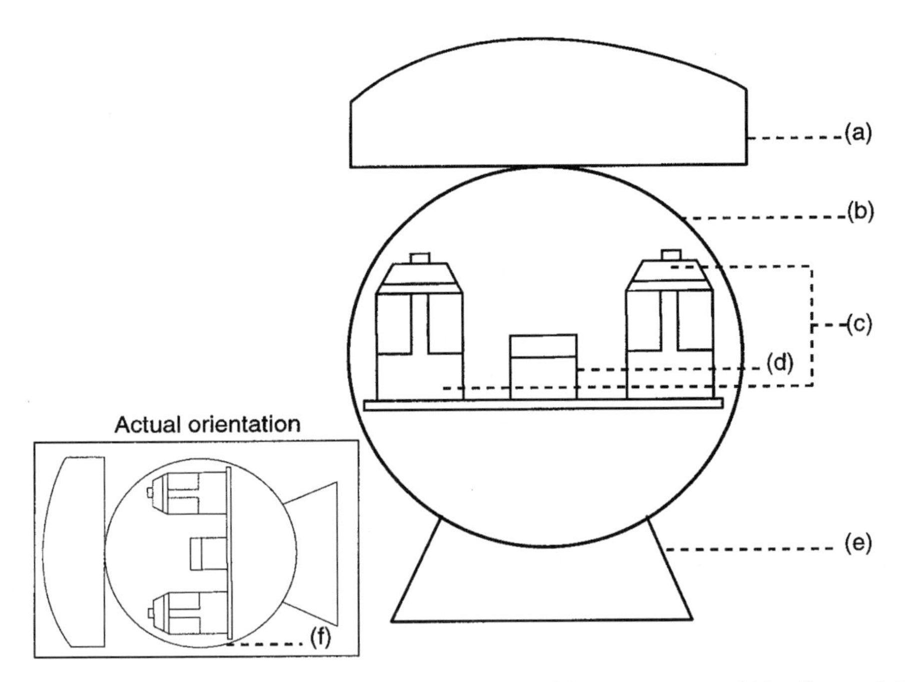

Cross section of Cosmos biosatellite: (a) instrument assembly compartment; (b) landing module; (c) monkey capsule; (d) rodent capsule; (e) energy storage unit; and (f) actual orientation just after landing. (Illustration: NASA, ARC, *Life into Space*)

Monkeys fly on Cosmos

The first Cosmos flight of the 1980s (Cosmos 1514) was markedly different from earlier flights, both in the scientific goals and the degree of cooperation it required. U.S. experiments went from passive, self-contained units to being integrated into the Soviet spacecraft. Two of these experiments – measuring cardiovascular and circadian rhythms – were integrated with onboard instrumentation, which required greater coordination of planning and engineering. Once again, experiments also studied the effects of weightlessness on animal physiology.

As with previous flights, the Soviets provided the animals. Ten pregnant Wistar rats joined two rhesus monkeys (*Macaca mulatta*) for the flight. On launch day, the rats were in day 13 of their 21-day gestation cycle. For the first time in the Cosmos/ Bion programme, monkeys were included in the payload. The monkeys, named Abrek and Bion, were about 3 years old and each weighed approximately 8.8 pounds.

The naming scheme for the monkeys on each of the Cosmos/Bion flights ran sequentially through the letters of the Russian alphabet (A, B, V, G, D, E, Zh, Z, I, K, L, M).

The Soviets had not flown monkeys before, having opted for dogs instead in their early rocket flights. Oleg Gazenko, the director of the IBMP, initiated the non-human primate programme for Cosmos in 1979. "It was so obvious to us that we had to fly monkeys if we were to resolve the big questions about manned space flight. But our

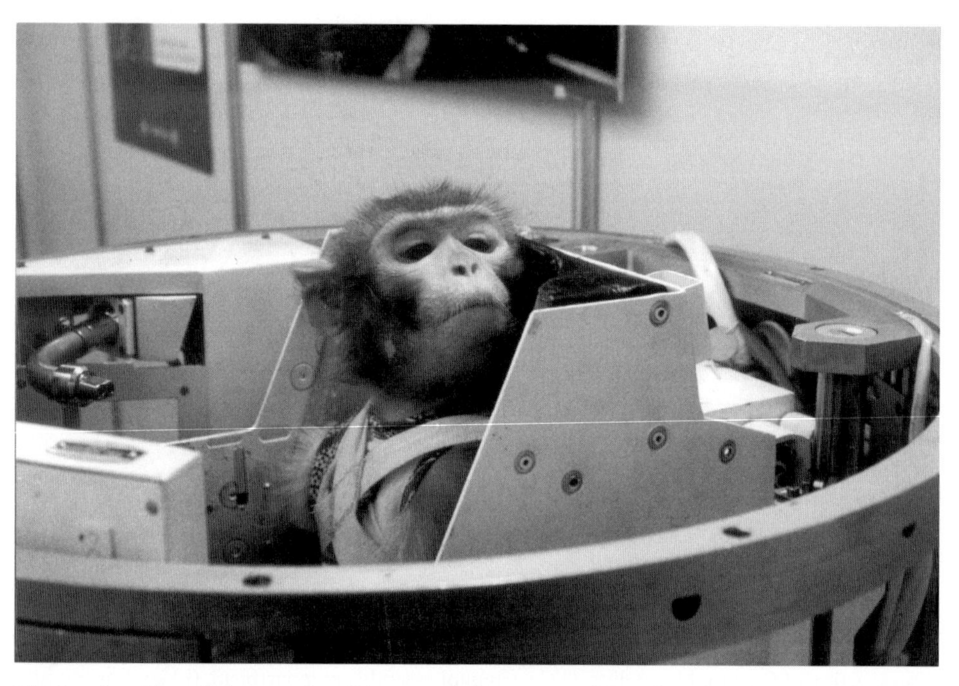

Monkey in a Bion flight capsule. This same design was used for all of the Bion monkey flights, Bion 6–12. (Photo: State Scientific Centre of Russian Federation, Institute for Biomedical Problems of the Russian Academy of Sciences)

expertise was with other animals, like mice and dogs, so we didn't dare fly monkeys for a long time.'' Gazenko credited close contact with American researchers for helping Soviet researchers overcome a ''mental barrier'' about using monkeys [10].

According to Bion project manager, Kenneth Souza, NASA shared basic information from its own experience working with monkeys, even lessons in what not to do, such as over-instrument the monkeys. ''The Russians took our lessons and learned from them'' [9].

Training for the Cosmos 1514 monkeys, which had begun 1 year in advance of the flight, involved familiarising them with the conditions of launch and re-entry and conditioning them to perform tasks for food rewards while in a restraint couch. Tasks involved pressing a lever with their feet, tracking a moving light with their eyes and a test of autonomic response to vertical oscillations [20].

In the months preceding launch, physiological sensors were implanted in the monkeys, including several brain and 15 subcutaneous electrodes, to take a wide array of physiological readings. In addition, a blood pressure cuff was surgically placed around the carotid artery of one of the flight monkeys and two of the control monkeys.

The two monkeys were secured in restraint couches in separate biological satellite BIOS capsules. The capsules were arranged so that the monkeys could see each other during the flight.

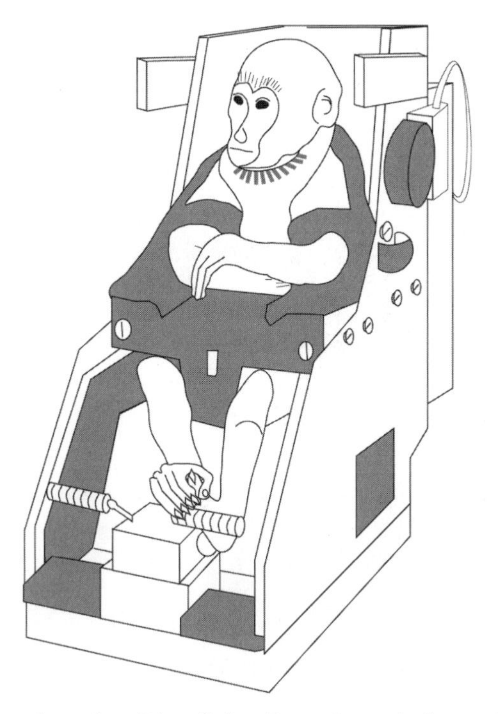

The animal restraint couch used on Bion flights. Restraint pads (in grey) held the animal in a comfortable position and cushioned against ground impact. (Illustration: NASA, ARC, *Life into Space*)

Two days into the flight, Bion's health began to deteriorate, forcing a decision to bring the flight back early. Nonetheless, Bion died 3 days after recovery. An autopsy indicated a strangled bowel as the cause of death, with no apparent connection to the flight or the implanted sensors [10].

Cosmos 1667 largely duplicated the animal payload from 1514, flying 10 rats and 2 rhesus monkeys, this time named Verny and Gordy. The U.S. flew only one experiment on this flight, a repeat of a cardiovascular experiment from Cosmos 1514, but with such improvements as more frequent sampling of in-flight data and a post-flight experiment 1 month after recovery using the same monkey.

Perhaps the most widely-reported event from Cosmos 1887 occurred when one of the monkeys, Yerosha, partially freed himself from his restraints and began to explore the capsule, necessitating an early return of the flight after $12\frac{1}{2}$ days. Touchdown occurred approximately 1,800 miles from the intended landing site, causing a delay in recovery and the loss of the some fish used in experiments, due to frigid weather [21].

Despite these setbacks, 1887 accomplished an ambitious schedule of experiments, 33 from the U.S. alone. A full complement of international partners also continued to contribute experiments. For this flight they included: Poland, Czechoslovakia, the German Democratic Republic, France, Romania, Bulgaria, Hungary and the European Space Agency.

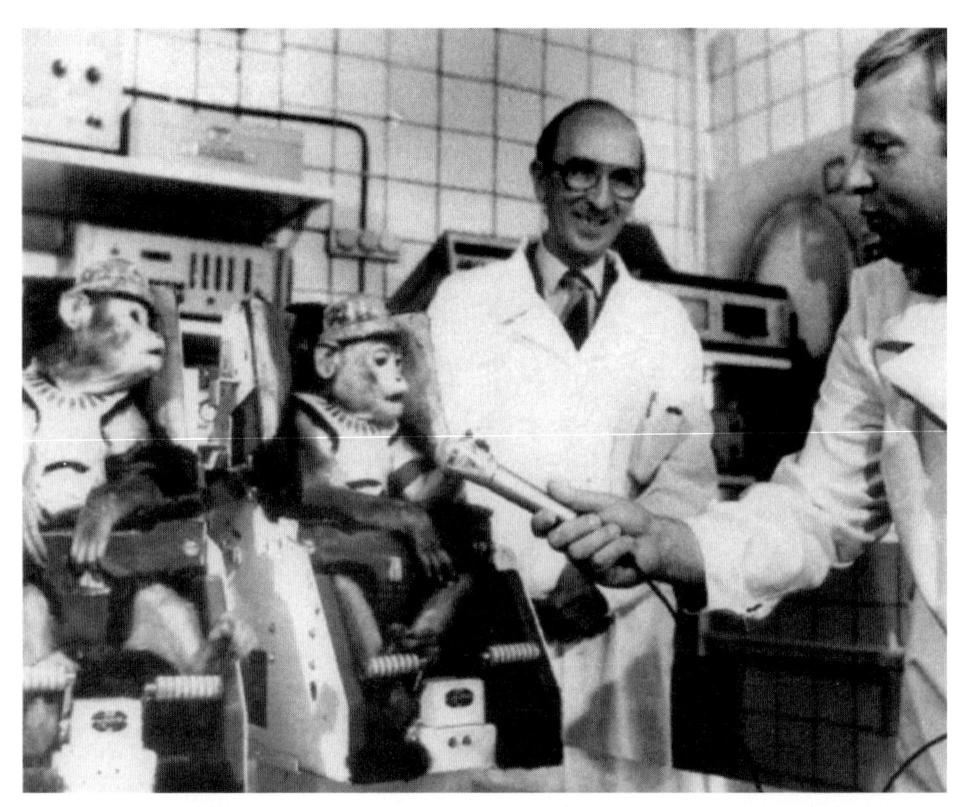

A Russian lab technician does a playful interview with rhesus monkeys Verny and Gordy before their flight aboard Cosmos 1667. Standing in the background is the director of the Institute of Biomedical Problems, Oleg Gazenko, a pioneer in the use of animals in space research. (Photo: authors' collections)

The final Cosmos mission of the 1980s, Cosmos 2044, launched on 29 September 1989 for a 14-day flight. Ten countries and the European Space Agency participated in this flight, which carried an extensive biological payload of rhesus monkeys, rats, fish, amphibians, insects, worms and other organisms. Among the 29 U.S. experiments were several conducted jointly with Soviet researchers.

By the end of the 1980s, researchers were able to draw some basic conclusions about the physiological reaction of animals to the weightless environment of space. Bones weakened and bone growth decreased; muscles atrophied and their biochemical composition changed; liver enzymes and red blood cells were also affected. Space flight also changed the behaviour and performance of flight animals [13].

The unprecedented opportunity offered with Bion was that new experiments could immediately be devised to test findings. Once physiological conditions were observed, new experiments could begin to study the physiological processes that caused those conditions. More sophisticated technology also allowed for more frequent physiological readings to better substantiate findings.

With his name emblazoned on his skullcap, Zhakonya is ready for flight aboard Cosmos 2044. The two monkeys on this flight provided researchers with a wealth of physiological data. (Photo: authors' collections)

POLITICS AND BIOSATELLITES IN THE 1990s

Just as the Soviet invasion of Afghanistan in 1979 cast a pall over international relations in the 1980s, the break-up of the Soviet Union more than a decade later became the defining political event in the early 1990s. Although neither event interrupted the schedule of Cosmos flights, the latter changed profoundly how they were carried out. But, rather than geopolitics, it would be the political storm surrounding animal experimentation that would bring about the end of the Cosmos missions.

By the time Cosmos 2229 launched in December 1992 for a 12-day flight, the Soviet Union had ceased to exist. The typically long list of countries providing experiments now included the Ukraine and Uzbekistan, new nations that had previously been Soviet republics.

The cast of international partners coordinated their efforts more closely on this flight than on any previous mission, especially in the development of hardware. The European Space Agency developed a Biobox facility, a fully automated, programmable incubator for biological experiments.

This greater coordination with the Russians exposed some of the fundamental differences between the two programmes. Michael Skidmore, Ames project manager for Bion 11, tells the story of working with the Russians during the final assembly of an enclosure cover. "The Russians found out that they were short three screws so someone on their team found a box of parts, dumped them out, fished through them until they found three screws that fit, and we were back in business. At NASA you can visualize a more 'resource intensive' environment where the screws would arrive in a certified container with a specific screw for each position and mounds of paperwork verifying each part's heritage back to the quarry" [22].

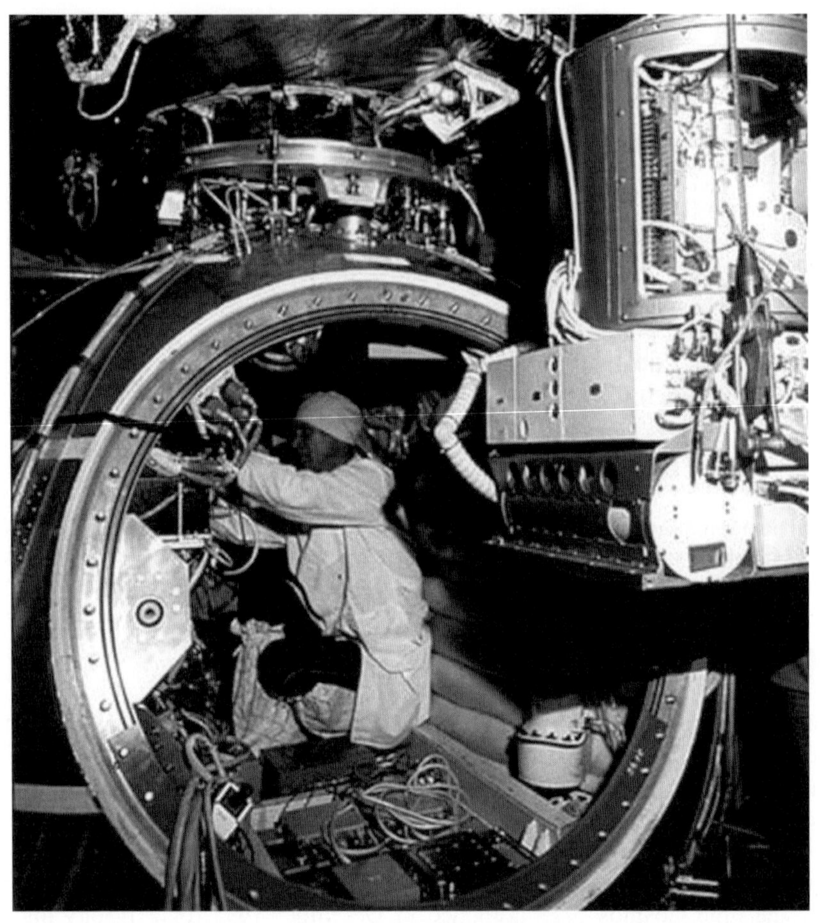

A Russian technician works inside the payload section of the Cosmos/Bion spacecraft. (Photo: authors' collections)

Despite these fundamental differences in approach, Skidmore was quick to point out that the teams managed to work effectively because U.S. and Russian technicians worked side by side to resolve problems as they arose, one unified team with the solitary goal of mission success.

Experiments on Cosmos 2229 studied the effects of microgravity on bones and muscles, including studies of bone strength, density, structure, biochemistry, calcium metabolism and neuromuscular function. Three studies examined the neurovestibular system, which controls the ability to perform coordinated movements. Experiments also investigated body temperature, circadian rhythms, plus immune and metabolic changes.

In addition to two rhesus monkeys named Ivasha and Krosh, specimens included Spanish newts, desert darkling beetles, fruit flies, silkworm larvae, clawed frog eggs, and animal and plant cell cultures.

The final Bion mission

The final flight in the Bion series (Bion 11) had no Cosmos designation. Unlike all previous flights, this was not a Soviet/Russian mission on which international partners flew experiments, but rather a true U.S.–Russian joint mission. In the past, the Soviet Union had assumed all costs of building, launching and recovering the satellites. With the demise of the Soviet Union, these costs would now be shared between NASA and the Russian Space Agency, amounting to about $15 million per partner, still a considerable bargain by NASA standards [13].

In fact, two missions were scheduled for the 1990s, Bion 11 and 12, in 1996 and 1998, both once again using two monkeys to further expand understanding of the effects of weightlessness. Bion 12 would largely repeat the experiments from Bion 11, to provide verification of the results.

A lot of science was planned for Bion 11. The number and complexity of the experiments run on each Bion mission had continued to increase and would reach its peak on Bion 11. What had begun for the U.S. as passive experiments and post-flight analysis of specimens had grown into an increasingly complex in-flight measurement of physiological performance. Equipment, instrumentation and data systems became more sophisticated and complex, while the procedures for handling of the experiments before, during and after the flights became more rigorous. The use of monkeys required lengthy and elaborate training and testing. International scientists went from being guest researchers to having full involvement in the planning, development and execution of all stages of the experiments [17].

Animal rights groups pressure NASA

NASA had to clear a few unexpected hurdles before Bion 11 flew. Animal rights groups in the U.S. had been applying increasing pressure on NASA to discontinue the use of monkeys in their research. That pressure contributed to NASA's decision in January 1994 to cancel a joint U.S.–French Shuttle project named the Rhesus Project [13]. The experiments planned for that mission were then transferred to the upcoming Bion 11 mission.

But, protests from the group People for the Ethical Treatment of Animals (PETA) intensified in 1995 and 1996 against the Bion 11 and 12 missions, as the organisation launched a public relations campaign and lobbied members of the U.S. Congress. An independent scientific review panel, convened at Ames Research Center in December 1995, praised the upcoming Bion missions and confirmed their scientific value [13]. But, PETA's efforts to influence Congress paid off when it succeeded in getting legislation passed to prohibit government funding for the Bion 11 mission.

Reacting quickly to this threat to its programme, NASA convened a Bion Task Force the following week to review the Bion 11 mission. Chaired by Dr. Ronald Merrill of the Yale University School of Medicine, the Task Force determined that the experiments planned for the mission were valuable and that the rhesus monkey was the proper species for the scientific objectives. Funding for Bion 11 and 12 was restored in the U.S. Senate, thanks in large part to a spirited defence of the programme by Senator John Glenn, the first U.S. astronaut to orbit the Earth [23].

Thwarted in its efforts to stop the programme, PETA staged a protest in early November 1996. Seven members of the group chained themselves together inside the Washington offices of NASA administrator, Daniel S. Goldin [24]. However, while this drama played out in the U.S., things stayed on schedule in Russia.

Preparing monkeys for Bion 11

In July 1995, the Institute of Medical Primatology, located near the Black Sea town of Adler, shipped 25 male rhesus monkeys to the IBMP in Moscow. For the next 18 months the animals trained to tolerate progressively longer periods of time spent in a flight safety seat and a capsule mock-up, to access food and drink from special dispensers, and to perform conditioned reflex motor tasks and psychomotor tests. Centrifuges introduced the monkeys to acceleration and deceleration forces, simulating launch and landing forces, respectively [25].

For one of the motor tasks, the monkeys had to touch a light stimulus that appeared on a panel before them. A second task required the monkeys to depress a foot pedal in response to a stimulus. The psychomotor test involved a series of graphic images displayed on a monitor. By manipulating a joystick, the monkeys could move a cursor to indicate a particular series of images.

Just as with engineering, differences between the U.S. and Russian "standard operating procedure" showed up as well in the area of animal care. According to Joseph Bielitzki, NASA's chief veterinary officer during Bion 11, the Russians also had a different attitude towards their animals, in particular the monkeys. "Their animal care programme was more personal than ours. Of course, you can't work with the same animals for almost two years, as they did, without becoming emotionally attached to them. They would come in after a weekend with a piece of vegetable or potato for them ... In some ways, they treated the monkeys almost like colleagues." Bielitzki found the Russians much more practical and their regulations less stringent, and yet they took good care of the animals and got good science from their experiments [26].

Six months prior to launch, 12 monkeys were selected to have numerous sensors and electrodes implanted and to undergo muscle and bone biopsies. Six monkeys made the final cut about 6 weeks before launch, two prime candidates (Lapik and Multik), two backups and two reserves. One month prior to launch, the monkeys were transported to the Plesetsk Cosmodrome launch site, where their health was carefully monitored. Just before launch, thermistors for recording skin temperature were attached to their heads and legs, brain electrodes positioned and bone growth marker injected [25]. Due to the complexity of the bioinstrumentation and onboard data systems, NASA personnel were present at the launch site to assist their Russian colleagues in last minute checkout of the equipment.

Bion 11 launched on 24 December 1996 in the midst of bitterly cold winter weather, with the temperature hitting $-37°F$. Once in orbit, the monkeys initially showed strikingly different reactions to their situation. Multik was active, while

Lapik was not. Later, they both performed their first motor task satisfactorily. But, then Multik damaged his display panel and could no longer respond to the task. Animal performance and response was otherwise unremarkable. In fact, both animals slept a lot. The ground crew monitored the monkeys via TV signals transmitted from the satellite.

When Bion 11 landed in Kazakhstan 14 days later, with its payload of monkeys, fruit flies, newts, molluscs and plants, the temperature was again around −37°F. For the first time in the Cosmos programme, NASA personnel were on hand for a satellite recovery. After the monkeys underwent a physical examination and a cleanup at the recovery site, and were found to be in good health, they were immediately returned to the IBMP in Moscow.

The tragedy of Bion 11

French and Russian veterinarians took part in the recovery. The American contingent included project veterinarian, Dr. John Fanton, as well as Dr. Bielitzki. Hired in 1995, Bielitzki was NASA's first ever chief veterinary officer. Over the years, vets had supported laboratory animal research at Ames, as well as on various NASA missions. Dr. William Britz had been on hand during the recovery of the chimpanzee Ham from his suborbital flight in 1961, but only on loan to NASA from the Air Force, which was also the case with Dr. Richard Simmonds on Cosmos 782 [27].

Because Bion 11 required fuller NASA involvement with all phases of the project, including the experimental animals, and because it had received such heated public criticism for its use of primates in the Bion programme, NASA wanted Bion 11 and Bion 12 to be model missions. Bielitzki, as the senior veterinary officer, was expected to provide close oversight of the animal experiments, to make decisions and to be accountable [26].

Back at the IBMP, 1 day following recovery, the Bion 11 monkeys underwent a battery of tests and biopsies, performed by the two French veterinarians. The Russian veterinarians were not present for this procedure. Bielitzki had already sedated the first monkey (Lapik), before a Russian anaesthesiologist arrived to sedate the other monkey, Multik [26].

During a 2006 interview, Bielitzki vividly remembered the scene. The procedures on the two monkeys had gone smoothly and been completed, the monkeys were coming out of anaesthesia, and the French vets were removing their scrubs. Bielitzki had just left the room to check on the status of the other animal when he was suddenly summoned back to find that one of the monkeys (Multik) had been physically sick and aspirated some vomitus. Frantic attempts to clear his airway and administer emergency drugs were ineffective. The condition quickly led to laboured breathing and heart arrhythmia, and finally cardiopulmonary arrest [25].

"The animal died in my hands," Bielitzki said, still frustrated by the unexpected turn of events, the loss of the animal and the enormous impact it would certainly have on the Bion programme [26].

Both Russia and the U.S. immediately launched reviews into the death of Multik. In the U.S., Dr. Ronald Merrill, who had chaired the review panel that recommended

NASA's participation in Bion 11, was called upon now to review the death of the monkey. The panel questioned Dr. Bielitzki at length, especially about the timing of the post-flight biopsies.

The biopsies themselves were a common procedure. Russian researchers had previously conducted similar biopsies on 10 monkeys after long space flights, following the same routine procedures used in scientific laboratories. But, they had never anaesthetised an animal on the 1st day after its return from space. Previously, it had never been done sooner than 7 days after landing. However, the Bion 11 protocol, which had been thoroughly reviewed by scientific panels, required that tissue be taken immediately from the monkeys in order to study how the body made the transition from weightlessness to gravity [28].

The Task Force determined there to be an unacceptable mortality risk in the use of anaesthesia for surgery on the day following return from space. It concluded that the physiological changes that occur in weightlessness had reduced the monkey's ability to tolerate the stress of anaesthesia [28]. Since the pending Bion 12 mission called for a repeat of the Bion 11 experiments, with the monkeys undergoing the same post-flight procedures, these monkeys would face the same "mortality risk".

"NASA is suspending its participation in primate research on the Bion 12 mission," read a NASA press release issued in April 1997, after Merrill's report [29]. Even though Merrill's report had not recommended the cancellation of involvement in Bion 12, NASA chose to cut its losses. The withdrawal of NASA's financial and technical support from Bion 12 was its death knell. The mission was subsequently cancelled, bringing an end to the Cosmos/Bion programme.

THE IMPACT OF BION

It is hard to overstate the value of the Bion programme and the impact it had on the understanding of how living organisms react to the weightless environment of space. The basic facts are impressive enough. Over the course of 23 years, 11 Cosmos/Bion satellites flew, with flight durations from 5 to 22 days. Experimental payloads included a variety of organisms, tissue cultures, plants and animals, though most experiments were carried out on Wistar rats and rhesus monkeys. Originally a Soviet programme, Bion transformed into a truly international collaboration, with investigators from many countries conducting onboard experiments and having access to specimens.

Collectively, the Bion missions represented the single largest source of data on how living organisms functioned in microgravity. With over 100 U.S. experiments, the missions accounted for one-half of all U.S. life sciences flight experiments with non-humans and resulted in more than 200 biomedical publications [17].

Hundreds of researchers, engineers and their students, in several countries, developed their skills and built careers around the Cosmos/Bion missions. Their expertise became a key resource for NASA and the Russian Space Agency as they conducted life science experiments aboard the Shuttle and the International Space Station.

It's fair to say that Bion was an enormously productive, cost-effective programme that did much to advance the study of life in space over two decades. Despite its abrupt termination in 1997, other avenues for life science research had begun to open with the U.S. Shuttle programme. The first veterinarian in space, Dr. Martin Fettman, flew as a payload specialist aboard STS-58 (18 October 1993), which carried Spacelab 2.

The flight of the shuttle *Columbia* (STS-78) in 1996 involved the Life Sciences and Microgravity Spacelab project. With studies sponsored by 10 nations and 5 space agencies, STS-78 was the first mission to combine both a full schedule of microgravity studies with a comprehensive life sciences payload. The flight included the veterinarian astronaut, Dr. Richard M. Linnehan.

Two years later, the shuttle *Columbia* (STS-90) carried aloft Neurolab (17 April to 3 May 1998), again with Dr. Linnehan aboard. The seven-person crew served as subjects of experiments as well as monitors of live biological experiments. Both of these missions were productive and served as models for future life science studies aboard the International Space Station.

Because of the circumstances surrounding the cancellation of the Bion programme, the non-human primate was essentially lost as a test subject for future space missions. About the time that the Bion 11 monkey died, NASA, in collaboration with the French space agency, CNES, had been training rhesus monkeys to fly on Spacelab Life Sciences 3, but that entire shuttle mission was cancelled.

Researchers studying the physiological effects of microgravity and space radiation were now forced to use either cells and microbes, rodents or humans, or to extrapolate results from laboratory experiments. This was a huge setback, according to Bielitzki. "That very same experiment should have been duplicated on Bion 12. They needed to determine whether that monkey's reaction was idiosyncratic or the result of space flight." The problem as he sees it is that this could be a huge problem for astronauts who might require emergency drugs or a surgical procedure post-flight [26].

For the potential problems it revealed, Kenneth Souza went so far as to say, "The death of that monkey was probably the most valuable piece of information learned in all of the Cosmos flights" [9]. Because of what that death suggested, namely that sensitivity to drugs is altered during and following spaceflight, NASA subsequently changed the procedures it would use to treat astronauts post-flight.

REFERENCES

[1] Asif Siddiqi, *Challenge to Apollo: The Soviet Union and the Space Race, 1945–1974*, NASA, Washington, D.C., 2000.

[2] V.V. Antipov, N.L. Delone, M.D. Nikitin, G.P. Parfyonov and P.P. Saxonov, "Some Results of Radiobiological Studies Performed on Cosmos 110 Biosatellite," *Life Science Space Research*, vol. 7, pp. 207–8, 1969.

[3] Raymond H. Anderson, "Soviet Dogs Lost Muscular Control in Space," *New York Times*, 17 May 1966.

[4] John B. West, "Physiology of a Microgravity Environment: Historical Perspectives: Physiology in Microgravity," *Journal of Applied Physiology*, vol. 89, no. 1, pp. 379–384, July 2000. Website: *http://jap.physiology.org/cgi/content/full/89/1/379*, accessed 29 March 2005.

[5] "Soviet Space Medicine, Session Two," Smithsonian Videohistory Programme transcript, interview with Oleg Gazenko, Abraham Genin and Evgenii Shepelev. Cathleen S. Lewis, interviewer, November 28 1989.

[6] Yevgeniy Ilyin interview conducted by Rex Hall and Bert Vis, Moscow, 17 April 2002, unpublished, used with permission.

[7] "Searching the Horizon: A History of Ames Research Center, 1940–1976, The Life Sciences Directorate," NASA SP-4304, NASA, Washington, D.C.

[8] Homer E. Newell, *Beyond the Atmosphere, Early Years of Space Science*, Chapter 16, NASA History Office, 1980. Website: *http://www.hq.nasa.gov/office/pao/History/SP-4211/ch16-1.htm*, accessed 4 March 2005.

[9] Kenneth Souza telephone interview conducted by Chris Dubbs, 8 March 2006.

[10] Kenneth Souza, Robert Hogan and Rodney Ballard (eds), *Life into Space: Space Life Sciences Experiments NASA Ames Research Center 1965–1990*, NASA, Washington, D.C., 1995. p. 42.

[11] President's Science Advisory Committee, Space Science and Technology Panel, "The Biomedical Foundations of Manned Space Flight," Washington, D.C., November 1969.

[12] Space Science Board, Space Biology, Washington: National Academy of Sciences, 1970; Space Science Board minutes, 13 January 1970.

[13] Kristen E. Edwards, "The US-Soviet/Russian Cosmos Biosatellite Program" (draft version), October 1999. NASA, Ames Research Center.

[14] Dr. Richard Simmonds telephone interview conducted by Chris Dubbs, 16 February 2006.

[15] "American Experiments on Cosmos 782," NASA Facts NF-77/10-77, Cosmos 782 Folder, Cosmos Biosatellite Files, Ames History Project Archive, Ames Research Center.

[16] S.N. Rosenszweig and K.A. Souza, *U.S. Experiments Flown on the Soviet Satellite Cosmos 782*, Final Reports, NASA TM-78525, NASA, Washington, D.C., September 1978.

[17] Richard E. Grindeland, Eugene A. Ilyin, Daniel C. Holley and Michael G. Skidmore, "International Collaboration on Russian Spacecraft and the Case for Free Flyer Biosatellites," in Gerald Sonnenberg (ed.), *Experimentation with Animal Models in Space*, Elsevier, North-Holland, 2005.

[18] M.R. Heinrich and K.A. Souza, *Final Reports of U.S. Rat Experiments Flown on the Soviet Satellite, Cosmos 1129*, NASA TM-81289, NASA, Washington, D.C., August 1981.

[19] "Bion Benefits People on Earth." Website: *http://lifesci.arc.nasa.gov/Bion_Ben.html*, accessed 15 December 2005.

[20] Tana M. Hoban-Higgins, Edward L. Robinson and Charles A. Fuller, "Primates in Space," in Gerald Sonnenberg, (ed.), *Experimentation with Animal Models in Space*, Elsevier, North-Holland, 2005.

[21] "A Brief History of Animals in Space," NASA History Division. Website: *http://history.nasa.gov/animals.html*, accessed 7 March 2006

[22] Michael Skidmore, "Three Screws Missing," Academy Sharing Knowledge, NASA. Website: *http://appl1.nasa.gov/ask/issues/5/stories/5_three_skidmore.html*, accessed 25 February 2006.

[23] "Senate Keeps Monkeys Floating in Space," *Human Events*, vol. 52, issue 43, p. 23, 15 November 1996, .

[24] "PETA Invades NASA," *Science*, vol. 274, issue 5290, p. 1085, 15 November 1996.

[25] Eugene A. Ilyin *et al.*, "Bion 11 Mission: Primate Experiments," *Journal of Gravitational Physiology*, vol. 7, no. 1, 2000.

[26] Joseph Bielitzki telephone interview conducted by Chris Dubbs, 27 February 2006.
[27] "Veterinary Medicine's Reach Out of this World," *dvm: The Newsmagazine of Veterinary Medicine*, 1 September 2002.
[28] Tony Reichhardt, "Spaceflight Monkey's Death 'Raises New Safety Concerns,'" *Nature*, vol. 387, issue 6628, p. 4, 1 May 1997.
[29] "NASA Concurs with Independent Review of Bion 11 Mission," NASA press release, NASA, Washington, D.C., 22 April 1997.

11

End of an era

At an ever-intensifying rate, animals, insects and other forms of biological life were fired into space in the 1960s and 1970s. Many perplexing questions needed to be resolved about survival in space, and these two decades would see creatures from nations such as China, France, the United States and the Soviet Union sent on missions of increasing complexity in order to answer these questions.

While the world may have been following the widely-reported manned race to the moon, rats, cats, spiders, mice, frogs, dogs and other animals were being routinely sent into space as biological test subjects. In 1968, two tortoises from the Soviet Union aboard a spacecraft named Zond became the very first creatures from Earth ever to fly around the moon.

The highly-publicised use of chimpanzees and monkeys in space flight was also moving increasingly into the spotlight, but for all the wrong reasons. Serious efforts would be made by animal support groups to save them from a lesser-known life of misery and degradation, eked out in untenable conditions well away from the public eye and awareness.

CHINA LOOKS TO THE FUTURE

The first launch of a Chinese T-7A sounding rocket, euphemistically designated the Peace-1, took place on 1 December 1963 from the Shijiedu launch facility, located on China's east coast. Aided in its development by Russian scientists, the two-stage rocket was an early element of China's forthcoming Long March series of indigenously-built, expendable launch boosters. Having reached an altitude of 72 miles, both the reusable upper stage of the T-7A and its detachable nose payload were deemed to have operated as planned. A successful parachute descent followed, and the components were eventually retrieved by the recovery team.

The T-7A rocket

Earlier that year, the China Academy of Sciences (CAS) Institute of Biophysics had proposed the use of the T-7A sounding rocket for what it designated as biological and high-altitude medical research. Some researchers into the Chinese space programme feel this decision may have actually been more largely influenced by the pending use of these rockets to gather high-altitude samples from China's top-secret atmospheric nuclear tests.

It was a well-tested rocket; an earlier version known as the T-7 had been fired as high as 36 miles to test winds and evaluate atmospheric temperature, pressure and density. In order to conduct further biological and geophysical experiments requiring far greater altitudes and the capability of carrying much heavier payloads, an advanced two-stage variant, designated the T-7A, was authorised in January 1962. Modifications to the T-7 were then carried out by the Shanghai Design Institute of Machine and Electricity.

Following these modifications, the T-7A could implement earlier CAS plans to launch several albino rats and mice on high-altitude ballistic flights. Each rocket was now capable of carrying 2 restrained albino rats, 3 free-floating albino rats, 4 free-floating albino mice and 12 test tubes containing fruit flies and other biological specimens. The two restrained rats were enclosed in a box-like container mounted on a board with springs attached to dampen the effects of vibration. The container itself was located within a copper wire screen that would shield it against random radio waves from the electrocardiogram telemetry system.

Re-designated the T-7A(S), the modified sounding rocket incorporated a newly-designed, 374-pound capsule mounted at the top of the rocket. Twelve inches in diameter, this capsule comprised the nose cone, photographic instrumentation, a sealed biological chamber with life-support capability and a telemetering system.

Mice such as these would be carried on Chinese T-7A rockets. (Photo: courtesy China Astronautic Publishing House)

A parachute recovery system had also been incorporated at the lower end of the separable nose capsule to allow for a post-flight retrieval. Altogether, the launch weight of the two-stage rocket and its payload was nearly 2,960 pounds [1].

Mission experiments

Several scientific and biological experiments were planned during the test rocket flight and following the recovery of the nose capsule. These would include:

- Transmission of electrocardiogram data from one of the restrained rats.
- Photographing the posture and attitudinal changes of the free-floating rats during the transition from high-g to zero-g. As part of a comparative study, the membranous labyrinth in the ear of one rodent had been cut, while that of the other rat had been left intact.
- Evaluation of the effects of solar radiation on the rats.
- The blood from the second restrained rat would undergo post-flight physical and chemical analysis.
- Following the recovery of the nose capsule, the albino mice would be dissected to study the effects of the high-altitude environment on their tissues and organs.
- The surviving rats, mice and fruit flies would be encouraged to breed to allow observation of the possible effects of the spaceflight environment and radiation on reproduction.

On three occasions a T-7A(S) rocket soared into the skies from the Shijiedu launch site carrying biological specimens. The first biological flight took place on 19 July 1964, and the second left the launch pad 11 months later, on 1 June 1965. The third launch in the series lifted off just 4 days later, on 5 June. Each of the rockets performed well, achieving altitudes ranging from 37 to 44 miles, and all three nose capsules were later recovered.

The success of the three-flight series encouraged the CAS Institute of Biophysics to propose a further series of rocket flights using the T-7A, this time upgraded to carry small dogs and other animals. In addition to a raft of physiological, behavioural and genetic tests on these live subjects, the flights would also be used to measure the intensity of ionising radiation that would be experienced in the biological cabin. A modified rocket, designated the T-7A(S2), was manufactured. This booster offered increased capacity in the nose section and could now carry 12 biological test tubes with fungus and actinomycin, in addition to a small dog and four albino rats. On board, lightweight 8-mm film cameras would record their reactions and behaviour under high-g forces and in brief moments of weightlessness.

According to British researcher Phillip S. Clark, in his 2004 report "The Development of China's Piloted Space Programme: From Sounding Rockets to Shen Zhou 5," published in the *Journal of the British Interplanetary Society*, several modifications had now taken place:

The nose section of T-7A(S2) comprised a nose cone, a biological cabin and a recovery system cabin. The magnetic tape recording system, nuclear sensitive emulsion block and electrical circuit control system were installed in the upper section of the cabin. The photographic and life-support systems, albino rats' box, biological test tubes and the dog's tray were installed in the lower cabin by four shock absorbers. In front of the tray there was a device for studying conditional reflex. A container was fixed behind the dog to retrieve its waste, and a safety-belt was provided on the tray ... There were three ventilation windows on the biological cabin which automatically opened at altitude to provide the animals with fresh air during descent [2].

Choosing the canine candidates

Given strict weight, mass and volume restrictions, the chosen dogs had to weigh no more than 14 pounds, and would have to be carefully trained to avoid any prospect of panic in a tightly-enclosed space. Thirty candidates were selected, and these underwent initial training and observation to assess their suitability and demeanour under stress. After a series of five evaluations the field was reduced to six dogs, and further tests narrowed the field to just two young dogs. One was a male dog named Xiao Bao (Tiny Leopard), and the second canine was a petite bitch called Shan Shan (Coral). These two dogs, which would be clad in protective garments for the high-altitude

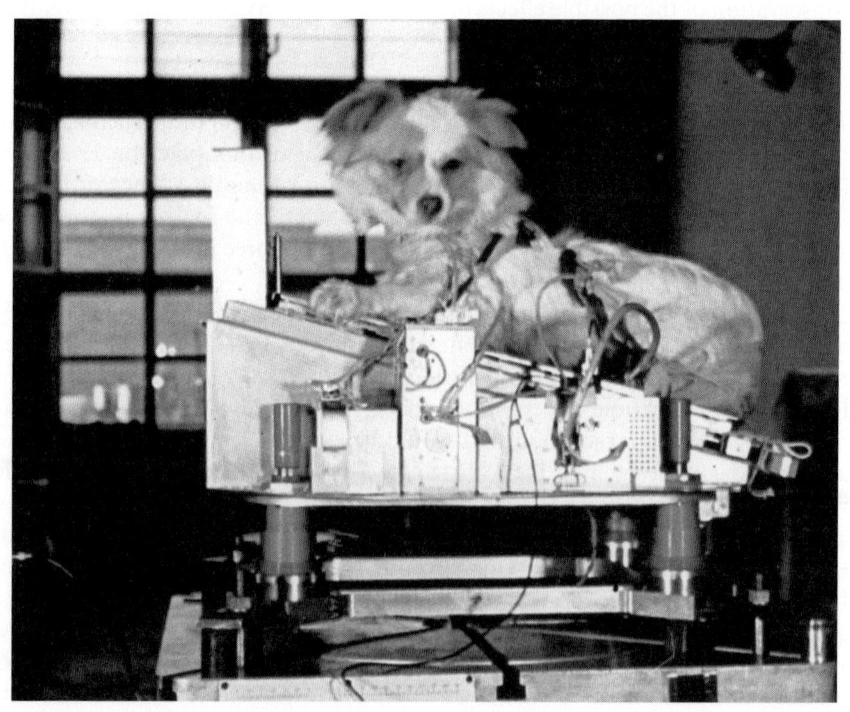

Chinese space dog, said to be Xiao Bao. (Photo: courtesy China Astronautic Publishing House)

ballistic flight, were carefully but rigorously trained to accept the unnatural conditions they might experience as part of an actual launch and recovery sequence enclosure in a confined space, noise, vibration, acceleration g-forces and the irritation of physiological sensors surgically implanted under their skin.

The first T-7A(S2) mission with Xiao Bao aboard took to the skies on the morning of 14 July 1966. Its trajectory was true, and the rocket executed the planned flight profile with a pleasing accuracy.

As the nose section returned beneath a billowing white parachute at a descent rate of around 33 feet per second, a circling Air Force helicopter took photographs. Once the nose section had touched down, recovery teams were quickly on the spot. They opened the biological cabin and released Xiao Bao, who seemed to be in good spirits. A medical examination held at the landing site confirmed that the frisky male was in perfect health. The rats that had accompanied him were quickly whisked away to laboratories for dissection and a meticulous examination of their internal organs.

The second and final flight of the T-7A(S2) took place just 2 weeks later, on 28 July 1966. Once again the biological cabin landed safely after an ideal launch and flight, and the dog Shan Shan was released into the care of her handlers, along with the albino rats and other animals [2].

An extended period without any biological rocket flights then seems to have ensued, and it would not be until 1990 – some 24 years later – that a new generation of animals rode into space atop Chinese rockets.

PROJECT GEMINI

NASA's Project Gemini was an intermediary, but crucial, manned spaceflight programme that would move America out of the one-man suborbital and orbital flights of Project Mercury, developing essential skills and procedures such as orbital rendezvous and extra-vehicular activity (EVA, or spacewalks), which were vital to the future success of the lunar-landing Project Apollo.

Early biological experiments

The first of the two-man Gemini–Titan missions, GT-3, carried astronauts Gus Grissom and John Young into orbit on 23 March 1965. Their primary task was to test the new craft's manoeuvrability during a three-orbit flight. NASA's Ames Research Center flew a single life sciences experiment on the mission, the object of which was to investigate the effects of microgravity on the fertilisation, cell division, differentiation and growth in a simple biological system; namely, eggs of the sea urchin *Arbacia punctulata* [3].

The urchin eggs were flown in a cylinder just over $6\frac{1}{2}$ inches long $\times$ 3 inches in diameter. The cylinder contained eight specimen chambers, each of which was divided into three compartments containing sperm, ova or fixative. The eggs could be fertilised, or fertilised eggs could be fixed, when the contents of the compartments were mixed by manually rotating a handle. Eggs in four of the chambers were fertilised just before launch, and the two astronauts were charged with fertilising the eggs in the other four chambers shortly after their spacecraft went into orbit.

Unfortunately for the experiment, there was a failure in the handle operating mechanism, as recalled by mission commander Grissom in his posthumously-released 1968 book, *Gemini*.

All I had to do was turn a knob, which would actuate a mechanism, which would fertilise some sea urchin eggs to test the effects of weightlessness on living cells. Maybe . . . I had too much adrenalin pumping, but I twisted that handle so hard I broke it off. (Later we learned that, by an odd coincidence, the ground controller who was duplicating our experiment, second for second, back on earth, had broken his handle in exactly the same fashion) [4].

Other life science experiments, involving embryonic frog egg packages (*Rana pipiens*), would later be carried on GT-8 and GT-12 in March and November of 1966, respectively. On these flights the cylinders would have four chambers, each containing five frog eggs [5].

The experiment was only partially completed on GT-8 due to the early termination of the flight, and a re-flight of the experiment on GT-12, while not conclusive, suggested that a gravitational field is not necessary for the eggs to divide normally. The tadpoles produced died for unknown reasons several hours after recovery of the spacecraft [6].

TORTOISES IN A RACE TO THE MOON

By late 1968, it seemed evident that the Soviet Union was slipping badly in the race to place a man on the moon, although Western observers knew from bitter experience that sheer audacity might get them over the line. However, the Soviet's heavy-lift N-1 booster, once envisaged as out-muscling NASA's massive Saturn V, was fast becoming an expensive disaster, involved in catastrophic post-launch failures at a time when America's astronauts were rigorously preparing to carry out the first manned Apollo mission, in an Earth-orbiting test of the new spacecraft. Russia's manned lunar programme, it seemed, was headed for oblivion.

The Zond programme

On 14 September 1968 there was great excitement in the Soviet Union, centred around a spacecraft heading for the moon after the successful night launch of a three-stage Proton rocket from Baikonur's Launch Complex 81. Zond 5 (officially designated 7K-L1) was one of a series of spacecraft intended to serve as precursory flights for later manned missions. It left the ground carrying the very first creatures and lower life forms ever to achieve a circumlunar flight around Earth's nearest celestial neighbour [7].

The biological payload transported in Zond 5's descent module comprised two Steppe tortoises (*Testudo horsfieldi*), some meal worms (*Tenebrio molitor*), hundreds of fruit fly (*Drosophila*) eggs, plants, seeds, bacteria and other living matter. The tortoises had been inserted into the spacecraft 12 days earlier, after which time they no longer received any food. As spaceflight researcher Asif Siddiqi noted in *Challenge to Apollo: The Soviet Union and the Space Race, 1945–1974*, "Physicians would study the deprivation of food until the recovery of the spaceship, to study the pathomorphological and histochemical changes in the animals over the course of several weeks" [8].

Following the launch, Zond 5 was initially inserted into a single "parking" orbit around the Earth while controllers checked the spacecraft's systems. They were satisfied the mission could proceed. As planned, the Block D third-stage engine fired once again, sending the spacecraft on a trajectory that would eventually loop it around the moon.

Trouble on the outbound journey

A major problem occurred on the way to the moon when the main stellar attitude control optical surface became contaminated by residue, rendering the control sensor unusable. Despite this, the spacecraft remained basically on course, eventually rounding the moon on 18 September. It was a wide orbit, extending some 1,200 miles above the lunar surface, but it was nevertheless a remarkable achievement for that time. Having circled the moon, Zond 5 began a return journey to Earth.

On the evening of 19–20 September, listeners tracking the flight of Zond 5 from the Jodrell Bank radio telescope in Cheshire, England, were surprised to suddenly hear the sound of a human voice emanating from the returning spacecraft. In the absence of information to the contrary, it was widely speculated that this was merely a recorded message that would allow Soviet controllers to monitor voice transmission from an inbound spacecraft. Some time later, opinions changed, and it is now believed that the voices were actually those of cosmonauts involved in the lunar-landing programme. They had been positioned at Soviet tracking stations and were transmitting reports via Zond 5 in order to practice their roles as part of an actual lunar crew.

Zond splashes down

During Zond 5's return journey a second attitude control sensor failed, ending hopes of a guided re-entry. Controllers now had to be satisfied with a direct ballistic entry into the atmosphere, which would result in a landing somewhere in the Indian Ocean.

On 21 September, having survived a faster-than-expected ballistic re-entry, the descent module was tracked heading for a splashdown off the coast of Madagascar. The craft's parachute was deployed and blossomed out some 4 miles above the backup landing area. The nearest Soviet craft equipped with tracking radar and searchlights was located some 100 miles away. Several hours after splashdown, Zond 5 was located bobbing around in heavy seas and successfully recovered by crewmembers from the Academy of Sciences ship, *Borovichi*. As the spacecraft had come down

The Zond 5 capsule floating in the Indian Ocean. It was recovered several hours after a heavy splashdown by the crew of an Academy of Sciences ship. (Photo: authors' collections)

in international waters, American Navy ships were also patrolling the area, ready to observe the recovery.

Once Zond 5 was safely on board the *Borovichi*, the module was quickly concealed beneath a large tarpaulin, at which time the American ships lost interest and sailed away. The spacecraft was later transferred to an oceanographic vessel, the *Vasily Golovnin*, which carried it to Bombay. On its arrival at Bombay on 2 October, Soviet officials packed Zond 5 into a container before it was driven to the city airport and transported back to Russia aboard an Antonov An-12.

The two Zond 5 tortoises are examined in Moscow following their historic circumlunar orbit, becoming part of the first payload of terrestrial creatures to fly to the moon and back. (Photo: authors' collections)

The heat-scarred capsule arrived in Moscow on 7 October, and was finally opened 4 days later, so checks on the condition of the biological cargo could begin. Among other results, it was later revealed that the two tortoises had lost around 10% of their body weight, but they were nonetheless healthy and displayed no loss of appetite [8].

More tortoises to the moon

As a precursor flight in the L-1 lunar programme, Zond 5 undoubtedly gave great encouragement to Soviet planners, allowing them to seriously contemplate a possible manned circumlunar flight of two cosmonauts in January 1969. However, this was contingent upon carrying out two more successful L-1 flights in the Zond series.

A number of different tortoises, worms, plants, seeds and other biological specimens would later be lofted into space on circumlunar flights (Zond 6 and 8), with the capsules and their cargoes successfully recovered for examination.

The Soviets had come tantalisingly close to achieving manned lunar missions, and America would send men to the moon by what was later revealed as a very narrow margin. But, while the Soviet Union's spacecraft and crews had been ready to proceed, catastrophic events had overtaken and curtailed the heavy-lift N-1 booster programme.

The maiden flight of the N-1 on 3 July 1969 – barely 2 weeks before the launch of Apollo 11 – ended in massive failure at the Baikonur Cosmodrome launch pad. Later investigations revealed that a loose bolt had been sucked into a fuel pump, causing the mighty booster to become erratic and explode moments after lift-off, destroying the booster and the launch pad. It would be the biggest explosion ever recorded in the history of rocketry [9]. The Soviet manned lunar space programme would never recover from these monumental setbacks, and while there were later attempts to continue the N-1 programme it was eventually abandoned following two further launch failures in 1971 and 1972.

While nothing is known of the eventual fate of the two tortoises, the Zond 5 descent module that triumphantly carried them around the moon is now on permanent display in Star City, near Moscow.

Several years later, along with two visiting crews of two cosmonauts, another two tortoises (again *Testudo horsfieldi*) and some zebrafish (*Danio rerio*) would occupy the Salyut 5 space station, which was actually the third and last of the Soviet Union's highly-secret Almaz military space stations [10]. It is not known if the biological cargo returned with the Soyuz 24 crew or together with the film canister that was purposefully jettisoned for recovery shortly after. The Salyut 5 space station itself was de-orbited on 8 August 1977.

THE FROG OTOLITH EXPERIMENT

At 1:00 a.m. on the morning of 9 November 1970, a Scout 1B four-stage, solid-propellant rocket took to the skies from Wallops Island, Virginia. It was on a mission undertaken by NASA's Office of Advanced Research and Technology (OART) to

Two Soviet N-1 moon rockets on the launch pads at Baikonur Cosmodrome in July 1969. (Photo: NASA)

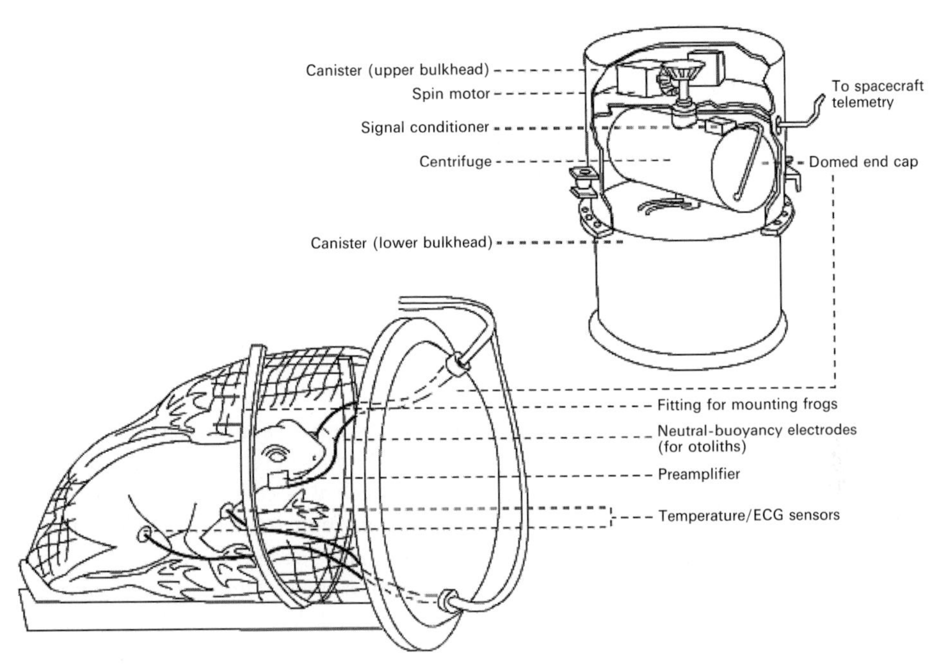

Frog Otolith Experiment Package, or FOEP. (Illustration: NASA, ARC, *Life into Space*)

achieve a better understanding of the function of the otolith and how it is affected by microgravity. The otolith is a sensory organ of the inner ear of vertebrates, which normally responds to changes in an animal's equilibrium and orientation within our planet's gravitational field.

Preparation for flight

The subjects used in this experiment were two male bullfrogs (*Rana catesbeiana*), a species with an inner ear labyrinth structure similar to that of humans. Initially known as the Frog Otolith Experiment (FOE), it was developed by Torquato Gualtierotti of the University of Milan, Italy, the later author of a book on the function and morphology of the vestibular system. He had been assigned to the Ames Research Center in California as a resident research associate, and his work was sponsored by the National Academy of Sciences.

The FOE would allow biologists and other researchers to gather neuro-physiological data on the response of the otolith to prolonged periods of weightlessness, such as that which would be encountered by astronauts during extended flights in micro-gravity.

Originally developed to fly during the Apollo Applications (later Skylab) Program, the Frog Otolith Experiment Package (FOEP) or what was now called the Orbiting Frog Otolith (OFO) programme was instead modified for a flight of 3–5 days on an unmanned satellite. The unmanned vehicle would better provide the low acceleration levels necessary to conduct the experiment.

Both animals would be housed in a compact, water-filled and self-contained centrifuge apparatus, which supplied the acceleration rotation required to partially simulate gravity during orbit. The bullfrogs would have ECG electrodes surgically implanted in their thoracic cavities and other electrodes connected to their vestibular nerves. These would measure the bioelectric action potential under conditions of weightlessness and during repeated simulated gravity stimulus obtained by activation of the centrifuge. The water in the apparatus would not only serve to cushion the vibration and acceleration associated with the launch phase, but would continually move carbon dioxide and heat away from each of the frogs. Data would be sent to ground receiving stations by telemetry. Unfortunately for the occupants, it was never planned to recover the satellite once the otolith experiment had ended.

The two amphibians were surgically prepared 12 hours before the expected launch time and then immersed in the water apparatus which was sealed inside the FOEP. They would breathe quite normally in the water through their skin. Meanwhile, a backup unit with similarly-prepared specimens was also readied in case of any problems. At around 10:00 p.m. the FOEP was installed in the satellite, and 3 hours later the Scout 1B (S178C) lifted off from Launch Complex LA3. Also on board was the Radiation Meteoroid spacecraft to demonstrate and evaluate improved instrumentation, and to gather near-Earth data of scientific interest. Soon after the OFO-A satellite had safely swept into orbit the centrifuge was activated.

The centrifuge now applied gravity stimuli in cycles, with each cycle lasting about 8 minutes. Each cycle comprised a 1-minute period free of acceleration, an 8-second period when rotation was slowly applied, and then 14 seconds of constant 0.6-g force. Over the following 8 seconds the rotation would slowly wind down to a complete stop, and then there would be a 6-minute period free of activity during which the after-effects of the rotation period could be examined. These cycles would be repeated every half-hour during the first 3 hours in orbit, but less frequently during the remainder of the flight.

Despite some heating and pressure malfunctions, the two frogs remained in good health throughout the mission, and successful vestibular recordings were transmitted to the ground. All the mission objectives were met, and data were collected for 6 days until the batteries eventually failed. The following day the satellite was de-orbited by ground controllers, and was quickly incinerated in the fiery heat of re-entry [11].

Following the successful OFO-A flight, interest in the experiment was maintained. Five years later, in 1975, the Vestibular Function Research Group investigated the possibility of flying further otolith experiments in an Earth-orbiting spacecraft, but work on this project was eventually discontinued.

OF ROCKETS AND POCKET MICE

When the final manned lunar voyage was launched towards Earth's near neighbour on 17 December 1972, the three Apollo astronauts had company. Just above the heads of

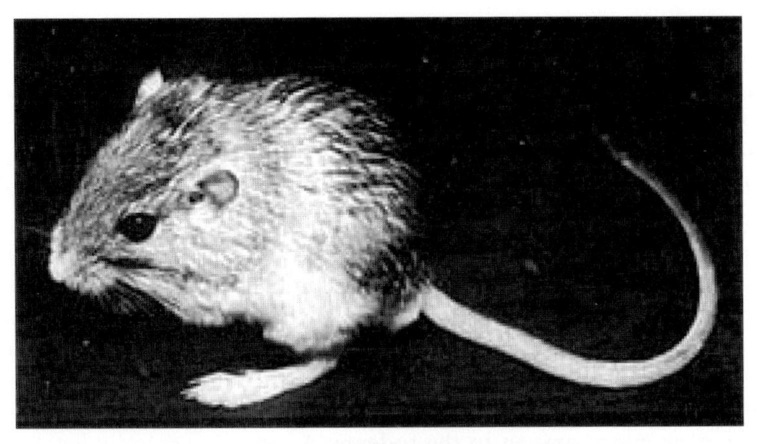

A moonshot candidate. (Photo: NASA)

Gene Cernan, Jack Schmitt and Ron Evans were five smaller space pioneers along for the ride in a sealed, self-sustaining canister. Unfortunately, they were never destined to share in the adulation bestowed upon their human counterparts.

Tiny space travellers

Some nematodes (microscopic roundworms) had been successfully carried aboard Apollo 16 eight months earlier, but these were structurally simple organisms that can be found by the thousands in any handful of soil. Apollo 17 had room for some more sophisticated life forms that would hopefully provide useful biological information as evaluative test subjects. The selected subjects were pocket mice *Perognathus longimembris*.

The pocket mouse is a small, lively nocturnal granivore inhabiting the more arid grassland regions of the western United States and northwestern Mexico. Although their common name suggests they are mice, these tiny desert-dwellers are actually heteromyid rodents. Essentially chosen for the long lunar flight because of their small size and weight (7–12 grams), they also possess an innate ability to lower their metabolic rate quite significantly when not active, and do not require drinking water. In other words, they are perfectly suited to space travel.

The five pocket mice obtained from the Californian desert were part of what was known as the Biological Cosmic Ray Experiment (or BIOCORE in NASA parlance). Conducted by the Ames Research Center, this experiment had been created to further assess whether microscopic lesions to the brain, eyes, skin and other tissues could result from exposure to heavy cosmic-ray particles on a lunar flight. In order to achieve this, each rodent had contoured, plastic cosmic-ray detectors, known as dosimeters, surgically implanted beneath its scalp several months in advance of the flight. Scalp tension held these dosimeters in place, and the mice displayed no related adverse effects in the lead-up weeks to the launch.

Two hermetically-sealed aluminium canisters designed to hold the pocket mice were specially prepared for the mission. One was intended for stowage in locker A-6 within the Apollo 17 Command Module *America*, while the other would be used to conduct parallel control studies on the ground at the Ames centre in California. The diminutive size and self-sufficient nature of the rodents meant that no human intervention or data recording would be required of the three-man crew on Apollo 17, and electricity was not required for lighting or any test facilities.

Seven perforated, cylindrical tubes were built into each of the two canisters, six of which formed a circle around the inside circumference. These tubes, all but one of which would house a pocket mouse during the flight, ran the full length of the 14-inch canister. They had been carefully designed to minimise tumbling for the occupants in weightlessness, but would also allow them to move around and eat a 30-gram supply of seed mixture without too much difficulty. One of the six tubes would remain empty to offer supplementary oxygen to the rodents. A seventh, centrally-located cylinder was manufactured from stainless steel and filled with granules of potassium superoxide, which would convert the animals' carbon dioxide exhalations into breathable oxygen. Also secured inside each of the canisters were two temperature-recording devices and pressure relief valves, while radiation detection units were mounted on the exterior of both canisters [12].

For purposes of formal identification, the pocket mice were given individual numbers – A3326, A3400, A3305, A3356 and A3352. Informally, it is known through an interview with participating astronaut Gene Cernan for the *Spaceflight News* magazine article "Last Man on the Moon" (July 1986), conducted by Nigel Macknight and Eddie Pugh, that the four male and one female subjects were given the whimsical pet names Fe, Fi, Fo, Fum and Phooey [13].

Bound for the moon

On 2 December, 2 weeks prior to the launch from the Cape, the pocket mice and their food supply were loaded into the aluminium tubes. Their canister was then sealed, flushed with oxygen and checked for leaks. Three days later the prime container with its moon-bound animals was placed into the locker, which in turn was inserted into *America*.

Altogether, the five pocket mice would spend 12 days and 13 hours in space. Following the safe recovery of the Apollo 17 Command Module, the BIOCORE canister was removed some 7 hours later and despatched to a tropical medical centre in Pago Pago, American Samoa [14].

Retired NASA scientist Dr. Delbert E. Philpott was the principal Ames Research Center investigator for the retina experiment, and he has many memories of that time, including the rushed post-flight trip to the Lyndon B. Johnson Medical Center on Pago Pago.

All the information I got from the astronauts was a note attached to the canister saying, "For what it's worth, I think I hear scratching on the inside." The trip from the airport to the LBJ hospital was in an ambulance because it was the only vehicle that could go faster than the very low speed limit on the island. This was to decrease the heat the aluminium canister would absorb during the trip. We feared it could kill the mice. I was a hero for having thought up this idea until 30 minutes after they arrived. Then the press called and said, "We understand that your mice have been taken to the hospital. What's wrong with them?" It took some fast talking to convince them that this was where the work had been planned all along [15].

Once the canister had been opened at the hospital, the male pocket mouse numbered A3352 was found to have died at some early stage of the lunar flight. Of the surviving four, two were reported to have been quite active and alert, while the remaining pair (including the female A-3326) were discovered hunched up, sluggish and uncoordinated.

Once exhaustive tests had been performed, the four surviving pocket mice were euthanised, together with all five from the ground control tests, to allow comparative autopsies to be carried out. Three days later the preserved bodies of the flown subjects were sent to the Ames Research Center where scientists would further assess the effects of prolonged cosmic radiation.

Unfortunately, the results of these post-mortem tests were relatively inconclusive, as the Apollo spacecraft had been so well shielded from the harmful effects of radiation it actually hindered the gathering of any significant results. Only circumstantial evidence was found of vulnerability to radiation from cosmic-ray particles, although high-energy (HZE) particles did seem to have minimally damaged the retinas of the flown rodents.

Veterinarian and BIOCORE project coordinator Richard Simmonds nevertheless found the results interesting. "The evidence was equivocal," he stated, "because you had to believe the predicted trajectory of the particles, and that the lesions seen followed that trajectory. Perhaps we may have euthanised the mice too early. I don't know if there would have been brain lesions too, if we had waited longer" [16].

As principal investigator for the retina experiment, Dr. Philpott told the authors he tends to disagree.

Our lab developed a very ingenious laser system to track the HZE particles, so I doubt that we could have been very far from their path. How long a damaged cell remains in the body is unknown, so I don't think waiting for a period of time after the flight would have helped. The body gets rid of what it doesn't need, so I doubt that it would be there very long.

We don't know if the nucleus has to be hit, and since the area around the HZE increases with the atomic number, the larger ones should be more damaging. In other words, a low atomic number HZE might not damage the cell, so we might be looking at one that was traversed and it would look fine [15].

More mice on Skylab

Seven months after the successful flight of Apollo 17, another cluster of six pocket mice was flown into orbit, together with the second Skylab crew (SL-3), Alan Bean, Jack Lousma and Dr. Owen Garriott.

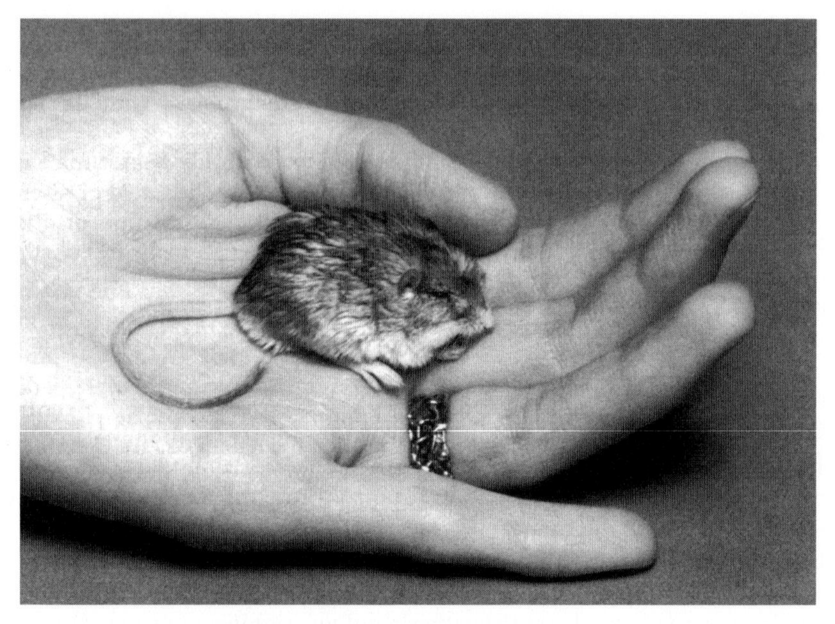

A Skylab pocket mouse. (Photo: NASA)

This time the pocket mice would be maintained in strictly controlled conditions of darkness and temperature as part of an experiment conducted by the University of California in Los Angeles to study the circadian system of a mammal during space flight. Primary data related to the circadian rhythm of body temperatures and activity of the pocket mice would be collected for later analysis, to assess whether these were affected by conditions of weightlessness over a 2-month period.

For the purpose of data collection, a total of 28 pocket mice were implanted with bio-telemetry devices for monitoring body temperature and movement a month prior to the flight. Six were later selected as the prime flight specimens and would be housed in individual circular cages. Another six would be used in parallel ground-based control exercises.

It was planned that a post-flight analysis and evaluation of the experiment would be derived from a master magnetic tape of all data retrieved from the Circadian Periodicity Experiment (CPE) units in conjunction with an examination of mission voice tapes. Average daily activity records would reference the time of arousal from torpor and would be correlated with temperature measurements.

According to NASA's report on the Skylab 3/Circadian Periodicity Experiment (CPE-1): "Baseline data was collected at the launch site in a specially designed holding unit, and flight and back-up CPE packages were loaded with six animals each" [17].

Concurrently, Stanford University researchers were engaged in studies related to the circadian rhythm of vinegar gnats (fruit flies or *Drosophila*), which would be carried out in conjunction with studies of the pocket mice. The objective was to study the stability of the circadian rhythm in these insects from pupa to adult state, and

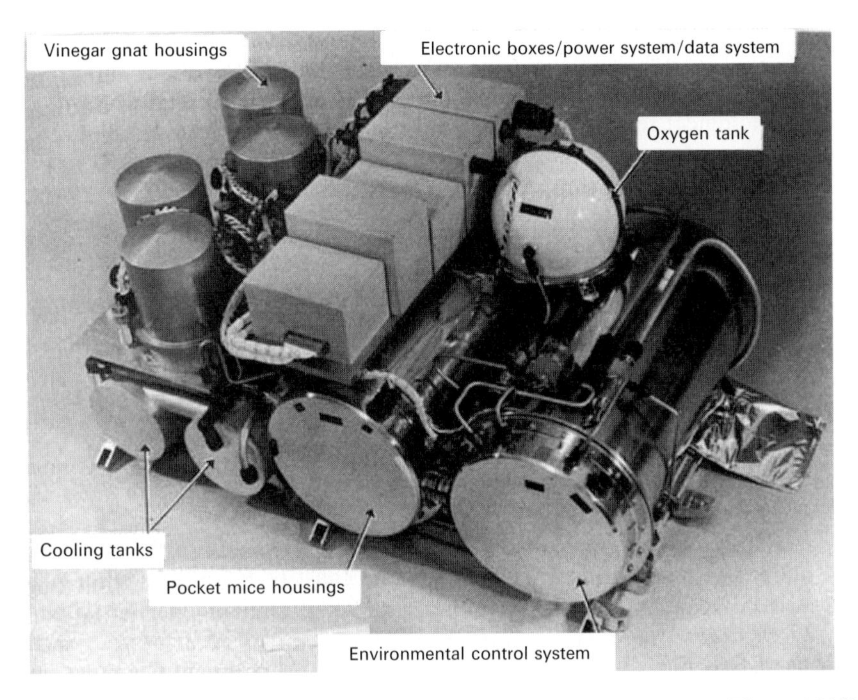

The CPE unit carried on SL-3 contained pocket mice and vinegar gnats. (Photo: NASA)

whether this daily rhythm was disturbed by microgravity or remained the same as on Earth. The pupae would be divided into four groups and flashes of light delivered to stimulate the emergence of adult flies at different times during the flight.

The three astronauts were launched to the Skylab space station on 28 July 1973 on a science mission that would last for 59 days. The CPE-1 experiment was simply one of a complex raft of scientific experiments to be carried out by the crew. In addition to the pocket mice and vinegar gnats, two Cross spiders (*Araneus diadematus*) named Arabella and Anita were launched into orbit as part of a unique experiment aboard Skylab.

WEAVING WEBS IN SPACE

The Skylab experiment involving spiders (ED-52, Web Formation in Zero Gravity) was conceived as part of the Skylab Student Project, a nationwide contest for Grade 7–12 students to help encourage their interest in science and engineering careers. Twenty-five students were selected as winners, and while many of the winners' experiments would be conducted during later space flights, the most highly-publicised idea was submitted by Judith Miles, a student at Lexington High School in Massachusetts. She had read a *National Geographic* magazine story about spiders creating webs and felt it would make a good study exercise aboard the space station [18].

Studying spiders

As the first arachnid study conducted in space (and the last for nearly three decades to come), biological researchers had expressed interest in the young student's novel concept and were keen to explore with her the web-spinning characteristics of spiders in a weightless environment. A comprehensive post-flight analysis of the experiment would be carried out by the Research Division of the North Carolina Department of Public Health [19]. Before the flight, crewmember Dr. Garriott met with Judy Miles to fully discuss the objectives of her experiment.

Three days prior to launch, the two spiders involved in the ED-52 experiment had been fed a house fly for food and were provided with a water-soaked sponge in their storage vials. *Araneus diadematus* is a hardy breed of spider, which can survive up to 3 weeks without food if an adequate water supply is made available.

First fish to fly

Also carried on the SL-3 flight were the first two fish to fly into space (mummichog minnows, *Fundular heteroclitus*), caught off the coast of Beaufort, North Carolina. The fish were added to the flight at the request of Garriott, who wanted to observe any disorientation they might experience when exposed to weightlessness. Another two mummichog fish would be carried on the follow-up SL-4 mission.

Garriott's fish were inside a clear plastic bag attached to a workshop wall for observation. Initially, the two minnows struggled to gain directional sense, swimming around and around in tight, irregular patterns. They finally seemed to visually adapt to the weightless environment and began swimming with their stomachs pointed towards the wall on which their plastic bag was hanging. It seemed to the crew that they were using the wall as a substitute river bed to orient themselves. Fifty minnow eggs had also been carried on the flight, and by comparison those that hatched quickly oriented themselves and swam without difficulty or confusion.

Unfortunately, experiments related to the circadian rhythms of the six pocket mice and vinegar gnats would prove unsuccessful. The facility housing the mice and pupae worked perfectly during the first part of the flight and some useful early data were obtained. A power supply malfunction 30 hours into the mission then resulted in the loss of both experiments.

Creating a tangled web

Once the three Skylab astronauts had settled into the space station as the second long-duration crew, Garriott had been tasked with releasing Arabella from her small vial into a lighted box resembling a window frame. This was equipped with a motion picture camera and a still camera positioned outside to record the spiders' activities and web construction. Until she was needed, Anita would remain enclosed in her launch vial. After some initial resistance to occupying her new home, Arabella was finally coaxed to enter the frame, although taking considerable time adjusting to weightlessness, and making what was described as "erratic swimming motions".

Dr. Owen Garriott filmed the web-making activities of spiders Anita and Arabella during Skylab mission SL-3. (Photo: NASA)

The following day, the spider produced her first small web in a corner, but the day after she had gained in confidence and completed an entire web. Weightlessness had obviously affected her web-weaving skills, as this effort was erratic in strand thickness, and not as geometric as those she and other spiders had daily spun on the ground. The strands were later noted to be thinner overall than those produced under normal 1-g conditions.

On 13 August, as part of the experiment, half of Arabella's web was deliberately destroyed to see if she would repair the damaged section, but she refrained. Garriott then fed the spiders tiny portions of rare fillet mignon and topped up their water supply. Following this, Arabella ingested the remaining half of her web and wove a whole new one, which was far more symmetrical and consistent than the first attempt. On 26 August, the spider was placed back in her launch vial, and Anita, similarly reluctant to emerge, was eventually coaxed into the observation cage. Her performance proved very similar to that of the other spider. "Their behaviour was the same," according to Garriott. "Scruffy web at first, then improving to very nice and symmetrical in a few more days" [20]. However, when he checked Anita's progress on 16 September, he found that the second spider had expired. Her remains were placed in the launch vial for examination after the flight.

Arabella spins a web in space. (Photo: NASA)

Return to Earth

Having occupied the Skylab station for a record-breaking 59 days, the three astronauts boarded their Apollo spacecraft and returned safely, splashing down in the Pacific on 25 September. They had clocked up an impressive 24,400,000 miles in travelling around the Earth 892 times. Arabella seems to have survived the orbital flight part of the mission, but she was also found dead after her launch vial was removed from the spacecraft after splashdown.

Meanwhile, the two spiders had created a good deal of public interest. "We were somewhat amazed, but pleased, to find all the publicity that Arabella and Anita had received on the ground," Dr. Garriott told the authors. "It seems everyone relates to spiders more easily than to solar flares or ergometer tests!

"Come to think of it," he added, "that seems pretty reasonable!" [20].

The dehydrated remains of Anita and Arabella are now part of the Smithsonian Institute's collection.

"SURPLUS TO REQUIREMENTS"

Once humans had travelled to the moon and returned safely, the once-vital role of the monkeys and chimpanzees in supporting this effort was virtually at an end. Colonies of these animals were being maintained at Holloman AFB, but now there was very little practical use for their services as aerospace test subjects.

On 1 July 1971, a decade after a cheeky chimpanzee called Enos paved the way for John Glenn and a succession of astronauts to safely orbit the Earth, Holloman's 6571st Aeromedical Research Laboratory closed down. The facility, which had been

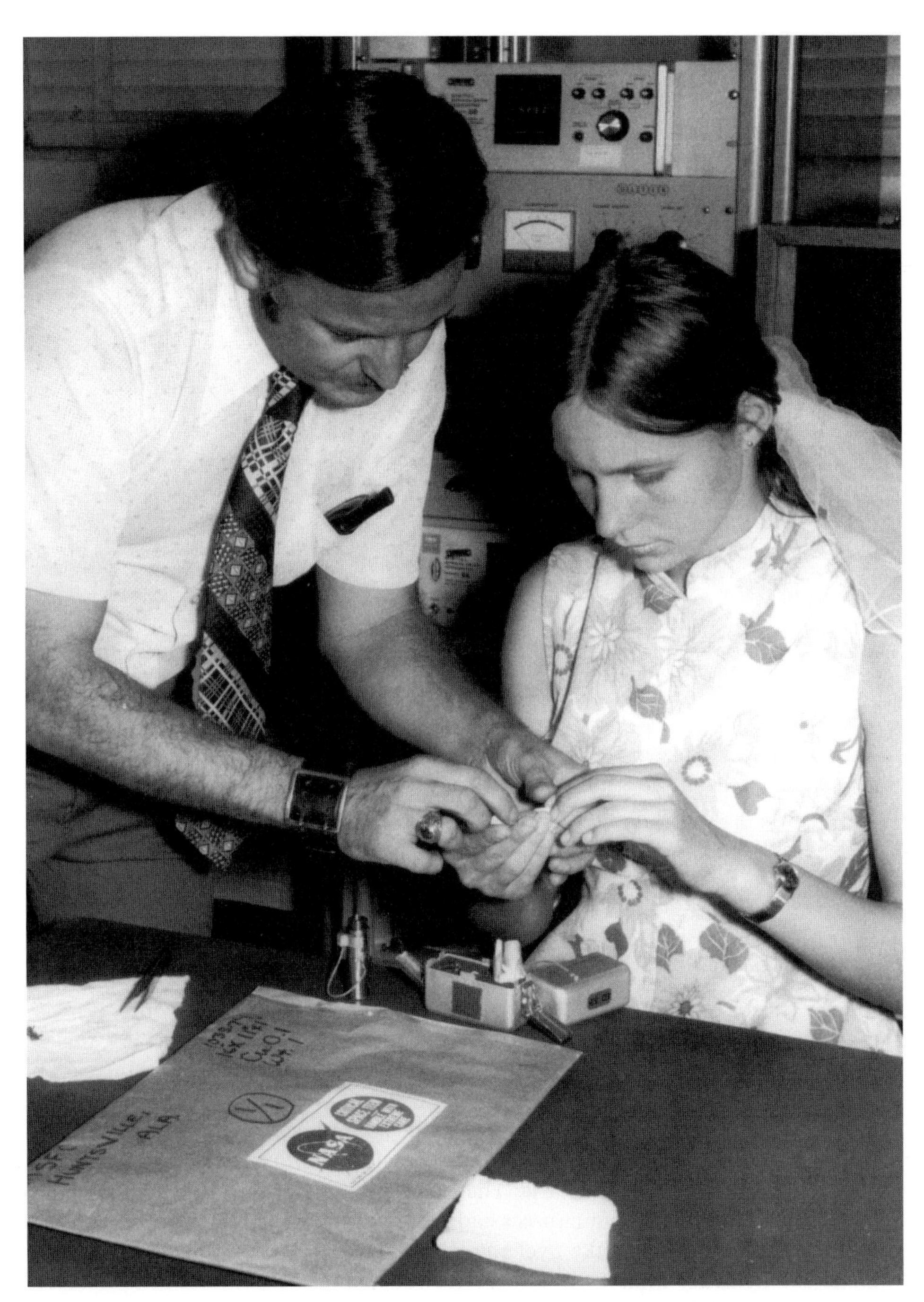

Judith Miles, student investigator for the Skylab spider experiment, together with her NASA science advisor, Dr. Raymond Gause. They are opening a package containing some of the web material returned from the SL-3 mission. (Photo: NASA, courtesy Jack Waite MSFC)

The chimp consortium at Holloman AFB in New Mexico comprised several acres surrounded by a high fence. (Photo: USAF, New Mexico Museum of Space History)

responsible for the upkeep and welfare of all animals housed at the base, no longer fell under the auspices of the Air Force. Instead it would be leased and operated by the Institute of Experimental Pathology and Toxicology (IEPT) of Albany Medical College, a privately-run affiliation of Union University in Albany, New York. They would assume full responsibility for the care and handling of the animals.

Coulston takes over

Following the closure, the chief of the Aerospace Medical Division Operating Location, Lt. Colonel Ralph Ziegler, issued a statement saying in part: "Officially, the lab was deactivated at the end of last year. We've just had a skeleton crew taking care of the animals until Albany could actively take over the program" [21].

The person who would now preside over the laboratory was Professor Frederick Coulston, an eminent toxicologist who had previously worked such important research areas as the control of malaria. The institute's director since 1963, his proposal to study the monkeys and chimpanzees had been accepted by the Air Force hierarchy, thus allowing the IEPT to take over the laboratory facilities and manage the large primate colony at Holloman. Ominously for the animals, Coulston openly foretold the future purpose he saw for them.

"As a medical institute we stress safety. For the past few years we have been working on safety in foods, chemicals, pesticides, additives, and drugs for veterinarian and human use. Thus far, we've been working with rhesus monkeys, but now we hope to graduate to chimpanzees.

"One of our major projects will be in the area of chimpanzee reproduction. We will study the basic principles of fertility and sterility among the members of the colony. We also plan to study the effects of drugs and chemicals on the chimps in the colony. I must stress that the chimps will be cared for in the most humane way. We're very conscious of the care and kindness needed to care for these animals."

Coulston also spoke of research into the areas of strong analgesia and hallucinogens. He said the laboratory could become "one of the great drug research centers in the nation."

He explained: "By having the chimp colony we can effectively measure the effect of these chemicals and drugs on them as total animals. Ones that are intact; non-anaesthetised. Since we can't experiment directly on humans, we have to rely on the chimp colony. And we can draw definite conclusions as to the effects of drugs on humans from the experimentation done on the chimps. We hope that the biological differences between the two species are so minimal that the effect of a drug on a chimp would be basically the same as it would be to a human" [21].

A new animal facility

In 1980 Frederick Coulston left Holloman and established a primate-testing toxicology laboratory called the White Sands Research Center, also based in Alamogordo.

Animal rights groups such as PETA (People for the Ethical Treatment of Animals) protested the use of monkeys in space research, especially those held at Holloman AFB. (Photo: PETA)

Meanwhile, management of the Holloman chimps had become the responsibility of the New Mexico State University, who would eventually build a $10.5 million taxpayer-funded animal research and housing facility on Holloman AFB.

The next quarter-century would prove to be a grossly unpleasant time for the space chimp colony at Holloman. Housed in steel cages, the chimps would be used to test everything from cancer-causing solvents in industrial cleaners to drugs for sexually-transmitted diseases. Many would be deliberately injected with HIV, allowing researchers to note any effects on the primates. However, despite their close DNA relationship to human beings, they would prove to be extremely poor test subjects in the study of HIV and AIDS. With one or two possible exceptions, none would come down with the full-blown disease.

In April 1993, Coulston took over the management of the Holloman chimpanzees from the New Mexico State University, which had decided to pull out of animal research. He subsequently formed The Coulston Foundation (TCF), which would operate two separate facilities – the White Sands Research Center and the chimpanzee laboratory at Holloman. The foundation had arranged to lease the Air Force chimpanzees for a nominal sum of $1 per year, and was able to advertise the animals' services for research, product-testing and insecticide development – a profitable undertaking. The National Institutes of Health also began funding foundation research, and their contributions would eventually exceed $40 million.

By October 1994, TCF, which had also been given full use of the former university research centre at Holloman, controlled around 500 chimpanzees from the New Mexico State University, the Air Force and Coulston's White Sands Research Center. It was now responsible for the largest captive chimpanzee colony in the world.

Awareness and protests grow

By the mid-1990s, however, the winds of political change were beginning to blow for the 144 Air Force chimps still held at Holloman AFB. Apart from an increased awareness of the plight of these animals, behaviourists' studies had begun to indicate their remarkable ability to create tools, communicate through the use of sign and oral language, and express to handlers a range of complex emotions.

Meanwhile, the National Research Council had become increasingly uncomfortable with the vast numbers of biological test chimps that were being held in primitive cages in Holloman and elsewhere across the country. For its part, the Air Force suddenly declared that the 144 chimpanzees at Holloman had become "surplus to requirements", and bids were sought on the primates and their housing. These bids were to be submitted by June 1998. The ambivalent attitude of the Air Force in selling off these animals to the highest bidder was probably best exemplified in the words of their spokesman, Colonel Jack Blackhurst, project manager for the divestment. "Chimps – right, wrong or otherwise – are basically personal possessions," he said. "They're just like a piece of equipment" [22].

Professor Coulston, who had been leasing the Holloman chimps for highly-profitable biomedical research experiments in AIDS, hepatitis A and hantavirus testing, wanted TCF to retain the use of the Air Force primates. However, he faced something of an uphill battle, as he had accumulated a long, documented history of serious animal neglect as the head of both the IEPT and TCF. In 1997, he was fined $40,000 by the U.S. Department of Agriculture (USDA) for serious breaches of the federal animal welfare act and for mistreating primates under his care. In January that same year an 11-year-old chimp named Jello had died after allegedly being anaesthetised in a negligent manner by an inexperienced veterinarian. The following month, a 2-year-old chimp named Echo had also died after botched surgery on an injured arm, again performed by an inexperienced veterinarian.

Preventable losses

In 1993, 4 years prior to the deaths of chimps Jello and Echo, Coulston's foundation had been investigated following the deaths of three chimpanzees named Robert, James and Raymond. These unfortunate animals had expired from heat prostration after the temperature in their tiny den one blistering summer day was reported to have soared to around 140°F. In December 1994, another four monkeys from the Holloman AFB colony had died from water deprivation after a malfunctioning supply hose had dried up and not been repaired for several days.

Lt. Colonel Denver Marlow, a veterinarian and then chief of the Air Force animal programme, cautiously admitted that TCF's "state of flux" in veterinary care had been "a concern of the USDA and a number of other parties" [22].

At this time, in total, Coulston had around 600 chimps housed at his Alamogordo-based foundation. He boasted to the press that he could raise chimpanzees like cattle, and his future plans included using the unfortunate animals as organ donors and blood banks for humans. "I don't think there's an excess," he said when asked about the vast number of primates held at his foundation. "I would like to have five thousand!" [22].

Several influential and concerned people, including a number of astronauts, had urged the Air Force to retire the animals to a wildlife sanctuary instead of subjecting them to further research. "It amazes me," said well-known chimpanzee researcher Jane Goodall, "that the Air Force would ignore the voices of so many who want these chimpanzees retired and, instead, sentence them to a life of fear, pain and loneliness in an establishment well known for its inability to care for chimpanzees" [23].

Formal charges laid

In March 1998 the USDA took the unprecedented step of filing a second set of formal charges against TCF for multiple violations of animal welfare, which included charges relating to "the negligent, entirely preventable deaths of two young, healthy chimpanzees (Jello and Echo) in 1997" [24].

Despite the Coulston Foundation's poor track record, and over the vigorous protests of animal welfare groups, the Air Force announced in June 1998 that it would be awarding its 111 surviving chimpanzees to the foundation, while another 30 healthy specimens (none of which had been used in medical tests) would be given to Primarily Primates, the operator of the country's largest primate sanctuary near San Antonio,

Minnie, the backup chimp for both Ham and Enos, with an unnamed animal handler in the early 1960s. (Photo: NASA)

where they would be allowed to roam freely within 15 acres of open space. At that time, they were the only animals the sanctuary could afford to purchase from the Air Force.

Although a total of about 65 chimps from the original colony had gone through at least some facet of spaceflight training, and two had flown into space, none of the

141 animals now in question had ever been launched into space. Minnie, who had passed away earlier that year, had come the closest, serving as backup candidate for both Ham and Enos. Aged 41, she had been the oldest chimp in the colony, while the youngest was her daughter, Li'l Mini, then aged 5.

Despite pending USDA charges against TCF and an official investigation into the foundation's activities, the Air Force officially signed over the remaining 111 chimpanzees on 8 October 1998, totally rejecting all calls to retire these animals to a sanctuary. Divestment project manager Colonel Blackhurst dismissed any criticism of the decision, saying TCF had been selected because of its record of animal care, said to be based on inspections by the USDA. "It did as well or better than the other chimpanzee facilities in the country," he stated [23]. One probable determinant in the Air Force's decision was the fact that only Coulston and TCF had access to the medical records of the entire chimp colony and other vital health information.

An "arbitrary and capricious" decision

Following the divestment announcement, the Center for Captive Chimpanzee Care (CCCC), an organisation founded and headed by anthropologist and project director Carole Noon, immediately filed a lawsuit against the Air Force over the fate of the chimpanzees, and for contravening its own chimpanzee divestment criteria. The suit alleged that the decision to award the chimps to TCF was "arbitrary and capricious, violating both federal law and the Air Force's own Request for Proposals." The court was asked to overturn the decision [25].

"From day one the Air Force has said its primary concern was the well-being of the chimpanzees," Dr. Noon contended. "Yet it gave them to a lab with the worst animal care record of any primate research facility in the country" [26]. The CCCC was not alone in its concerns; on 23 September a letter signed by several dozen members of Congress had been sent to the House Government Reform and Oversight Committee, requesting hearings be held to investigate what might have been an improper divestment of the animals by the Air Force.

Sadly, the chimpanzee death toll continued to mount at TCF, as did the number of charges, alleged violations and investigations. In January 1998 a chimpanzee named Holly died from the side-effects of a drug being tested on her. Five months later another two chimps named Terrance and Muffin passed away during tests of the same drug, and in May 1999 a chimp named Eason died while undergoing a spinal disc replacement experiment. Then, according to concerned whistleblowers at the facility, a chimpanzee named Gina died in June 2000 after being locked outside without shelter for several hours in the relentless New Mexico summer heat. A 10-year-old chimp called Ray also succumbed after veterinarians failed to treat him for several days after he became ill. A female chimp named Donna would also die with pus in her peritoneal cavity and a ruptured uterus after carrying a dead foetus in her womb for up to 2 months.

In October 1999, as the result of a yearlong lawsuit against them, the Air Force handed over a further 21 of their chimps to the CCCC. Eventually, these animals would be retired to a newly-constructed and permanent, 200-acre sanctuary set on a former citrus grove in Fort Pierce, South Florida, which had cost $2.5 million to purchase and build. Here, they would live in family groups on man-made islands, taking shelter in comfortable, chimp-friendly accommodations and enjoying a number of recreational facilities.

In April 2001, after undergoing a socialisation process in two separate groups, the chimps were finally released, unhindered, into the sanctuary. There would be no more medical experiments for them; no more cages or stuffy, overheated, windowless cells. "They're off limits," said Dr. Noon. "They don't work any more" [27].

By now, with many companies following sensitive global trends and disassociating themselves from any animal research involving their products, Coulston and the TCF had begun experiencing severe financial difficulties, and were taking in less than 50% of what was needed to run the facility. The foundation was also being financially ravaged by a barrage of prosecutions for safety violations. These suits not only involved the animals, but were filed on behalf of disgruntled foundation employees who had to work in what they called unsafe or unsanitary conditions.

End of an anthropoid era

In June 2001, after several years of complaints, prosecutions, testimony and unsavoury public exposure, the National Institutes of Health finally withdrew the Coulston Foundation's animal welfare assurance, which immediately made the lab ineligible for federal funding. It also marked the beginning of the end. That December, the First National Bank of Alamogordo filed a foreclosure lawsuit against TCF in an endeavour to collect over a million dollars in unpaid loans. Bills for unpaid payroll taxes also rolled in, causing Coulston to sell his personal belongings in a frantic attempt to stay in business. Then his employees threatened to walk out unless they were paid overdue wages.

In desperation, with personal and foundation debts totalling several million dollars now hanging over his head, Coulston contacted the CCCC and began negotiations to sell off the lab's buildings, equipment and land. The CCCC agreed, on condition that all the remaining chimpanzees and monkeys would be immediately handed over to them. Thanks to a generous grant from the Michigan-based Arcus Foundation, a settlement involving $3.7 million was reached and TCF soon ceased to exist as a research laboratory.

Finally, in September 2002, the long, sorry saga of the Air Force chimp colony began drawing to a close when TCF started the process of transferring the remaining 266 chimps and 61 monkeys from its research facilities to the care of the CCCC. The deal also included 22 chimps and 29 monkeys who were out on lease or loan to other institutions. "We are no longer interested in biomedical research on non-human primates," said Don McKinney, a TCF spokesman. "We are moving on" [28].

Under Carole Noon's instructions, the animals were immediately taken off their bland food intake and placed on a new and welcome diet of fresh fruit and vegetables. As with the earlier intake of animals, the males would undergo vasectomies prior to being placed into mixed sex social groups for rehabilitation purposes and eventual transfer to the Florida sanctuary to see out their lives.

In December 2003, 2 weeks after his 89th birthday, Dr. Frederick Coulston died in Alamogordo, New Mexico. In his eulogy, long-time friend and colleague Don McKinney pointed out that Coulston had once told reporters he felt his finest personal achievement was in his earlier studies of the malaria parasite. "Dr. Coulston was, and remains actually, the last of the scientific giants from the early 20th century," McKinney stated. "He was one of the people who laid the groundwork for our medical research today" [29].

There is still more work to be done at the Fort Pierce sanctuary in South Florida. It is currently undergoing further expansion and will one day house the remaining animals still being cared for in New Mexico. Eleven additional 3-acre islands, complete with jungle gymnasium equipment for fun and recreation, are now being constructed. Each of the islands is linked by a land bridge to indoor accommodations in which the animals are fed their meals and sleep without any kind of restraint.

It had been a long and hard-fought battle for many individuals and organisations to save the descendants of a colony established by the Air Force in the 1950s to test the effects of space travel. But now, according to Carole Noon, the hard work is behind them and their beloved primate charges are adapting well to an existence in which they are no longer confined in "a place full of dirty cages with no light [where] the smell was so bad it made your eyes water."

"The main thing we're focusing on," she emphasised, "is improving their lives today" [28].

REFERENCES

[1] "T-7A," Mark Wade's *Encyclopedia Astronautica.* Website: *http://www.astronautix.com*

[2] Phillip S. Clark, "The Development of China's Piloted Space Programme: From Sounding Rockets to Shen Zhou 5," *Journal of the British Interplanetary Society*, vol. 57, no. 11/12, November/December 2004.

[3] "Programs, Missions, and Payloads: Gemini 3," *Life into Space: Space Life Sciences Experiments, NASA Ames Research Center, 1965–1990*, NASA, Washington, D.C., 1995. Website: *http://lis.arc.nasa.gov/lis/index.html*

[4] Virgil "Gus" Grissom, *Gemini: A Personal Account of Man's Venture into Space*, Macmillan, Toronto, Ontario, 1969.

[5] R.S. Young and J.W. Tremor, "Session III – The Effect of Weightlessness on the Dividing Egg of *Rana pipiens*, *Bioscience*, vol. 18, no. 6, June 1968, pp. 609–615.

[6] "Experiment Information, Frog Egg Growth," NASA Life Sciences Data Archive, JSC (S003 2/2). Website: *http://lsda.jsc.nasa.gov/scripts/experiment/exper.cfm?exp_index=232*

[7] "Zond 5," *NSSDC Master Catalog Display: Spacecraft* (NSSDC ID: 1968-076A), NASA Goddard Space Flight Center, Greenbelt, MD, 2006.

[8] Asif Siddiqi, *Challenge to Apollo: The Soviet Union and the Space Race, 1945–1974*, NASA Publication SP-2000-4408, NASA, Washington, D.C., 2000, pp. 655–656.

[9] "N-1 Rocket," Wikipedia free online dictionary. Website: *http://en.wikipedia.org/wiki/N-1_rocket*

[10] "Lunar L-1," Mark Wade's *Encyclopedia Astronautica.* Website: *http://www.astronautix.com*

[11] *Orbiting Frog Otolith Satellite Mission Performance Report*, Contract NAS6-1637 (No. 1333-032), December 1970.
[12] "Biological Cosmic Ray Experiment (BIOCORE)", *NSSDC Master Catalog: Experiment* (NSSDC ID: 1972-096A-11), NASA Goddard Space Flight Center, Greenbelt, MD, 2005.
[13] Nigel Macknight and Eddie Pugh, "Last Man on the Moon," *Spaceflight News*, 7 July 1986, pp. 24–37.
[14] "Programs, Missions and Payloads: Apollo 17," *Life into Space: Life Sciences Experiments, NASA Ames Research Center, 1965–1990*, NASA, Washington, D.C., 1995.
[15] Delbert E. Philpott, letter to Colin Burgess, 14 November 2005.
[16] Webb Haymaker, Bonne C. Look, Eugene V. Benton and Richard N. Simmonds, "Biomedical Results of Apollo: The Apollo 17 Pocket Mouse Experiment (BIOCORE)," Chapter 14, Section IV, NASA JSC, 2002.
[17] M.K. Fairchild and B.A. Hartmann, Skylab S071/S072, "Circadian Periodicity Experiment: Experimental Design and Checkout of Hardware." Contract NAS2-6897 (NORT 73-320), November 1973.
[18] Jack Waite (Marshall Space Flight Center) letter to Colin Burgess, 8 December 2005.
[19] Lee B. Summerlin (ed.), "Skylab, Classroom in Space," NASA Marshall Space Flight Center, NASA Publication SP-401 (Chapter 3), NASA, Washington, D.C., 1977. Website: *http://history.nasa.gov/SP-401/Ch3.htm*
[20] Owen K. Garriott, email correspondence with Colin Burgess, 21 November and 20 December 2005.
[21] "Aero Med Closes, Albany College Opens Lab," *The Forty-Niner* (Holloman Air Force Base newspaper), vol. 1, no. 27, 8 July 1971.
[22] Geraldine Brooks, "In Chimp Sell-Off, Military Finds It Has Monkey on its Back," *The Wall Street Journal*, 30 December 1997.
[23] Paul Recer, "Most Air Force 'astro chimps' to be used in further research," *The Associated Press*, 16 August 1998
[24] Eric Kleiman (Research Director, In Defense of Animals), letter to Secretary of Defense, William Cohen, 23 July 1998.
[25] "Doris Day Animal League Supports Space Chimpanzee Lawsuit Against Air Force," *Business Wire*, 8 October 1998.
[26] "Air Force Faces Legal Challenge over Fate of Space Chimps," *PRNewswire*, 8 October 1998.
[27] "Space chimps go free," Melbourne *Herald Sun* newspaper, 22 August 2001, p. 30.
[28] Kathy A. Svitil, "Late retirement for the Space Chimps," *Discover* magazine, January 2003.
[29] "Frederick Coulston Dead at 89," *The Associated Press*, 16 December 2003.

12

Shuttling into space

The advent of the space shuttle, which flew for the first time in 1981, presented researchers with an ideal platform for the study of animal and plant life in space, as hundreds of biological experiments could be carried aloft, examined or pursued in relative comfort.

THE ERA OF THE SPACE SHUTTLE

Nothing more biologically ambitious than pine, oat and mung bean seedlings, a few moths, house flies and a small number of honeybees were carried on Space Transportation System 3 (STS-3), and the bulk of the crew's duties were confined to a routine observation of the experiments. But these biological passengers were the forerunners of many thousands of other creatures that would ride the rockets with the shuttle astronauts.

Creating suitable habitats

Jack Lousma was the commander of space shuttle *Columbia* on STS-3, together with Gordon Fullerton. The flight, which lifted off from the Kennedy Space Center on 22 March 1982, was an 8-day test flight of the shuttle's systems, so there was plenty of time for observing experiments and the Earth below. At one stage Lousma wrote in his flight log:

It was beautiful and sunny when we pulled up the window shade and looked out ... halfway through the pass the sun went down, and now it's black as night. I guess we'll roll over and go back to bed.

The moths are very lively, the bees have all gotten stationary, and the flies took to walking. They decided there wasn't any use flapping their wings and going out of control, so they just float and wiggle their legs. As far as the moths go, they like to stand on the side of the Plexiglas cage too, but a lot more of them seem to fly, and some have adapted to flying from one place to another [1].

Fourteen months later, six white rats would become the forerunners of many other spacefaring rodents on the eighth shuttle mission, STS-8.

According to a NASA information brochure *Animals in Orbit*, taking living creatures into space on the shuttle requires special considerations.

Due to the housing needs and the practicalities of space travel, the lowest form of life is most suitable for space travel. Often, experiment results using snails and fish can be applied to human conditions: inner ear exams can be done in a snail rather than a highly evolved mammal, and genetic studies can be conducted in fish. While there is not a one-to-one transfer, the similarities are enough to gain necessary information [2].

All of the animal experiments flown on space shuttle missions have either been housed in the mid-deck area, or within a laboratory research module specifically configured to fit within the shuttle's cargo bay. The preferred and most frequently used option for numerous rodent experiments on NASA's orbiters, first used on STS-8, was the mid-deck. Forty-two lockers are located on the mid-deck, which can be utilised for payloads and experiments. Whenever a rodent experiment was flown, up to three lockers were fitted with a purpose-built animal enclosure module, or AEM, which would be integrated into the mid-deck the day before launch. The AEM, manufactured by the Convair Division of the General Dynamics Corporation, was managed by the NASA Ames Research Center in California. These small, portable, self-contained habitats had been commissioned by NASA and created by biologists to maximise the results and the comfort of these unwitting space travellers. Normally, five to eight rodents would be loaded into each AEM prior to their insertion into the mid-deck lockers, but this number would depend on several factors, such as the species and weight of the animals involved and the duration of the mission.

Solving the problems

In the 1950s the first rodents fired into space had relied on potatoes for sustenance, but on shuttle missions compressed nutrient food bars, sufficient for each flight, would be supplied. These were generally attached to the side of the AEM habitats, while pressurised and replenishable water-sipper systems also made life a lot more bearable for the occupants. Dr. Daniel Holley, a biology professor at San Jose State University was one of those who helped design these facilities. "If they're going into space," he explained, "I want them to be in a comfortable, healthy environment with adequate lighting, water and feeding systems. Rats are sentient beings – they feel. If you squeeze its paw it will withdraw" [3].

For the sake of the crews, it had also been seen as necessary to devise an elaborate ventilation system in order to minimise the smell resulting from the rodents' bodily functions. A laminar airflow was developed, which would suck all the faeces and urine

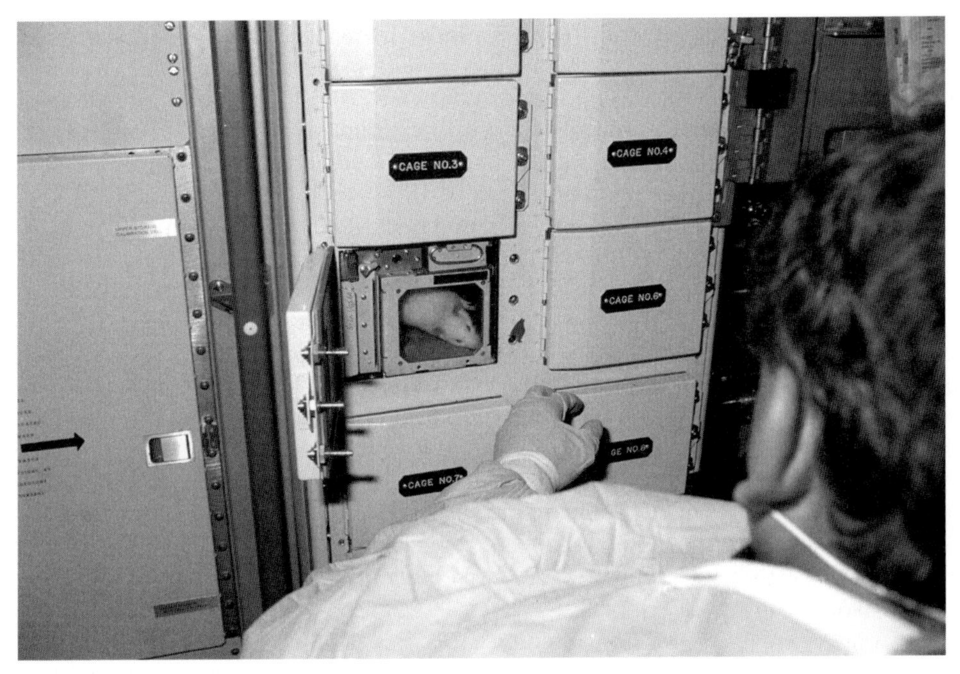

Taken during STS 51-B, this photograph shows how the animal enclosure modules (AEMs) were positioned inside lockers on the shuttle's mid-deck. Mission specialist Norm Thagard is checking on the rodents in one AEM. (Photo: NASA)

into an exhaust system. "They didn't really want to sit there and smell rodent poop all day," observed former NASA chief veterinarian Joseph Bielitzki. In the same 1998 interview he told *Metro* magazine's Cecily Barnes that flying rats in space had helped determine how to prevent a slew of ailments experienced by humans in space, including calcium deficiency, cardiovascular deconditioning and nausea. "Rats have probably been the single most important research animal that we have flown to date," he emphasised. "We have flown probably 700 of them" [3].

Laura Lewis, a member of NASA Ames Institutional Animal Care and Use Committee was asked how well animals adapted to living in microgravity. "Amazingly, they adapt very quickly," she responded. "Within five minutes, mice are floating in their living spaces, grooming themselves and eating, just as they would on Earth. Good science sets up hypotheses for an experiment, but sometimes that result is not what you expect. While we test our projects on the ground and in simulators, once we get into space we are sometimes surprised by what we learn" [3].

Shuttle life science begins

The diversity of animal payloads flown on shuttle missions has allowed seminal flight experiments that have significantly advanced researchers' understanding of gravitational physiology, while investigations involving space-borne plant and animal tissues have likewise contributed to improvements in gravitational biology.

As part of a student-sponsored experiment on STS 41-C, a total of 3,400 honeybees (*Apis mellifera*) and a queen were sent into orbit on shuttle *Challenger* in an aluminium container known as the bee enclosure module (BEM). It was part of a student experiment called "A Comparison of Honeycomb Structures built by *Apis mellifera* (SE82-17)," which sought to determine whether honeycombs could be produced in microgravity. A ground-controlled colony of bees was kept in an identical container for the purpose of comparison with the comb-building activities of those in orbit.

In their log book, the crew noted that "by Day 7, comb well developed, bees seemed to adapt to 0-g pretty well. No longer trying to fly against top of box. Many actually fly from place to place." It was felt that this adaptation might indicate a certain "learning" capacity on the part of the bees [4].

Weightlessness did not seem to greatly upset the worker bees in their task; they constructed a perfect 30-square-inch hive just as they would have done on Earth, although they did tend to cling to the sides of their container. The queen bee managed to fill the comb with 35 eggs, but none of these would hatch when returned to Earth.

3,400 bees were carried on board STS 41-C. Crewmember James van Hoften observes their honeycomb-building efforts in space. (Photo: NASA)

SOME SERIOUS MONKEY BUSINESS

On 29 April 1985, 2 squirrel monkeys and 24 rats were launched aboard shuttle *Challenger* as part of a life sciences payload. While monkeys were the original space travellers for the United States, and American astronauts had been rocketing into orbit for nearly a quarter of a century, it was the first time that humans and monkeys had flown into space together.

One of the two squirrel monkeys who flew on STS 51-B. (Photo: NASA)

Destined for space: these five rats were flown on STS 51-B. (Photo: NASA)

Spacelab experiments

The shuttle flight, designated STS 51-B/Spacelab 3, was crewed by seven astronauts. Many experiments were conducted in the large European Space Agency Spacelab, which nestled in *Challenger*'s large payload bay. The animals on this flight were part of an experimental programme for NASA's Ames Research Center in California, and were housed in special windowed cages within the unique Research Animal Holding Facility (RAHF), which maintained them in vivarium-like conditions. The monkeys were actually last-minute passengers on the flight; the four animals specifically trained for the mission were found to have a disease pre-flight, which could have been passed on to the astronauts, and healthy ones were substituted.

The two squirrel monkeys from the Ames centre lived in individual cages. It was the task of mission specialists William Thornton and Norman Thagard, both medical doctors, to keep watch over the animals throughout the mission. Thornton, as primary animal handler on the mission, had been working with the research team at Ames for several years before the flight.

Food and water consumption, animal activity, and environmental parameters were monitored continuously throughout the flight. Meanwhile the biotelemetry system (BTS), implanted in four of the rats, provided continuous data on their heart rate and deep body temperature throughout the flight.

Mission specialist Dr. Bill Thornton monitors the condition of one of the two monkeys on STS 51-B. (Photo: NASA)

The no-name monkeys

Unlike earlier rocket flights, these monkeys were not given names. Instead, they were simply known by the numeric denomination given to them at the Ames Research Center – 3165 and 384-80 – which the crew would simplify for their own use. "I don't think the American public would have been too concerned if one of the rats had died," explained mission commander Robert Overmyer. "But pictures of the monkeys got about and we received lots of mail as a result of the media coverage. I would not let the crew name the monkeys – not even unofficially. We called them 'Specimen No. 1' and 'Specimen No. 2'. If something should happen to them I didn't want the public to come down heavily upon us because some personalised monkey friend of ours had died. 'Specimen No. 1' and 'Specimen No. 2' may sound inhuman, but at the same time it kept the attachment from ever being there" [5].

The two medical mission specialists, Thornton and Thagard, were not supposed to hand-feed the monkeys. Their job was simply to keep an eye on their activities while the monkeys fed themselves on banana-flavoured pellets. This changed when one of the animals became sick and seemed to be growing worse as the flight continued. The specialists at Ames sent word that the animals could be hand-fed proper food items if it was felt this would help, and a search was made for the remains of a banana one of the crew had eaten earlier. Half of this, the last banana on board, had been thrust into a rubbish container, but the crew could not locate it despite a careful search.

Overmyer said the crew, and the researchers at Ames, were worried that the monkey might die. "The best thing that happened to us," he declared, "is that Bill was able to hand the banana pellets to the monkey through the window, and when he did that the monkey seemed to perk up."

Thornton explained further. "Once the ground-based experts decided to let me manually feed him, with a brief period of play, cajoling, putting human scent on the food (rolling in palms) and hand feeding, and with the cage door open he began to gorge himself. A second bout of play was required the next day after which he behaved normally" [5].

The monkeys and rats all survived the flight, although at one time some of their droppings floated free from the cages, and the astronauts had to quickly vacuum them up. The rats did provide pilot Fred Gregory with a few bad moments, when it was his turn to see how they were doing. When he peeped through the glass into their module, all he could see was motionless rats floating around. He thought they were dead, and expressed his concern to mission commander Bob Overmyer. The two men discussed the unwanted option of relaying this news to the ground, but first of all Gregory broke the news to Bill Thornton, who seemed quite non-plussed. "Have you tried tapping on the unit?" he suggested. Gregory did just that, and received an immediate, animated response from inside the module. He would later say that the rodents were so much at ease in weightlessness that he was almost jealous.

One of the Ames mission researchers, Bonnie Dalton, also told of an interesting experiment the crew tried on the rats. "They squirted water at them to see what they would do, and the rats reached out with their little paws and brought the droplets of water to their mouth. So they're extremely adaptable" [6].

After 7 days, and 111 orbits, *Challenger* touched down smoothly at Edwards Air Force Base in California, after which the two squirrel monkeys returned to the animal care facility at Ames for tests and post-flight evaluation. Monkey 3165 still lives at the facility, while the other has since died of natural causes.

TRAGEDY, AND A LENGTHY HIATUS

It was fully intended to send more monkeys and rats on subsequent Spacelab missions. Then, on 28 January 1986, space shuttle *Challenger* and her crew of seven were lost in a tragic explosion during ascent. As a result, many such experiments had to be postponed for several years.

Back to business again

Manned shuttle flights resumed in September 1988, with *Discovery* successfully completing a 4-day "return to flight" mission designated STS-26, operated by a crew of five astronauts. Two flights later, on STS-29, the crewmembers were accompanied by four Long Evans rats inside an AEM. Each of the rodents had had a small piece of bone surgically removed before the flight to see how well it grew back during the mission. Tissues from all of the flight rats and ground control rats, as well as some others that had been placed into simulated microgravity, were examined and compared post-flight by light and electron microscopy. The results of these tests seemed to indicate that the healing of fractures was delayed in those rodents maintained in actual and simulated microgravity.

Rats and the meaning of life

On mission STS-41, shuttle *Discovery* lifted off from the Kennedy Space Center on 6 October 1990. Located in the orbiter's mid-deck was a payload known as the physiological systems experiment (PSE), sponsored by the Pennsylvania State University's Center for Cell Research. The corporate affiliate leading this investigation was Genentech, Inc., a San Francisco-based biotechnology company engaged in the research, development, manufacture and marketing of recombinant DNA-based pharmaceuticals. NASA's Ames Research Center would provide payload and mission integration support. The venture between the three groups would result in the first commercial space research project in the life sciences.

According to the STS-41 press kit issued by NASA:

Research previously conducted by investigators at NASA, Penn State and other institutions has revealed that in the process of adapting to near weightlessness, or microgravity, animals and humans experience a variety of physiological changes including loss of bone and lean body tissue, some decreased immune cell function, change in hormone secretion and cardiac deconditioning, among others. These changes occur in space-bound animals and people soon after leaving Earth's gravitational field. Therefore, exposure to conditions of microgravity during the course of space flight might serve as a useful and expedient means of testing potential therapies for bone and muscle wasting, organ tissue regeneration and immune system disorders [7].

In this experiment, eight healthy male Wistar rats destined for the flight would receive one of the natural proteins developed by Genentech, while another eight would accompany them without receiving the proteins, providing a comparative study at flight's end 4 days later. Post-flight analyses were conducted by Genentech, while additional studies were carried out by investigators from Penn State's Center for Cell Research.

ONE GIANT LEAP FOR AMPHIBIANS

Meanwhile, biological studies continued on other manned missions. When Russia's Soyuz TM-11 flew into orbit and docked with the Mir space station in December 1990, one prominent member of the three-man crew was 48-year-old Japanese media reporter Toyohiro Akiyama.

The very reluctant astronaut

A former Vietnam War correspondent, his passage to Mir had been paid for by the Tokyo Broadcasting Service (TBS) to celebrate the 40th anniversary of Japan's biggest television company. Akiyama would become both the first Japanese citizen and the first full-time journalist to fly into space.

There were many critics of this paid space junket, especially for a passenger described in the British newspaper *Today* as "a chain-smoking, whisky-drinking Japanese executive." Akiyama, who suffered from space sickness early on in his mission, did live up to some of this criticism. Normally, a four-packs-a-day smoker, he began moaning during live televised conferences from orbit that he could not wait to get back home to resume his habit. "I'm gasping for a cigarette!" was one of his more quotable comments.

Despite the criticism, Akiyama did carry out some Japanese science experiments during his 6-day tenure aboard the Russian space station. One such experiment for the Japanese Institute of Space and Astronautical Science would assess the adaptability of six green Japanese tree frogs (*Hyla japonica*) to weightlessness, as well as their readaptation to 1 g once they had returned to Earth. Selected from 1,500 candidates, they would be the first frogs to fly into space on a manned spacecraft, although there had been an earlier attempt to hatch frogs from eggs carried into orbit. These timid, skittish frogs, growing less than an inch long, are biologically equipped with small suction cups on their legs to enable them to grip most surfaces in the wild.

According to a spokesman for TBS, the frogs were given "certain stimuli" early in the flight to gauge their reaction, while the same experiment was carried out on a control group of frogs located in a Japanese laboratory. The experiment, recorded by Akiyama in space, was repeated several days later and the data recorded for comparative analysis. The flight was not without its moments of humour: during a live broadcast back to Earth, one of the frogs leapt up and went sailing out of sight until he was recovered. Akiyama wryly observed: "Well, there you are. Frogs are frogs, everywhere" [8].

Froggie he did ride

Arboreal species such as the green tree frog sometimes fall or jump to the ground from branches with all four limbs outstretched and their abdomens inflated, ready to right themselves quickly on landing to escape any predators. It was noted by researchers that the typical posture of these frogs when floating in weightlessness was similar to that of frogs in "free-fall" back on Earth.

It was also discovered that most of the Mir frogs would bend their heads right back when perched on surfaces, increasing their grip by keeping their hind legs partially extended and pressing their stomachs against the contact surface. It was a posture they would hold for many seconds, and was not normal behaviour for this species in the 1-g environment except during retching and vomiting. At these times, a hyperextension of the head by the frogs was believed to increase abdominal pressure and compress the stomach to aid in the ejection of stomach contents. It raised the distinct possibility that the frogs may have experienced symptoms of motion sickness or even space adaptation sickness, which would lead to future experimentation with other amphibians exposed to microgravity.

On 10 December Akiyama returned to Earth aboard the Soyuz TM-10 spacecraft, together with all six frogs. Eventually, following a multitude of tests, four were dissected by the Molecular Endocrinology Laboratory in Rouen, France, in order to study neuropeptide secretion by glands in the hearts and brains of the frogs [9].

SOME SURPRISING DEVELOPMENTS

Frog eggs would prove to be an ideal study medium in space, shedding light on the importance of gravity to animal life. According to Dr. Emily Holton, branch chief of the Gravitational Research Branch at NASA's Ames centre, the "most elegant and definitive developmental biology experiment in space used the amphibian as a model." There would be many surprises associated with one space flight in 1995, including a totally unexpected result associated with egg fertilisation in the virtual absence of gravity.

Unexpected behaviour

As Dr. Holton wrote in her report, *The Impact of Gravity on Life*, "On Earth, the fertilized frog egg rotates upon sperm penetration, and this rotation is thought to be essential for normal development. Upon fertilisation, the egg begins to divide and form the embryo that, after an appropriate time, emerges from the jelly-like egg as a tadpole.

"Female frogs were sent into space and induced to shed eggs that were artificially inseminated. The eggs did not rotate and yet, surprisingly, the tadpoles emerged and appeared normal. After return to Earth within 2–3 days of hatching, the tadpoles metamorphosed and matured into normal frogs.

"This developmental study produced multiple important findings. It showed that vertebrates can be induced to ovulate in space and that rotation of fertilised eggs is not required for normal development in space" [10].

SPACELAB AND LIFE SCIENCES

On 5 June 1991, shuttle *Columbia* was launched on mission STS-40. This was the fifth Spacelab mission and the first dedicated life sciences mission to make use of the 7-metre habitable Spacelab module, which was secured in the shuttle's payload bay.

Parabolic flights on NASA's KC-135 aircraft can simulate microgravity for short periods. On this flight, scientists were training an adult frog for a shuttle experiment that investigated the development of frog eggs in weightless conditions. (Photo: NASA)

The first Spacelab life sciences mission

The mission was co-designated Spacelab Life Sciences-1, or SLS-1, and would feature the most detailed physiological measurements taken in space since the days of the Skylab missions. While many of these experiments would involve tests conducted by (and on) the 4-man, 3-woman crew, other experiments were centred on 30 male Harlan (Sprague/Dawley) rats and 2,478 tiny jellyfish polyps (*Aurelia aurita*) encased in flasks and plastic bags filled with artificial sea water packed inside an incubator.

Many of the developing structures of jellyfish resemble structures of humans, according to principal researcher, Dr. Dorothy B. Spangenberg, although they are less complicated. It was felt that jellyfish might be used to predict events that can occur in embryos of more complex life forms during space flight. Early in the flight, *Columbia*'s crewmembers injected thyroxine or iodine into the jellyfish containers, to induce the polyps to divide and metamorphose into free-swimming larvae, or ephyrae. These ephyrae developed during the flight and were able to pulse and swim despite the weightless environment. Studies were also made by Dr. Spangenberg's research team to determine whether microgravity causes a decrease in the calcium content of the jellyfish [11].

A question of muscular atrophy

The physiological and anatomical rodent experiment (PARE-1), carried on STS-48 in September 1991, was the first in a series of planned experiments that focused on physiological and developmental adaptation to microgravity. All of the payload and integration support was provided by NASA Ames. The PARE-1 experiment was designed to examine changes caused by exposure to microgravity in anti-gravity

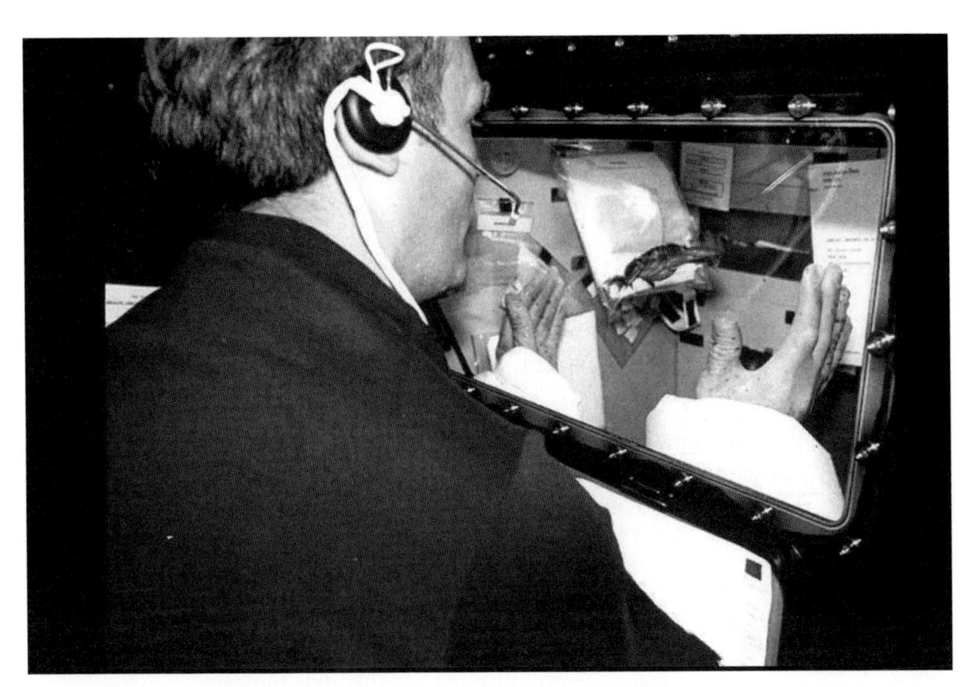

STS-47/Spacelab-J crewmember Mark Lee works with an adult frog using a glove box to protect the frog during test procedures. Female frogs carried aboard SL-J were induced to ovulate and shed eggs, which were then artificially fertilised. (Photo: NASA)

muscles (those used for movement) and in tissues not involved in movement. Previous flight experience using adult rats had indicated that muscle atrophy resulting from exposure to the weightless conditions was of serious concern, particularly in planning any future long-duration missions. Another major objective of the PARE experiments was to determine whether microgravity caused an increase in glucose uptake in the presence or absence of insulin.

Eight young, healthy Sprague-Dawley rats were carried on board shuttle *Discovery*, housed in a single AEM occupying a locker on the orbiter's mid-deck. Another group of identical control rats in similar housing and conditions (apart from weightlessness) would be studied on the ground for comparative analysis. Following the mission, the rat tissues would be examined by principal investigator, Dr. Marc Tischler of the College of Medicine, University of Arizona.

Spacelab flies again

STS-58 was the second Spacelab mission dedicated to life sciences and at 14 days the longest shuttle flight to that time. Experiments carried out on this flight, launched on 18 October 1993, would involve investigations of the cardiovascular, cardiopulmonary, regulatory, neurovascular and musculoskeletal systems. To aid in these studies, a batch of 48 rats would be loaded into the SLS-2 Spacelab module mounted in

Columbia's payload bay. The facility actually used to contain the rodents was described in the pre-flight STS-58 mission report, issued by NASA Headquarters [12].

Research Animal Holding Facility

NASA Ames Research Center, Moffett Field, Calif.

The rodent Research Animal Holding Facility (RAHF) is a general-use facility for housing rodents in life sciences experiments in the Spacelab. It is a self-contained unit providing food, water, temperature and air-flow control, waste management and lighting for the animals on board. It can accommodate twenty-four 400-gram rodents.

The rodent RAHF contains twelve cages that are removable for easy access to the animals. A cage can contain up to two animals, one in each of two compartments measuring 4 by $4\frac{1}{4}$ by 10 inches. Each cage contains a waste management system plus individual feeders and watering lixits. Food and water are available ad lib.

Additional control can be exercised over temperature and light/dark cycles. Protection against cross-contamination between crew and animal is provided through bacteriological isolation. An environmental control system is mounted on the back of each cage module to circulate conditioned air through the cages.

Cage temperature, animal activity, lighting, humidity and water consumption can be monitored by the ground crew and by the astronaut crew onboard. Food consumption on orbit is monitored by the crew. The rodent RAHF flew successfully on the 1991 Spacelab Life Sciences-1 mission. Two rodent RAHFs will fly on the SLS-2 mission.

On previous flights the subject rats had been dissected post-flight, but it had been discovered that their organs had already begun to re-adapt to gravity. As a result, one of the payload specialists on this crew was a veterinarian, Martin Fettman, D.V.M., Ph.D, whose research and teaching interests had focused on selected aspects of the pathophysiology of nutritional and metabolic diseases. The rodents had been dosed with tracers so that the process of adaptation to microgravity could be examined in greater detail, and Dr. Fettman would be responsible for decapitating six of the rats with the use of a small guillotine before dissecting them to preserve their organs in a space-adapted state. This way, their tissue and bone could be sampled in conditions of weightlessness. Several of the others would be sacrificed immediately after landing, in order to isolate the effects of descent, while the remaining rats would be killed at intervals in order to fully gauge the process of re-adaptation.

Post-flight analyses of the bone marrow, spleen and thymus from flight and ground-control rats would provide additional material on the effects of microgravity on the blood system. Another crucial experiment would focus on the effects of space flight on the muscles from the rats' hindquarters. It had previously been noted that after 2 weeks in orbit, pathological changes had occurred in the skeletal muscles of rat subjects; of principal concern was the fact that the rodent's muscles had atrophied markedly through lack of use in weightlessness. It was feared that similar problems could potentially cause irreversible loss of muscle strength in astronauts on long-duration space flights, which could even prove fatal.

This flight experiment would compare the atrophy rates of muscles used primarily to oppose gravity with muscles used for movement. Tissues would be examined for physical and chemical changes that might also be related to the stress of launch, microgravity, re-entry and re-adaptation to Earth's gravity. These experiments would result in developing in-flight countermeasures to prevent damage to the muscular system. This research would also have other applications in helping to counter muscle deterioration for humans who might be bed-ridden or hospitalised for extended periods.

RODENTS LEAD THE WAY IN RESEARCH

Further experiments on the vestibular and muscular systems of Norwegian/Sprague-Dawley rats would take place on STS-66 (launched 3 November 1994), STS-70 (13 July 1995), STS-72 (11 January 1996) and STS-77 (19 May 1996).

Astronauts and AstroNewts

Aquatic creatures had first been introduced to space flight during the Skylab missions, and the testing and observation of these subjects would continue on shuttle mission STS-65, which featured the flight of the second international microgravity laboratory (IML-2), a major Spacelab sub-programme. The first of these laboratories had been carried on STS-42 in January 1992.

One research facility carried aboard the IML-2 was the Aquatic Animal Experiment Unit (AAEU), which would be used in a programme known as the Fertilisation and Embryonic Development of Japanese Newt in Space – known as AstroNewt for short. As suggested by the longer title, this was an experiment on the effects of gravity on the early developmental process of fertilised eggs, using a unique aquatic animal, the Japanese red-bellied newt (*Cynops pyrrhogaster*). The experiment, according to principal investigator Michael Wiederhold of the University of Texas Health Science Center in San Antonio, was carried out to determine whether babies born in space could adapt to Earth's gravity on return. It was also thought that the experiment might also point the way towards development of food production for long space flights. The tiny newts have a unique breeding characteristic: after mating, the females will release their eggs only if given a hormonal signal (usually coming as the days lengthen into spring) [13].

The AAEU cassettes carried four newts. Two of the newts had been treated by controlled hormone injections on the ground to induce egg-laying, producing 144 eggs which were then fertilised. The other two newts would be treated on orbit by the crew. More eggs were successfully obtained on the mission and two baby newts were hatched, beginning on Flight Day 5, although one of the adult female newts died early

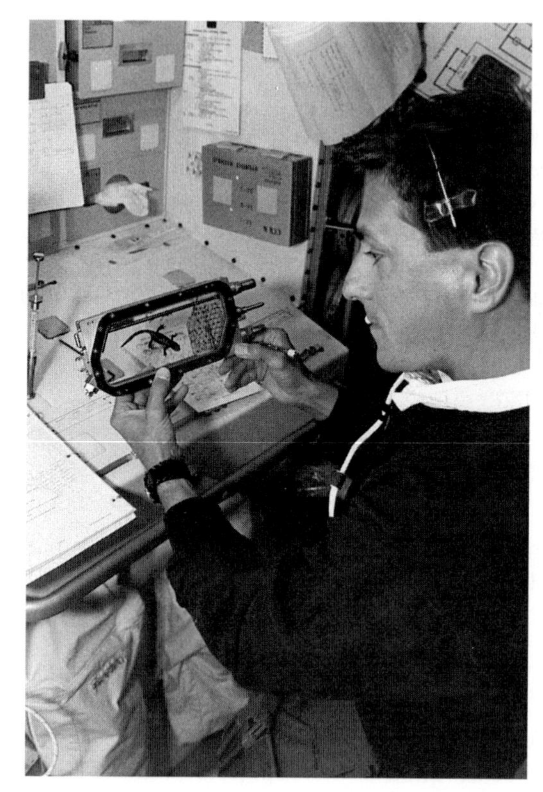

STS-65 crewmember Don Thomas conducting the AstroNewt experiment at the Aquatic Animal Experiment Unit (AAEU) inside the IML-2 science module. (Photo: NASA)

in the flight. Mission specialist Dr. Don Thomas, noting the success of the test, said, "The baby newts are pretty active. They are swimming around the little egg chambers pretty good." Later, on Flight Day 11, he reported that a second female adult newt had died, after earlier producing eggs. The following day Dr. Thomas reported that the remaining newts, including those hatched on the mission, appeared to be doing well [14].

Tanks, tests and transparent fish

Two pairs of pre-fertilised medaka (*Oryzias latipes*), a small freshwater fish found in Japan's ponds and rivers, were also carried in a special aquarium in the AAEU for principal investigator Dr. Kenichi Ijiri of Tokyo, Japan. The mission would last 14 days, and as medaka eggs only take 8 days to hatch the crewmembers could observe the entire mating and hatching process, and make post-hatching observations. In fact, these tiny fish would hold the distinction of being the first vertebrate creatures ever known to have successfully mated in space.

Tiny, transparent medaka fish flew on several shuttle missions. (Photo: NASA)

Unlike the fish carried into orbit with the second Skylab crew, which went a little frantic in weightlessness, initially looping wildly, the Japanese found that the medaka would remain quite sedate throughout the experience, and the females would orient their bodies for mating based on crewmembers periodically opening a window on their tank, allowing artificial light to flood in and simulate daylight.

In viewing downlinked videos made of the mating process, Dr. Ijiri noted that initial attempts at mating by the two pairs of fish failed, and the females would often attack the other courting couple with short pecking jabs. On the third day of the flight, videotaping crewmembers caught the first mating between the little golden fish, and Ijiri was delighted as he watched the live downlink. "I could see with my own eyes this dramatic scene," he later reported. Eventually eight tiny offspring were born in orbit." According to an exuberant Ijiri, the four medaka pioneers "having overcome ... weightlessness, and the complexity of love and hatred," lived long enough to become great-grandparents [15].

The fish, eggs and fry are all mostly transparent, so the crew could also record the development of the body and internal organs on video. Crewmembers reported that some medaka fry had hatched during the mission. Video footage was later delivered to Dr. Ijiri, who was able to study behavioural aspects of the adult and fry fish.

Following *Columbia*'s safe landing at the Kennedy Space Center after a record-breaking 14-day, 18-hour mission, the two surviving adult newts would be dissected by researchers. Some pathological changes were discovered in several organs of the adult newts that returned alive from their space flight. The medaka would later be shipped back to Japan to live out their lives and mate under scientific scrutiny as biologists looked for any lingering effects of space flight.

STS-66 carried a National Institutes of Health payload (NIH-R-1), which allowed for the study of effects of space flight on developing rats over an 11-day flight. Two AEMs were loaded, each of which contained five pregnant rats. Experiments were carried out on behalf of nine U.S. principal investigators, as well as one from France and another from Russia.

A second National Institutes of Health payload (NIH-R-2) would fly on STS-72. Dozens of neonatal rats, all between 5 and 15 days old, were sent aloft to allow scientists to study critical periods in their neurological development. An advantage to using rats is that they develop far more quickly than humans in certain areas; in fact, a rat develops a nervous system in 3 weeks – something that would take years in a human.

Flying fish and hornworms

On the STS-70 mission, NASA and the Walter Reed Army Institute also collaborated on the space tissue loss-B experiment, which investigated the effects of microgravity on embryogenesis, and once again the medaka egg was used as the biology model. Additionally, Dr. Marc Tischler of the University of Arizona's College of Medicine had a student experiment on board which involved an examination of the hormonal system of the tobacco hornworm (*Manduca sexta*).

In continuing biological research, 12 Norwegian rats were carried on *Columbia*'s mid-deck on STS-78, a space flight lasting nearly 17 days. The rats were euthanised following the flight. Medaka embryos were also carried on this flight in a continuation of studies of development conditions for other vertebrates, such as humans.

STS-90 NEUROLAB

When shuttle *Columbia* soared into space from Pad 39B at KSC on Friday, 17 April 1998, the STS-90 crew of seven was not alone in orbit. Their spacecraft carried a vast, fully enclosed menagerie of mice, adult and baby rats, two kinds of fish and thousands of crickets on a 2-week mission dedicated to science. The crew had been given a challenging and exhaustive mandate – to conduct intense neurological research and observe changes in the nervous system in the microgravity of space – and these animals were a crucial part of those studies.

A veritable raft of experiments

Neurolab was the name officially given to this important science mission, carrying an international suite of 26 complex experiments that would be conducted during the flight. According to a book containing the results from the mission, edited by crewmember Jay C. Buckey, Jr., M.D. and mission scientist Jerry Homick, Ph.D., the goal was to understand how the nervous system helps humans adapt to weightlessness. There were five main areas of interest: balance, sensory integration, development of the nervous system, blood pressure control, sleep and circadian rhythms [16].

The seven-member crew included two medical doctors, a veterinarian and a physiologist who served as both subjects and operators in carrying out several of the 26 experiments. These ranged from studies of the inner ear and sleep patterns, to a simple eye–hand coordination experiment to see how well astronauts could catch a tossed ball in microgravity.

The STS-90/Neurolab crew. Standing (from left): James Pawelczyk, Richard Linnehan, Kay Hire, Dave Williams, Jay Buckey. Seated (from left): Scott Altman, mission commander Rick Searfoss. (Photo: NASA)

Before the flight, physician Dafydd (Dave) Williams gave news reporters a few details of their science objectives and concluded by saying, "One of the big goals of this mission is exploring inner space, or the innermost workings of the human nervous system. There's so little that we really understand about the detailed functioning of the nervous system. We're going into outer space to understand inner space" [17].

Spacelab, for the final time

Nestled into *Columbia*'s cargo bay was a facility known as Spacelab, a bus-sized, pressurised laboratory developed by ESA and built by European industry. On this mission it housed the Neurolab payload. The innovative Spacelab had proved a useful working laboratory in the weightlessness of space, allowing science to be carried out in a comfortable "shirtsleeve" environment. This would be Spacelab's final scheduled flight after 15 years of valuable service and would also prove to be the first and only Neurolab mission.

The Spacelab module containing the Neurolab payload being lowered into *Columbia*'s payload bay at the Kennedy Space Center. (Photo: NASA)

In the days leading up to the mission (which would suffer a 24-hour delay to allow a faulty mid-deck network data processor to be replaced), scientists sorted through 10,000 garden-variety crickets to locate the ones they specifically wanted on the flight. A total of 1,514 crickets were selected for placement in a sealed botany experiment incubator known as BOTEX. This number would comprise 824 babies at 3 distinct ages and 690 eggs that should hatch a few days into the flight. A centrifuge was installed in the incubator to provide artificial gravity for a control group of crickets.

There was a special reason crickets were chosen for this flight, according to German neurobiologist and cricket expert Eberhard Horn. "The crickets have an external gravity sensor," he explained. "So you can see immediately what happens in space with such an animal." When asked if the chirping of the crickets would prove a nuisance to the crew, Horn laughed, and said that crickets chirped by rubbing their wings, and none of them would be old enough to have the wings necessary for serenading. "If they start to sing," he added with a smile, "then we have a result – an unexpected one!" [18]. The crickets were accompanied by 18 pregnant mice, 135 water snails, 96 rats, 223 fish and some rootless water plants called hornweed. Suitable habitats had been provided to contain and sustain the animals. The fish, snails and crickets would not be removed from their habitats during the flight.

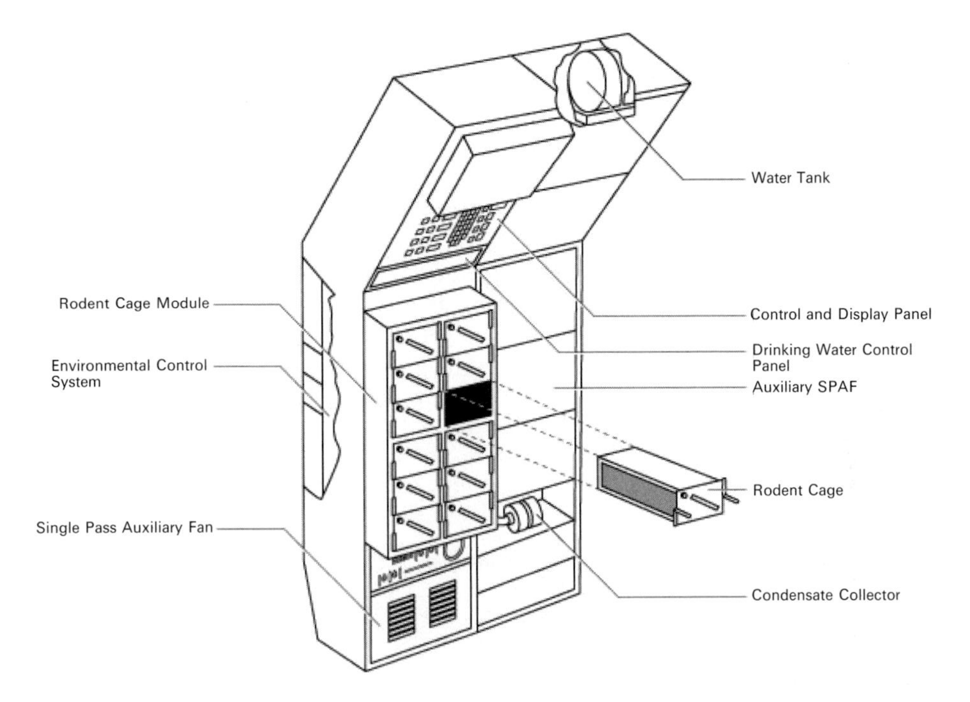

The Neurolab RAHF module. (Illustration: NASA)

Rats in hiding

The rats, on the other hand, were housed in two facilities – the research animal holding facility (or RAHF) and two animal enclosure modules (or AEMs). The RAHF, with temperature and humidity environmental control, consisted of 12 cages that could house 2 adult Fischer rats or 1 female and 8 young rats, 8 days old at launch. Water outlets in the cages known as lixits would operate when touched by the rats, while food bars were dispensed through a mechanical system.

Once they had achieved orbit, the crew quickly began preparing Spacelab for the work ahead. Examinations – many of them demanding – were scheduled to be carried out on both the crewmembers and the animals.

The crewmembers would later find that the greatest disadvantage to the rodents' RAHF was an inability to see the animals properly, as the viewing windows offered only a partial view of the front cage, and these windows became further obscured over time as particles coated the windows. The rats would also tend to spend most of their time at the back of their cages.

One major setback occurred when it was discovered on Flight Day 8, the first activity day involving the young rodents, that nearly half of the youngest rats were dead – a surprisingly high mortality rate. Forty-five rodents had died from what was later thought to be maternal neglect, related to the baby rats failing to nurse in weightlessness, and complicated by low blood sugar, low body temperature and dehydration. It came as a severe blow to the crew, especially after telemetry data had indicated normal food and water consumption and RAHF operation According to NASA's chief veterinarian Joe Bielitzki, it was 5 to 10 times the number scientists expected to die. "Their surrogate mother rats weren't drinking enough water," he explained, "and either shunned the young animals or simply did not produce enough milk for nursing" [19].

Despite the losses, it was felt that there were probably enough rats left to conduct the mission's research into nervous systems development in microgravity, according to project scientist Louis Ostrach. Over the first few days of the flight, the young rats were supposed to learn to walk in reduced gravity, following which they would be operated on or dissected to see how their nerves and weight-bearing muscles had developed. However, it was felt that, with the greatly reduced number, other experiments would have to be dropped, while some would require sharing parts of the brains of the same rats.

Weightlessness and the development of muscles

The youngest rodents were part of a study to understand how weightlessness would affect the development of muscles. To continue the study, crewmen Jay Buckey and Dave Williams picked six healthy animals for leg surgery. They made a small incision in both hind legs, injected a fluorescent dye, and then sealed the wounds using surgical glue. It was the first surgery in space on creatures meant to survive. Following post-flight dissection, the fluorescent dye would allow researchers to see how muscles and nerves had developed in weightlessness. "Everything went well," Buckey reported of the surgery [20].

Several more of the youngest rodents died or had to be euthanised by the crew, although Richard Linnehan, the veterinarian aboard *Columbia*, said that several sick rats had improved after being fed a mixture of water and Gatorade[1] and warmed with heated fluid bags. "They all seemed to perk up quite a bit," he told scientists on the ground. "We feel we're over the hump" [20].

Spacemen and specimens

Meanwhile, the science crew was also monitoring the effects of microgravity on the nervous systems of oyster toadfish and swordtail fish. The toadfish were enclosed in the Vestibular Function Experimental Unit (VFEU), while the swordfish and water snails were contained in the Closed Equilibrated Biological Aquatic System (CEBAS) mini-module, a locker-sized fresh water habitat located on the shuttle's mid-deck. The CEBAS was designed to allow the controlled incubation of aquatic species in a self-stabilising, artificial ecosystem. The much larger toadfish was an excellent model for

[1] Gatorade is a non-carbonated rehydrating sports drink used to replenish carbohydrates and electrolytes depleted during exercise.

The oyster toadfish carried on STS90/Neurolab. (Photo: NASA)

an examination of the vestibular function, as the architecture of its inner and middle ear are very similar to those of mammals with respect to the vestibular apparatus [21].

Animal rights groups soon learned of the in-flight deaths of the young rodents. "NASA has an appalling record," declared Mary Beth Sweetland, the director of research, investigations and rescue department for PETA (People for the Ethical Treatment of Animals). "It can't keep animals alive on the ground or in space" [22].

Meanwhile, Joe Bielitzki said *Columbia*'s crewmembers were doing everything humanly possible to prevent more deaths. "The crew has really done yeoman's work in this case," he responded. "Anybody that has tried to rear an orphaned animal ... understands the number of hours and the effort that has to go into saving a single animal, let alone 45 or 50 of them" [22].

A dwindling population

The adult rats and the rats that were 14 days or older at launch all did well throughout the flight. Most of the surviving youngest rats improved as the mission wore on, although the attrition rate continued. By Flight Day 9 there had been 44 rats remaining; the following day 4 more had died or been euthanised, and by Flight Day 13 only 38 of the youngest rats remained, although it appeared the situation had stabilised appreciably due to the crew's efforts. Packs of gel containing water had been added to the cages in addition to the lixits, and this helped. There was still some evidence that a few mothers were neglecting the young rats, so these were removed and placed in cages where other young rodents seemed to be doing much better.

At the end of its 16-day mission, *Columbia* and its crew swooped through a clear noontime sky and returned to Earth, ending 2 weeks of intensive laboratory work.

Although the flight was at an end, the experiments were not over. Within an hour of touching down, the crew were hustled off to medical tests that lasted several days. Six of the seven astronauts left the shuttle on stretchers because doctors wanted to preserve their weightless state as long as possible. NASA also rushed to offload the surviving animals so scientists could begin studying the few dozen baby rats that were still alive, as well as the nearly 2,000 fish, snails, crickets and older rodents. The sooner the astronauts and animals could be examined, they knew, the greater the likelihood of observing space-induced changes in their nervous systems.

Post-flight solutions

Several factors had contributed to the in-flight deaths of so many young rodents in the RAHF, when rats of the same age had flown successfully prior to Neurolab. Later investigations revealed that inadequate housing and an inability to provide the crew with effective monitoring of the rodents probably contributed to the high mortality rate.

In a book published after the flight (*The Neurolab Spacelab Mission: Neuroscience Research in Space*), crewmembers Buckey and Linnehan and ground-based alternative flight veterinarian Alexander Dunlap wrote that, "The problem on Neurolab might have been averted or minimised if the crew or ground had had some indication of problems within the cages. As it turned out, however, none of the monitoring systems were adequate to detect the problems. Lixit counts remained normal in all cages. The food bar measurements were consistent with previous experience. The windows were not adequate to allow the crew to see the problems in the cages, and the activity monitoring system had been shown prior to Neurolab to be unreliable" [23].

It was recommended that future cage designs should incorporate internal surfaces which could be more easily grasped by the rats, allowing for better three-dimensional navigation in microgravity, as well as more effective and reliable ways of monitoring the animals.

Despite problems that arose during the mission, the majority of the animal experiments went well and produced a series of interesting findings and published articles.

FLIGHTLESS BIRDS AND AVIAN EXPERIMENTS

With protracted experiment time becoming available aboard Russia's Mir station, scientists were keen to study the potential effects of weightlessness on embryo development, and whether it might induce abnormalities later in the life cycles of animals. Obviously, this would have important ramifications for the long-term future of humans in space. It was also felt that many of the protocols developed during this

The quail incubator was used to transport fertilised quail eggs to the Mir space station. (Photo: NASA)

investigation could prove useful in the treatment of diseases such as cancer, heart disease, genetically inherited syndromes and congenital defects.

The Russian quail egg story

In 1995, the Life Sciences Division at NASA's Ames Research Center was able to fly a group of quail eggs to Mir aboard a Russian Progress transport vehicle, with a second group scheduled to fly later that year. The eggs to be used in the first round of experiments were collected from a quail colony at the Institute of Biomedical Problems in Moscow and were hand-carried to the launch site. Reporting on the upcoming mission, Dr. Gary Jahns, the Shuttle/Mir payload manager at the Ames centre, explained that: "The first group will include three females and one male quail. During the initial stages of the research, astronauts will put the eggs that we anticipate will be fertilised in space into an on-board incubator. Scientists will study the eggs at various stages of development" [24].

On arrival at the space station, the eggs were transferred aboard by the Mir-18 extended-stay crew, which included U.S. astronaut Dr. Norman Thagard, and placed into the incubator. This crew, and the next to occupy Mir, would conduct experiments to help investigators determine the effects of microgravity, not only on

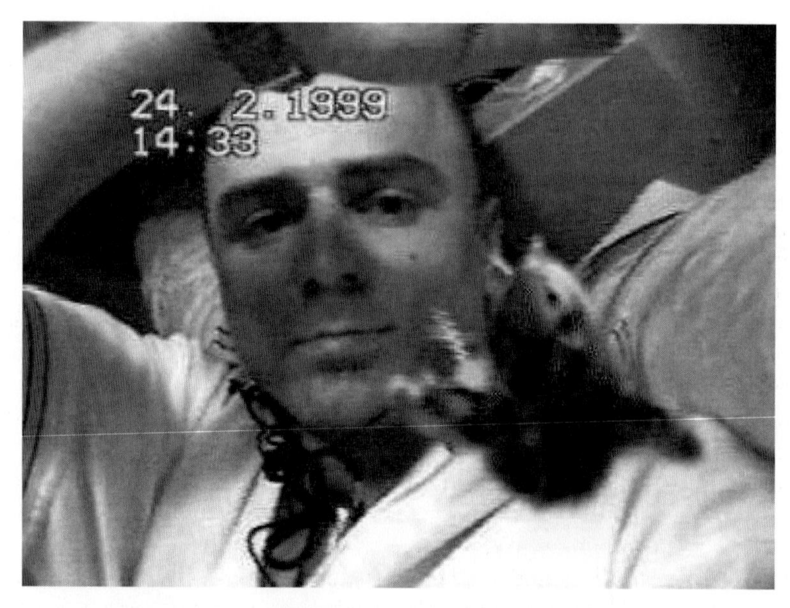

Quail eggs would fly to the Mir space station on several occasions. In 1999, Soyuz TM-29 delivered a 3-man crew and 60 quail eggs to Mir. The birds began hatching a few days after the docking. Here the first Slovak cosmonaut, Ivan Bella, watches a newly-hatched quail fledgling. Only three of the birds would survive to return to Earth. (Photo: NASA)

the physiological processes, but on the functional and structural development of vestibular receptors in the hatched birds. The quail would then be returned to Earth aboard shuttle ferry flights.

Unfortunately, many of the quail embryos failed to develop as expected, and most did not survive past 5 days of incubation.

On the following Mir-19 flight, in-flight development progressed to a further stage than noted in the earlier group, with 10 embryos developing past 7 days of incubation and 1 developing to 16 days. No abnormalities in gross morphology were noted in the embryos.

Beatles in orbit?

STS-93, launched 23 July 1999, was the first ever U.S. space mission launched under the command of a female pilot, Eileen Collins. One of the experiments loaded on board involved a familiar and much-beloved winged creature often regarded as a symbol of good luck: the ladybug (also known as the ladybird). In nature, ladybugs provide gardeners and farmers with a wonderful service – they love to eat the tiny but destructive aphids. But aphids are cunning; when they spot a ladybug they instantly fall down from the plant they were chewing to escape being devoured.

In the STARS (Space Technology and Research Students) programme funded by Spacehab, Inc., schools from around the globe were invited to participate in research, and to view on-line data of their experiments. A group of students from the all-girl Chile-based Liceo No. 1 Javiera Carrera High School developed an experiment that asked a simple question: Place some wheat plants, ladybugs and aphids into a container, take away gravity, and what happens to the aphids?

STS-93 pilot Jeff Ashby discussed the experiment prior to his delayed flight aboard shuttle *Columbia*, and said he was personally interested in finding out how the aphids' escape mechanisms would work in weightlessness, and what would happen to the relationship between predator and prey. "One of the things that extra time has allowed us to do," he told a press gathering, "is to come up with names for the four ladybugs we have. I think they have been very appropriately named after the Beatles: John, Paul, Ringo and George. We're taking these ladybugs up and we're going to release them and see what they do" [25].

Also included in the STARS experiment was a study by students from Albany, Georgia, on the metamorphosis of pupae larvae from a chrysalis (cocoon) to the maturation stage of painted lady butterflies in microgravity.

Following the flight, it was determined that gravity had little effect on the voraciousness of the ladybugs. Without the ability to jump or drop from the leaves, the aphids became easy prey and were consumed once the ladybugs were introduced into their container. As well, the Georgia High School student experiment was a complete success, with a butterfly emerging from its chrysalis to become the first such creature in space. The results of this experiment are now on permanent display at the Smithsonian Air and Space Museum in Washington, D.C. [26].

The STAR programme proved so successful that a new round of student experiments would be carried on board shuttle *Columbia*, on the ill-fated STS-107 flight.

CHINA RESUMES BIOLOGICAL FLIGHTS

In his 1998 book, *The Chinese Space Programme: From Conception to Future Capabilities*, author Brian Harvey reveals that at least three recoverable satellites known as the Fanhui Shei Weixing (FSW) carried animals into space in the 1990s. Adapted from former military reconnaissance satellites, these vehicles carried animals on three recorded flights.

Shenzhou and state secrecy

According to Harvey's research, these flights were:

- FSW-1/3. Launched 5 October 1990, carrying guinea pigs. Although they were never recovered, the guinea pigs became the first Chinese animals in orbit.
- FSW-2/2. Launched 3 July 1994, carrying an unspecified number of animals.
- FSW-2/3. Launched 20 October 1996, carrying an array of small animals for the Shanghai Institute for Theoretical Physics. On this third launch in the FSW series, China became the third nation, after the Soviet Union and the United States, to

intentionally launch animals into orbit and safely recover them after their flight [27].

In order to be recognised as a major space player, China's scientists and policy-makers were keen to launch a human into space and to enjoy the prestige associated with becoming the third nation to do so, ahead even of Europe and Japan. A manned space programme would also draw some attention away from the nation's military space activities. This was cited in a July 2002 report published by the U.S. Department of Defense, which stated that: "While one of the strongest immediate motivations for this appears to be political prestige, China's manned space efforts almost certainly will contribute to improved military space systems in the 2010–2020 time frame" [28].

The technology applied to this effort to challenge what the Chinese refer to as "the fourth frontier" had eventually resulted in a Soyuz-based spacecraft originally referred to as Project 921 and later as Shenzhou, or Divine Vessel.

Talking of taikonauts

The first unmanned Shenzhou flight took to the skies at 6:30 a.m. on the morning on 20 November 1999. Twenty-one hours later, having completed 14 orbits, the re-entry module returned to Earth, landing safely. The spacecraft had been launched from the Jiuquan Satellite Launch Centre (JSLC) in Gansu Province, northwestern China, atop a Long March CZ-2F launch vehicle. It swept into orbit 10 minutes later.

Shenzhou was an electronic experimental prototype without any payloads, life-support equipment or emergency escape system. Despite this, the successful flight of the first Shenzhou was seen as a significant milestone in China's plans to finally launch human candidates, known alternatively as *taikonauts* or (as the Chinese prefer) *yuhangyuans*, into space. Photographs of the re-entry capsule after touchdown were quickly released amid a flurry of self-congratulatory euphoria and state propaganda.

The second, more complex flight came 14 months later, but this time the latter part of the flight was very quickly shrouded in secrecy. Five years later, the second test flight of the Shenzhou spacecraft still remains an impenetrable question for researchers. The Chinese state media reported that Shenzhou-2 carried "various life forms", including animals, plants, aquatic creatures, microbes and cells, but did not go into any specifics. Rumours emanating from an industry source at the time suggested that the live payload actually included a monkey, a dog, a rabbit and some snails, but this has never been confirmed.

A second Shenzhou

Shenzhou-2 was launched on 9 January 2001. As with the first spacecraft, the launch vehicle was a CZ-2F, which lifted off the pad at JSLC at 9:00 a.m. Beijing time. The Chinese defence department later described the payload as "a sophisticated space science laboratory, with a total of 64 science experiment payloads, including 15 in the re-entry module, 12 in the orbital module and 37 on the forward external pallet." They added that these included "a micro-gravity crystal growing device, life sciences experiments with 19 species of animals and plants, cosmic ray and particle detectors, and China's first gamma ray burst detectors" [29].

The spacecraft assembly would make three orbit-raising manoeuvres during its 7-day flight, reaching a 206 × 216-mile orbit by the completion of the initial phase of the mission. The ability of controllers at the Beijing Space Command and Control Centre to manoeuvre the spacecraft's orbital module certainly took many Western observers by surprise. After 108 orbits, the descent and service modules were separated from the forward orbital module and external pallet. Contemporary reports from China stated that after braking rockets were fired on the service module, the re-entry module separated and landed amid grasslands in Inner Mongolia's Autonomous Region at 5:22 p.m. Beijing time on 16 January.

Then, as observers waited for further news, the Chinese media unexpectedly became ambiguous and low-key when reporting on the landing, merely stating that the flight had successfully concluded. More than 2 weeks after the reported return of the re-entry module, the state-controlled Xinhua News Agency was still maintaining a stoic silence about the mission, prompting space officials to deny mounting rumours that crucial systems had failed during the module's descent through the atmosphere. Zhang Xiaodong, a spokesman for China Aerospace Science and Technical Consortium (CASC), which had been formed in 1999, fended off queries by stating quite adamantly that "nothing went wrong". Dr. Liu Yongding, the life sciences payload manager for the mission, also refused to comment when asked about the animals that were said to have been aboard the re-entry module [30].

There seems little doubt that the mission ended in failure. Unlike the previous operation, no photographs have ever been released of the returned Shenzhou-2 capsule. Despite Chinese denials, and statements alluding to "a perfect mission", it does appear that an unknown but obviously major problem occurred at some vital stage of the re-entry procedure.

A programme shrouded in mystery

When the unmanned Shenzhou-3 touched down in Inner Mongolia on 1 April 2001, a few hours shy of a full week in space, photographs of the returned re-entry module were released within a couple of hours. By comparison, no photos have ever been released of what appears to have been a catastrophic end to the earlier mission.

No animals were carried on the flight, although a human dummy rigged up with sensors was carried as part of a test of the life-support system.

"Unlike decades ago when many technical difficulties existed, making a dummy is now simpler," said Zhuang Fenggan, the chairman of CASC, when discussing the flight. "We could, say, measure simulated blood pressure using a dummy on a mission. If a monkey is launched into space, unless it is tightly secured and made immobile, the monkey could touch anything on board and cause troubles.

"Speaking from another perspective, we want to guarantee the safety of the *yuhangyuans* [astronauts] much as with the animals. Otherwise animal protection groups would protest" [31].

Recovery of the Shenzhou-3 capsule. (AP/Xinhua News Agency)

It is highly unlikely that animal welfare would actually have been a major concern for the CASC, but immediate worldwide condemnation would have followed the loss of the rumoured monkey, dog and rabbit aboard Shenzhou-2. This spectre of bad publicity probably resulted in a state-imposed news blackout following the apparent loss of the re-entry module on that flight.

TRAGEDY STRIKES AGAIN

On 1 February 2003, space shuttle *Columbia*, on mission STS-107, broke up in the searing heat of re-entry into the Earth's atmosphere, and all seven crewmembers perished.

Sole survivors

Incredibly, hundreds of worms that were part of an experiment aboard the doomed shuttle were found alive in debris recovered from a crash site in Texas – the sole survivors of the tragedy. *Caenorhabditis elegans* worms, which live in soil or among rotting plants, are tiny creatures about the size of a printed comma in a newspaper, and they have two sexes – male and hermaphrodite. These worms were part of an experiment testing a synthetic nutrient solution. They were found in six canisters, each holding eight Petri dishes, within a 9-pound locker from the shuttle's mid-deck that

The Biological Research in Canisters (BRIC) container in which surviving roundworms were discovered after shuttle *Columbia* broke up during the landing phase of mission STS-107. The container, from the shuttle's mid-deck, was found amid debris recovered in east Texas. (Photo: NASA)

had somehow survived the fiery plunge to Earth. "To my knowledge, these are the only live experiments that have been located and identified," said Bruce Buckingham, a NASA spokesman at the Kennedy Space Center. Along with the worms, the locker was found to contain traces of moss from another biological experiment [32].

There were several scientific experiments on board *Columbia* involving animals, including insects, spiders, fish, bees and silkworms. Thirteen Sprague-Dawley rats had also been loaded onto *Columbia* for this flight. In one experiment involving eight of these rodents, Ames researchers were looking at how the process of adapting to microgravity changes the distribution of fluids in the body, including the brain, kidneys and lungs. In all of these organs, a family of proteins called aquaporins transport water, and tests on the eight rats would have helped investigate the effect of space flight on these aquaporins. Post-flight, it had been planned to remove the brains, pituitary glands, kidneys and lungs of the rodents for processing and analysis using conventional electron microscopy and immunocytochemistry detection. All of the rats perished in the fiery break-up of *Columbia*.

Not a place for stressed-out scorpions

At the time of researching and writing this book, the most recent biological space flight to carry living creatures had been launched and recovered by Russia.

On 31 May 2005 an unmanned space laboratory called Foton-M2 had been flawlessly launched aboard a three-stage Soyuz LV booster from the Baikonur Cosmodrome in Kazakhstan. Aboard were Spanish ribbed newts, Cuban fresh water crayfish, Georgian snails, thorn-back tritons (marine gastropods), scorpions, geckos and lizards and other creatures and micro-organisms that would spend more than 2 weeks circling the globe.

Also on board the 1,200-pound laboratory was a diverse suite of more than 39 science experiments provided by several nations, mostly European. They included such fields as physical and material sciences, biology and exobiology (adapting Earth's life forms to alien environments). Many of these experiments were being reflown after the Foton-M1 capsule was destroyed on October 2002, when the Soyuz carrier rocket exploded just seconds after lift-off.

The chief of the European Space Agency's research programmes in Russia, Christian Feichtinger, reported that Russian scientists were pursuing 4 biological experiments and around 20 experiments in physics, chemistry and space biotechnology in Foton. He said that an evaluation of the effects of stress on the nervous system of scorpions and experiments with fluids in conditions of weightlessness were among the most interesting lines of work. In particular, lizard studies could assist scientists in developing methods for treating osteoporosis, or bone fragility caused by a deficiency in calcium, while tritons might provide vital clues about organ regeneration.

Parachuting safely back on Earth 16 days later on 16 June, the Foton's spherical entry module touched down about 90 miles southeast of the town of Kostanay in Kazakhstan, close to the Russian border [33].

Shenzhou flights continue

China had announced plans to carry some 14 grams of semen from pedigree pigs on the nation's second manned space voyage, Shenzhou-6, which was launched on 12 October 2005. This flight carried two taikonauts, Fei Junlong and Nie Haisheng, on a low Earth orbit mission lasting 5 days. China's state media had said this involved a study to determine whether exposure to outer space, and particularly the impact of microgravity and cosmic radiation, alters the genetic make-up of the sperm.

According to Wang Jinyong, head of the Chongqing Academy of Animal Husbandry Science (CAAHS), the sperm would come from two carefully selected Rongchang pigs, named after Rongchang province in southwest China's Chongqing municipality. When the samples were returned to Earth they would be used to fertilise eggs in test tubes by CAAHS researchers, and the results of the tests would be released approximately 2 years later.

The day before the launch, however, it was revealed that there would be no animal experiments carried on the flight after all. Liu Luxiang, director of the Centre for Space Breeding at the Chinese Academy of Agricultural Sciences, said the sole focus of Shenzhou-6 would be to determine the physical reactions of the crew to the environment of space [34]. The next manned Shenzhou flight, manifested to carry three taikonauts, will probably not take place until 2008.

It is now recognised that travelling beyond the known protection of Earth's atmosphere can expose a spacecraft and its occupants to potentially dangerous amounts of cosmic radiation. Fortunately, for manned space flights conducted to date (including the Apollo lunar missions), radiation doses received by astronauts have been quite low and of no real clinical significance. However, as missions increase in duration and we advance on to lunar bases and later travel to Mars, the insidious dangers of radiation for human space explorers will become increasingly evident.

It remains a difficult and worrying field to investigate, according to Frank Cucinotta of NASA's Space Radiation Health Project at JSC. "It's a question of radiation," he admits. "We know how much radiation is out there, waiting for us between Earth and Mars, but we're not sure how the human body is going to react to it" [35].

In January 1971, and halfway to the moon, the crew of Apollo 14 became aware of curious light flashes, especially noted in the darker recesses of their spacecraft and during scheduled sleep periods. Because it presented nothing more than a curious biological phenomenon, they only reported these flashes once they had returned to Earth. Investigators soon realised that cosmic radiation was responsible and created the Biostack experimental packages that were carried on the two final Apollo missions. Five pocket mice travelled on such a package aboard Apollo 17. These studies helped to reveal the flux and energy levels of these electronic particles that can pass straight through spacecraft walls and to a degree demonstrated the biologic consequences when cosmic radiation "hits" occurred in susceptible tissues.

In 1973, Skylab astronaut Bill Pogue was asked to record retinal light flashes as the space station passed through a hazardous zone of intense radiation known as the South Atlantic Anomoly (SAA). Here, roughly located over the eastern Brazilian coastline, the close-to-spherical electromagnetic radiation belt that extends to around 400 miles beyond our planet is pulled downwards to within 60 miles of the Earth's surface by a still-mysterious force. Like Skylab, today's high-orbit spacecraft routinely pass through this area of intense radiation several times each day. At one stage during the 84-day orbital mission, Pogue set himself to record each light flash he experienced by pressing or "keying" his microphone. To the shock of many, at times he could not key fast enough to record all the hits.

Former NASA scientist-astronaut Duane Graveline, now involved in biomedical research at KSC, is deeply concerned about the effects of cosmic radiation on future long-duration crews.

"Life as we know it developed within the protective electronic shield of the Earth," he told the authors. "Raining down on this electromagnetic shield are ions of both solar and Galactic origin. Protection from these extraordinarily energetic articles will provide an immense challenge, as they can pass through the walls of a spacecraft as if they did not exist.

"To get to the moon we must venture beyond Earth's shield, out to where we are subject to the full impact of cosmic radiation, which is lethal to life as we know it."

Graveline also remarked that during a 6-month stint aboard the Mir space station, NASA astronaut Jerry Linenger reported an inability to sleep when the Russian space station swept through the SAA zone. "He even tried moving his sleeping platform so as to reposition his head behind lead storage batteries and other places offering thicker spacecraft walls, but little benefit was obtained" [36].

Humans have now been travelling into space for more than four decades, yet we are still largely unappreciative of the potentially lethal dangers that lurk out there in deep space.

The difficulties of establishing long-duration bases on the surface of the moon and subsequently travelling on to Mars are not only confined to the development of new and advanced spacecraft and boosters, fuel and life-support systems. If we are to ever routinely travel to other worlds, ways must be found to protect space travellers on long-duration missions, during which they will be bombarded by lethal rays that could potentially destroy the crew's minds and bodies.

Unpalatable as it may seem, many feel the answer is to consider a renewal of what project scientists routinely did in the post-war years – sending animal emissaries out into deep space on protracted flights before committing any humans to such missions.

A whole new frontier of exploration and knowledge is waiting to be crossed, and once again animals might help to prove that these problems, and others yet to be realised, can be overcome with persistence and ingenuity.

REFERENCES

[1] Jack Lousma, diary notes, STS-3 NASA Media resource Kit. Website: *http://www.jsc.nasa.gov/history/shuttle_pk/mrk/FLIGHT_003_STS-003_MRK.pdf*

[2] "Animals in Orbit," published by NASAexplores, 25 April 2002. No author given.

[3] Daniel Holley quote from "Rat Here, Rat There" by Cecily Barnes. *Metro* magazine, 19–25 February 1998.

[4] "A Comparison of Honeycomb Structures Built by *Apis mellifera* (SE82-17)," NASA Life Sciences Data Archive, NASA JSC, Houston, TX, 2004.

[5] "Overmyer's Ark," *Space Flight News* magazine, No. 13, January 1987, pp. 34–47, unnamed author.

[6] Bonnie P. Dalton, NASA Oral History, interview conducted by Rebecca Wright, 23 April 2002.

[7] NASA Press Kit STS-41, issued by NASA, October 1990.

[8] "Japan's First Spacefarer," *Space Flight News* magazine, No. 63, February 1991, pp. 28–35, unnamed author.

[9] Masamichi Yamashita and Tomio Naitoh, undated report "Comparative Approaches to the Study of the Gravitational Biology of Frogs: Adaptation and Diversity in Gravitational Responses," Shimane University, Matsue, Japan.

[10] Dr. Emily R. Morey-Holton, "The Impact of Gravity on Life," Chapter 9 in Lynn Rothschild and Adrian Lister (eds), *Evolution on Planet Earth: The Impact of the Physical Environment*, Academic Press, New York, 2003.

[11] "From Undersea to Outer Space," *Lift-off to Learning*, Educational Product EV-1997-07-004-HQ, NASA, Washington, D.C.

[12] STS-58 Mission Report, NASA, Washington, DC, 1993.

[13] "Early Development of a Gravity-Receptor Organ in Microgravity (8913083 1/3)," Life Sciences Data Archive, NASA JSC, Houston, TX, 2006. Website: *http://lsda.jsc.nasa.gov/scripts/experiment/exper.cfm?exp_index=668*

[14] STS-65 Mission Report, NASA, Washington, D.C., 1994.
[15] Kathy Sawyer, "Sexual Revolution in Zero Gravity," *Washington Post* newspaper, 15 September 2000.
[16] Jay C. Buckey, Jr. and J.L. Homick, "The Neurolab Spacelab Mission: Neurospace Research in Space" (NASA SP-2003-535), NASA JSC, Houston, TX, 2003.
[17] Marcia Dunn, "Columbia mission will explore brain," *Florida Today* newspaper, 18 April 1998.
[18] Marcia Dunn, "Jiminy! Columbia to fly with 1,500 crickets," *Florida Today* newspaper, 14 April 1998.
[19] "Space Race proves too tough for rat race," *The Australian* newspaper, 27 April 1998.
[20] "Shuttle rats gets leg surgery," Melbourne *Herald Sun* newspaper, 1 May 1998.
[21] A. Christopher Maese, Louis H. Ostrach and Bonnie P. Dalton, "Neurolab Document ID 20020073543," NASA Center for Aerospace Information (CASI), February 2002.
[22] Pauline Arrillage, "NASA to probe shuttle rat deaths," *The Associated Press*, 29 April 1998.
[23] Jay C. Buckey, Jr., Richard M. Linnehan and Alexander Dunlap, "Animal Care on Neurolab," *The Neurolab Spacelab Mission: Neuroscience Research in Space* (NASA SP-2003-535), NASA JSC, Houston, TX, 2003.
[24] "Avian development studied on Mir space station," NASA news release 95-57 reproduced in *Marsbugs: The Electronic Exobilogy Newsletter*, vol. 2, no. 5, 5 May 1995. Website: *http://www.lyon.edu/projects/marsbugs/1995/19950505.txt*
[25] "Ladybug or Lady Beetle," Enchanted Learning, 2006. Website: *http://www.enchantedlearning.com/subjects/insects/Ladybug.shtml*
[26] Daniel Sorid, "Rumble on the Shuttle: Ladybugs vs. Aphids," Space.com Website: *http://www.space.com/scienceastronomy/planetearth/shuttle_stars.html*
[27] Brian Harvey, *China's Space Program: From Conception to Manned Space Flight*, Springer/Praxis, Chichester, UK, 2004.
[28] Leonard David, "Pentagon Report: China's Space Warfare Tactics Aimed at U.S. Supremacy," Space.com, 1 August 2003. Website: *http://www.space.com/news/china_dod_030801.html*
[29] "Shenzhou 1, 2, 3, 4 Unmanned Spaceflight Mission," *China Defence Today*. Website: *http://www.sinodefence.com/strategic/mannedspace/shenzhou1234.asp*
[30] Wei Long, "China Says Shenzhou Okay: Orbital Module Operational," *Space Daily*, 2 February 2001.
[31] Wei Long, "Shenzhou Design Changes Reason for Launch Delay," *Space Daily*, 13 March 2002.
[32] "Worms survived Columbia disaster," *BBC News*, 1 May 2003. Website: *http://news.-bbc.co.uk/go/pr/fr/-/2/hi/science/nature/2992123.stm*
[33] "Successful Conclusion of Foton-M2 Mission," RedNova.com. Website: *http://www.rednova.com/news/display/?id=156536&source=r_space*
[34] "Shenzhou 6," Wikipedia online encyclopedia. Website: *http://en.wikipedia.org/wiki/Shenzhou_6*
[35] Dr. Tony Phillips,"Can People Go to Mars?" *Science@NASA*, 17 February 2004. Website: *http://science.nasa.gov/headlines/y2004/17feb_radiation.htm*
[36] Dr. Duane Graveline, interview with Colin Burgess, Florida, 16 May 2005.

Epilogue

There was a particularly intense decade in the use of animals in the development of space travel, from 1951 to 1961. It was a period of transition, from the first successful suborbital flight of the Soviet dogs Tsygan and Dezik to the orbital flight of the U.S. chimpanzee Enos.

In that formative decade, when the development of powerful new rockets took man incrementally closer to the long-held dream of space travel, animals played a critical role in answering the most fundamental questions of space flight: Could living beings survive rocket flight, weightlessness and exposure to cosmic radiation? And, further, what sort of equipment was necessary to keep terrestrial inhabitants safe in the unforgiving environment of space? Life-support systems, medical monitoring and telemetry systems came of age in this period, thanks in part to the use of animals.

During these pioneering years, the dogs of the Soviet space programme and the monkeys and chimpanzees of the U.S. space effort were treated very much like astronauts, undergoing vigorous training, subjected to dangers and garnering lavish media attention, just as would their human counterparts in the years to come. They became objects of national pride and pawns in the ongoing political gamesmanship of the Cold War Space Race.

The advent of manned space flight changed all of that. Although humans could quite clearly function in space for extended periods of time, the weightlessness and radiation they encountered caused structural and functional changes in the body that might endanger the health of astronauts and compromise the success of future missions. Once again, animals came into service, to study these effects and answer this new set of fundamental questions.

In the U.S. Biosatellite and the USSR/Russia Bion programmes, monkeys and rats especially proved their worth in ongoing studies of microgravity, radiation and artificial gravity. But their role had radically shifted from that of their pioneering brethren. They were no longer animal astronauts pioneering space travel, but rather laboratory animals, playing a smaller but still important role.

In the investigation of the physiological changes caused by prolonged stays in space, the use of animals, as opposed to humans, offered two key advantages. For one, the population of animals studied was more uniform than the human population. The Wistar rat, for example, subject of extensive study in space, is a strain specifically developed for biological and medical research. Its physiology is so well known that changes are more easily noted and take on greater significance.

The second advantage is that highly invasive studies could be conducted, utilising organs and extensive tissue samples, revealing elemental, biological changes. Also, experimental conditions could be more completely controlled in ways that would not be possible with humans.

On board the various biosatellites and the space shuttle, the use of experimental animals demonstrated the existence of a "deconditioning syndrome" that affected living beings, causing comprehensive structural, functional and metabolic changes in bones, muscles and many of the body's physiological systems. The impact of these significant changes was mapped in considerable detail in the behaviour and the living tissue of experimental animals.

Whether the immediate future of human space exploration remains stuck in low Earth orbit or ventures on to long-duration flights and visits to the moon or Mars, animals will probably continue to play a critical role. In laboratories and on board spacecraft, it is likely that basic research with animals will continue to explore the mechanisms behind the physiological changes affecting astronauts. They will also play their role in directed research that more closely supports future missions. Current emphasis seems to favour this latter approach.

One such project, set for launch in the near future – if funding materialises – is the Mars Gravity Biosatellite Project. Working under NASA contract, the Massachusetts Institute of Technology and Georgia Institute of Technology will launch 15 mice into low Earth orbit in 2010. For 5 weeks the spinning satellite will simulate Martian gravity– which is about one-third that of Earth's. This will be the first time that mammals are used to explore the effects of partial gravity. This directed research, investigating the long-term physiological response to Martian gravity, supports preparation for manned missions to the Red Planet.

This is the role that animals seem destined to play in the future development of space travel, a role very similar to one that they have played in the past. At each new juncture in the advance of our knowledge of space, at each new frontier where unknown dangers might lurk, these intrepid space travellers will continue to lead the way, making it safer for the humans that follow.

Appendix A

U.S. monkey research flights

WSPG White Sands Proving Ground
CC Cape Canaveral
WIMTC Wallops Island Missile Test Center

Launch date	Launch site	Launch vehicle	Monkey/ape (genus/species)	Status
11.06.1948	WSPG	V-2 No. 37 (Blossom No. 3)	Albert I Rhesus macaque (*Macaca mulatta*)	Did not survive (a)
14.06.1949	WSPG	V-2 No. 47 (Blossom No. 4B)	Albert II Rhesus macaque (*Macaca mulatta*)	Did not survive (a)
16.09.1949	WSPG	V-2 No. 32 (Blossom No. 4C)	Albert III Philippine macaque (*Macaca fascicularis*)	Did not survive (a)
08.12.1949	WSPG	V-2 No. 31	Albert IV Rhesus macaque (*Macaca mulatta*)	Did not survive (a)
18.04.1951	WSPG	Aerobee USAF-12	Albert V Rhesus macaque (*Macaca mulatta*)	Did not survive (a)

Launch date	Launch site	Launch vehicle	Monkey/ape (genus/species)	Status
20.09.1951	WSPG	Aerobee USAF-19	Albert VI (a.k.a. Yorick) Rhesus macaque (*Macaca mulatta*)	Did not survive (b)
21.05.1952	WSPG	Aerobee USAF-26	Patricia and Michael, both Philippine macaques (*Macaca fascicularis*)	Both survived
13.12.1958	CC	Jupiter AM-13	Old Reliable (a.k.a. Gordo) Squirrel monkey (*Saimiri sciureus*)	Did not survive (c)
28.05.1959	CC	Jupiter AM-18	Able Rhesus macaque (*Macaca mulatta*) Baker Squirrel monkey (*Saimiri sciureus*)	Both survived
04.12.1959	WIMTC	Little Joe LJ-2	Sam Rhesus macaque (*Macaca mulatta*)	Survived
21.01.1960	WIMTC	Little Joe LJ-1B	Miss Sam Rhesus macaque (*Macaca mulatta*)	Survived
31.01.1961	CC	Redstone/ MR-2	Ham Chimpanzee (*Pan troglodytes*)	Survived
10.11.1961	CC	Atlas 32E	Goliath Squirrel monkey (*Saimiri sciureus*)	Did not survive (d)
29.11.1961	CC	Atlas/MA-5	Enos Chimpanzee (*Pan troglodytes*)	Survived
20.12.1961	CC	Atlas 6F	Scatback Rhesus macaque (*Macaca mulatta*)	Did not survive (c)

Launch date	Launch site	Launch vehicle	Monkey/ape (genus/species)	Status
29.06.1969	CC	Thor Delta N/ Biosatellite 3	Bonnie Pig-tailed macaque (*Macaca nemestrina*)	Died post-flight (e)
29.04.1985	CC	Space shuttle STS 51-B	Specimen No. 3165 and Specimen No. 384-80 both squirrel monkeys (*Saimiri sciureus*)	Both survived

(a) Impact with ground.
(b) Heat prostration following recovery.
(c) Presumed drowned after splashdown.
(d) Rocket off-course, destroyed.
(e) Dehydration following recovery.

Appendix B

Soviet space dog programme

This list is wrong. Quite bluntly, that is the only way to begin any credible list of the launches in the Soviet space dog programme. Incomplete record-keeping, conflicting "official" lists, Soviet inclinations towards secrecy and covering up shortcomings, problems with transliteration of Cyrillic to English, the Russian penchant for pet names, the practice of changing the names of some dogs pre-flight – all of these have contributed to confusion over the years.

Below is our attempted listing of all the suborbital and orbital dog flights of the Soviet space dog programme, between the years 1951 and 1966. Despite labouring long and hard to assemble some sort of definitive list, we must assume (and acknowledge) that it contains errors. The four principle factors contributing to the difficulty of assembling an accurate list are given below.

THE RUSSIAN LANGUAGE

The Russian language allows for so many derivatives to names that it is difficult to keep them straight in any historical record of the dog programme. Anyone who has had the experience of reading a Russian novel, thinking it has 20 characters in it when in fact it has only 5, has experienced this first hand. There are so many diminutives and nicknames associated with most common Russian names that you eventually lose track of who's who. The same thing occurs with the names of the dogs that have flown in the Soviet space programme. To gather all the names that have appeared over the years in books, articles, diaries and other sources, you would think that a hundred dogs had participated in the flights; whereas, in reality, the number is only a few dozen. Sometimes dogs' names are given in full forms, sometimes in diminutives, sometimes changed to neuter or masculine forms, sometimes made into adjectives. For instance, the dog Smelaya (feminine adjective form) is sometimes listed as Smyeliy (default neuter/masculine form) and Ryzhaya as Ryzhik. The name Zhemchuzhnina can appear as Zhemchuzhnaya. Zhemchuzhnina is the actual thing (pearl), while Zhemchuzhnaya is an adjective form (pearly, pearl-like, pearl-coloured)

TRANSLITERATION

Problems transliterating Cyrillic words into English have resulted in multiple spellings for some dog names. The Russian language has a number of sounds that cannot be represented by a single letter in the English alphabet. How these sounds are rendered from Russian to English varies from translator to translator. Take as an example one of the canine subjects to fly in the first dog flight. The dog's name is usually given as Tsygan. Russian has one letter for "ts" – as does German (z) – whereas English must use two. Over the years, translators struggling to give the best representation of this Russian name in English have come up with such variations as Tsigan, Zhegan, Zigan. Yet they are all the same dog. A similar transliteration problem can be seen in the dog named Dzhoyna. The Russian letter represented by the "Dzh" sound is similar to the sound of the English letter "J" as in John. So, the name of the dog Dzhoyna will sometimes appear as Joyna.

RENAMING DOGS

For various reasons, dogs were sometimes renamed before their flights. Oleg Gazenko has referred to this practice with the space dog that lived with him for 12 years, Zhulka. According to a 2001 interview in the Russian magazine, *Nauka I Zhizn*, Gazenko claimed that Zhulka (which means "Mutt") flew under different names on two suborbital flights and one orbital attempt. For some flights she was named Zhemchuzhnaya and for another Pushinka. This does not quite square with the list below. Pushok, a variation of Pushinka, made a suborbital flight on 21 February 1958. But, Zhemchuzhnaya also made two suborbital flights. Forty years after the fact, did Gazenko's memory slip, or is the record wrong?

The orbital attempt to which he refers, 22 December 1960, involved additional name-changing, compounding the confusion. Some sources (Nikolai Kamanin) list Zhemchuzhnaya on this flight; however, the name given for the other dog is Zhulka. What to make of that? Most sources list the dogs on this flight as Kometa and Shutka. It is reasonable to assume that one of these dogs is a renamed Zhulka. The motivation for a name change on this occasion may well be that it seemed more appropriate to have a dog on this historical flight named Comet (Kometa) rather than Mutt (Zhulka).

CONFLICTING LISTS

If any list of dogs and rocket flights could be said to form the basis of the list below it would be that published in Vladimir Yazdovskiy's 1996 book *On the Trail of the Universe*, and yet we have deviated from that list on numerous points. Yazdovskiy's list is occasionally in conflict with other lists or with information from other sources. To the best of our ability, we have tried to reconcile those discrepancies. We have added flights missing from his list or changed dates when substantiated by other sources. We have attempted to sort out the inherent confusion with the names, including the habit of occasionally reusing certain names after a dog has died. On the rare occasion when two historical sources were in irreconcilable conflict, we have made an educated guess.

The authors would appreciate hearing from anyone who has impeccable evidence conflicting with this list.

* Died

Launch date	Dogs	Rocket	Altitude (miles)
22-7-51	Tsygan (Gypsy), Dezik	R-1V	62
29-7-51	Lisa* (Fox), Dezik*	R-1B	62, failed
15-8-51	Mishka, Chizhik	R1-B	62
19-8-51	Smelaya (Courageous), Ryzhik (Ginger)	R1-V	62
28-8-51	Mishka*, Chizhik*	R1-B	62, failed
3-9-51	Neputevyy (Screw Up), ZIB (acronym: substitute for Missing Dog Bobik)	R1-B	62
26-6-54	Lisa-2, Ryzhik (Ginger)	R-1D	62
2-7-54	Damka (Little Lady), Mishka*	R-1D	62
7-7-54	Damka Ryzhik*	R-1D	62
25-1-55	Rita*, Lisa-2	R-1E	62
5-2-55	Lisa-2*, Bulba*	R-1E	62, failed
4-11-55	Malyshka (Little One), Knopka	R-1E	62
31-5-56	Malyshka, Linda	R-1E	62
7-6-56	Albina (Whitie), Kozyavka (Little Gnat)	R-1E	62
14-6-56	Albina, Kozyavka	R-1E	62
16-5-57	Ryzhaya, Damka	R-2A	132
24-5-57	Ryzhaya*, Dzhoyna*	R-2A	132

Launch date	Dogs	Rocket	Altitude (miles)
25-8-57	Belka (Squirrel), Modnitsa (Fashionable) (Belka anaesthetised)	R-2A	132
31-8-57	Damka, Belka (Belka anaesthetised)	R-2A	132
6-9-57	Belka, Modnitsa (Modnitsa anaesthetised)	R-2A	132
3-11-57	Laika* (Barker)	R-7	Orbital
21-2-58	Palma, Pushok	R-5A	280
2-8-58	Kusachka, Palma	R-2A	132
13-8-58	Kusachka, Palma	R-2A	132
27-8-58	Belyanka (Whitie), Pestraya	R-5A	280
31-10-58	Zhulba*, Knopka*	R-5A	280, failed
2-7-59	Otvazhnaya (Brave One), Snezhinka (Snow Flake)	R-2A	132
10-7-59	Otvazhnaya, Zhemchuzhnaya (Pearly)	R-2A	132
15-6-60	Otvazhnaya, Malek	R-2A	132
24-6-60	Otvazhnaya, Zhemchuzhnaya	R-2A	132
28-7-60	Chaika*, Lisichka* (Little Fox)	R-7	Was to be orbital, failed
19-8-60	Belka, Strelka (Little Arrow)	R-7	Orbital
16-9-60	Palma, Malek	R-2A	132
1-12-60	Pchelka* (Little Bee), Mushka* (Little Fly)	R-7	Orbital, failed
22-12-60	Kometa (Comet), Shutka (Joke)	R-7	Orbital
9-3-61	Chernushka (Blackie)	R-7	Orbital
25-3-61	Zvezdochka (Little Star)	R-7	Orbital
22-2-66	Ugolyok (Little Piece of Coal), Veterok (Little Wind)	Soyuz	Orbital

* Died

Appendix C

U.S. biological rocket flights 1946–1960

V-2 ROCKET BIOLOGICAL FLIGHTS

Launch vehicle	Launch date	Launch site	Biological payload	Approx. altitude (miles) and result summary
V-2 No. 7	09.07.1946	WSPG	Specially developed strains of seeds	83; samples not recovered
V-2 No. 8	19.07.1946	WSPG	Specially developed strains of seeds	4; samples not recovered
V-2 No. 9	30.07.1946	WSPG	Ordinary corn seeds	104; seeds recovered
V-2 No. 12	10.10.1946	WSPG	Rye seeds	112; seeds recovered
V-2 No. 17	17.12.1946	WSPG	Fungus spores	117; spores not recovered
V-2 No. 20	20.02.1947	WSPG	Fruit flies	68; flies recovered alive
V-2 No. 21	07.03.1947	WSPG	Rye seeds, corn seeds, fruit flies	101; biological results were not recorded
V-2 No. 29	10.07.1947	WSPG	Rye seeds, corn seeds, fruit flies	10; biological results were not recorded
V-2 No. 37 Blossom 3	11.06.1948	WSPG	Monkey Albert	39; parachute failure; animal died
V-2 No. 44	18.11.1948	WSPG	Cotton seeds	90; seeds recovered
V-2 No. 50	11.04.1949	WSPG	Not known	54; biological results were not recorded

V-2 ROCKET BIOLOGICAL FLIGHTS

Launch vehicle	Launch date	Launch site	Biological payload	Approx. altitude (miles) and result summary
V-2 No. 47 Blossom 4B	14.06.1949	WSPG	Monkey Albert II	83; parachute failure; animal died
V-2 No. 32 Blossom 4C	16.09.1949	WSPG	Monkey Albert III	3; rocket exploded; animal died
V-2 No. 31	08.12.1949	WSPG	Monkey Albert IV	82; animal died on impact
V-2 No. 51	31.08.1950	WSPG	Mouse	84; animal died on impact

AEROBEE BIOLOGICAL ROCKET FLIGHTS

Aerobee No.	Launch date	Launch site	Biological payload	Approx. altitude (miles) and result summary
USAF-12	18.04.1951	WSPG	Monkey Albert V, several mice	38; parachute failure; animals died
USAF-19	20.09.1951	WSPG	Monkey Albert VI, 11 mice	44; all animals recovered, monkey died 2 hrs after impact
USAF-20	21.05.1952	WSPG	Two monkeys Patrick & Michael, two mice	39; all animals safely recovered

PROJECT MIA (MOUSE-IN-ABLE)

Launch vehicle	Launch date	Launch site	Biological payload	Approx. altitude (miles) and result summary
Thor 116 IRBM/ Aerojet 1040/ Able	23.04.1958	CC	Mouse Mia-1 (a.k.a. Minnie)	Not known; rocket exploded, nose cone not recovered
Thor 118 IRBM/ Aerojet 1040/ Able	09.07.1958	CC	Mouse Mia-II (a.k.a. Laska)	1,400; nose cone lost
Thor 119 IRBM/ Aerojet 1040/ Able	23.7.1958	CC	Mouse Wickie (a.k.a. Benji)	1,400; nose cone lost at sea

JUPITER BIOLOGICAL ROCKET FLIGHTS

Launch vehicle	Launch date	Launch site	Biological payload	Approx. altitude (miles) and result summary
Jupiter IRBM AM-13/ Bio-flight No. 1	13.12.1958	CC	Monkey Old Reliable (a.k.a. Gordo) and *Neurospora*	290 altitude, 1,300 distance; nose cone lost at sea
Jupiter IRBM AM-18/ Bio-flight No. 2 (2A and 2B)	28.05.1959	CC	Monkeys Able and Baker, *Neurospora*, seeds, pupae, sea urchin eggs, sperm	300 altitude, 1,500 distance; nose cone and both animals safely recovered
Jupiter IRBM AM-23	16.09.1959	CC	Two frogs, 12 pregnant mice, seeds, pupae, urchin eggs and sperm	Missile destroyed by range safety officer after flawed launch

"LITTLE JOE" – PROJECT HERMES FLIGHTS

Launch vehicle	Launch date	Launch site	Biological payload	Approx. altitude (miles) and result summary
Little Joe LJ-2	04.12.1959	WIMTC	Monkey Sam, *Neurospora*, beetle eggs, *Escherichia coli*, cancer cells, seeds	53 miles, 194 downrange; flight successful, monkey safely recovered
Little Joe LJ-1B	21.01.1960	WIMTC	Monkey Miss Sam	9 miles, 12 downrange; flight successful, monkey safely recovered

ATLAS MISSILE FLIGHT

Launch vehicle	Launch date	Launch site	Biological payload	Approx. altitude (miles) and result summary
Atlas D 71-D	13.10.1960	CC	Three black mice; Sally, Amy and Moe	650 miles, 5,000 downrange; nose cone and mice safely recovered

WSPG White Sands Proving Ground, New Mexico
CC Cape Canaveral, Florida
WIMTC Wallops Island Missile Test Center
IRBM Intermediate Range Ballistic Missile

Appendix D

French biological rocket flights, 1961–1967

Table of French medical–biological, experimental ballistic flights carried out between 1963 and 1967 by CERMA (Research and Studies Centre of Aerospace Medicine) using Veronique and Vesta rockets. All launches took place at the Hammaguir test range in Algeria.

Date	Vehicle	Animal	Result (miles)
22.2.1961	Veronique AG124	Rat RC139: Hector	69; recovered
15.10.1962	Veronique AG137	Rat RC271: Castor	75; recovered
18.10.1962	Veronique AG136	Rat RC268: Pollux	69; not recovered
18.10.1963	Veronique AG147	Cat CC341: Felicette	97; recovered
24.10.1963	Veronique AG150	Cat CC333 (Unnamed)	55; not recovered
7.3.1967	Vesta 04	Monkey: Martine	152; recovered
13.3.1967	Vesta 05	Monkey: Pierrette	146; recovered

Appendix E

Chinese T-7 sounding rocket launches

(data incomplete)

Launch date	Launch vehicle	Launch site
1960, February 19	T-7M	Laogang
1960, September	T-7	Shijiedu
1963, December 1	T-7A	Shijiedu
1964, July 19	T-7A (S)	Shijiedu
1965, June 1	T-7A (S)	Shijiedu
1965, June 5	T-7A (S)	Shijiedu
1965	T-7	Jiuquan
1966, July 14	T-7A (S2) (a)	Shijiedu
1966, July 28	T-7A (S2) (b)	Shijiedu
1968, August 8	T-7/GF-01A	Jiuquan
1968, August 20	T-7/GF-01A	Jiuquan
1969	T-7	Jiuquan
1969, June	T-7A	Jiuquan
1969, July	T-7A	Jiuquan

(a) Carried dog Xiao Bao.
(b) Carried dog Shanshan.

Appendix F

Bion research flights

Bion number	Cosmos number	Launch date	Primates carried	Flight duration (days)
Bion 1	605	31.10.1973	None	21.5
Bion 2	690	22.10.1974	None	20.5
Bion 3	782	25.11.1975	None	19.5
Bion 4	936	3.08.1977	None	18.6
Bion 5	1129	25.09.1979	None	18.5
Bion 6	1514	14.12.1983	Abrek and Bion	5.0
Bion 7	1667	10.07.1985	Verny and Gordy	6.9
Bion 8	1887	29.09.1987	Yerosha and Dryoma	13.0
Bion 9	2044	15.09.1989	Zhakonya and Zabiyaka	14.0
Bion 10	2229	29.12.1992	Krosh and Ivasha	12.0
Bion 11	None given	24.12.1996	Lapik and Multik	15.0

Appendix G

Space shuttle life science orbital flights

LARGE ANIMALS

Mission	Launch date	Animals carried	Duration (days)	Shuttle
STS 51-B Spacelab 3	29.04.1985	Squirrel monkeys No. 3165 (*Saimiri sciureus*) No. 384-80 (*Saimiri sciureus*)	7	Challenger

SMALL ANIMALS

Mission	Launch date	Animals carried	Duration (days)	Shuttle
STS-8	30.08.1983	6 Lewis Wistar rats (*Rattus norvegicus*)	6	Columbia
STS 41-B	03.02.1984	6 Lewis Wistar rats (*Rattus norvegicus*)	8	Challenger
STS 51-B Spacelab 3	29.04.1985	24 male albino rats (*Rattus norvegicus*)	7	Challenger
STS-29	13.03.1989	4 Long-Evans rats (*Rattus norvegicus*)	5	Discovery
STS-41	6.10.1990	16 Sprague-Dawley rats (*Rattus norvegicus*)	4	Discovery

SMALL ANIMALS (*cont.*)

Mission	Launch date	Animals carried	Duration (days)	Shuttle
STS-40 Spacelab SLS-1	05.06.1991	29 Sprague-Dawley rats (*Rattus norvegicus*); Jellyfish (*Aurelia aurita*)	9	Columbia
STS-48	12.09.1991	8 female albino rats (*Rattus norvegicus*)	5	Discovery
STS-47 Spacelab J1	12.09.1992	2 Japanese koi (carp) (*Cyprinus carpio*); 4 frogs (*Xenopus laevis*)	8	Endeavour
STS-52	22.10.1992	12 male albino rats (*Rattus norvegicus*)	9	Columbia
STS-54	13.01.1993	6 male albino rats (*Rattus norvegicus*)	6	Endeavour
STS-56	08.04.1993	16 Sprague-Dawley rats (*Rattus norvegicus*)	9	Discovery
STS-57	21.06.1993	12 Fisher-344 rats (*Rattus norvegicus*)	10	Endeavour
STS-58 Spacelab SLS-2	18.10.1993	48 Sprague-Dawley rats (*Rattus norvegicus*)	14	Columbia
STS-60	03.02.1994	12 Sprague-Dawley rats (*Rattus norvegicus*)	8	Discovery
STS-62	04.03.1994	12 Fisher-344 rats (*Rattus norvegicus*)	14	Columbia
STS-65 Spacelab IML-2	08.07.1994	Newt (*Cynopus pyrrhogaster*); Medaka fish (*Oryzias latipes*); Moon jellyfish (*Aurelia aurita*)	15	Columbia
STS-66	03.11.1994	10 Sprague-Dawley rats (*Rattus norvegicus*)	11	Atlantis
STS-63	03.02.1995	12 Sprague-Dawley rats (*Rattus norvegicus*)	8	Discovery
STS-70	13.07.1995	10 Sprague-Dawley rats (*Rattus norvegicus*); Tobacco hornworms (*Manduca sexta*)	9	Discovery

SMALL ANIMALS (*cont.*)

Mission	Launch date	Animals carried	Duration (days)	Shuttle
STS-72	11.01.1996	6 adult and 60 neonatal rats (*Rattus norvegicus*)	9	Endeavour
STS-77	19.05.1996	12 Sprague-Dawley rats (*Rattus norvegicus*); Tobacco hornworm pupa (*Manduca sexta*)	10	Endeavour
STS-78	20.06.1996	12 Sprague-Dawley rats (*Rattus norvegicus*)	17	Columbia
STS-80	19.11.1996	14 Sprague-Dawley rats (*Rattus norvegicus*)	18	Columbia
STS-89	22.01.1998	Snails (*Biomphalaria glabrata*)	9	Endeavour
STS-90	17.04.1998	152 rats (*Rattus norvegicus*); 18 mice (*Mus musculus*); 135 snails (*Biomphalaria glabrata*); 4 oyster toadfish (*Opsanus tau*); 229 swordtail fish (*Xiphophorus helleri*); 1,500 crickets (*Acheta domesticus*)	16	Columbia
STS-93	23.07.1997	Fruit flies (*Drosophila melanogaster*); 4 ladybugs (*Coccinella septempunctata*); Caterpillars (*Vanessa cardui*)	5	Columbia
STS-95	29.10.1998	Oyster toadfish (*Opsanus tau*)	8	Discovery
STS-106	08.09.2000	Fruit flies (*Drosophila melanogaster*)	12	Atlantis
STS-107	16.01.2003	13 Sprague-Dawley rats (*Rattus norvegicus*); 8 orb weaver spiders (*Eriophora biapicata*); 3 carpenter bees (*Xylocopa virginica*); 15 harvester ants (*Pogononyrmex* spp.); 5 silkworms and 3 cocoons (*Bombyx mori*); Fruit flies (*Drosophila melanogaster*) 4 medaka fish eggs (*Oryzias latipes*); snails (*Biomphalaria glabrata*)	16	Columbia

Index

Italicized page numbers refer to figures or figure captions in text

Printing: Mercedes-Druck, Berlin
Binding: Stein+Lehmann, Berlin